ALGEBRAIC NUMBERS AND HARMONIC ANALYSIS

North-Holland Mathematical Library

VOLUME 2

NORTH-HOLLAND PUBLISHING COMPANY – AMSTERDAM · LONDON

AMERICAN ELSEVIER PUBLISHING COMPANY, INC. – NEW YORK

Algebraic Numbers and Harmonic Analysis

YVES MEYER
Faculté des Sciences d'Orsay,
Université de Paris-Sud, France

1972
NORTH-HOLLAND PUBLISHING COMPANY – AMSTERDAM · LONDON
AMERICAN ELSEVIER PUBLISHING COMPANY, INC. – NEW YORK

Library of Congress Catalog Card Number: 77–183278
North-Holland ISBN: 0 7204 2452 6
American Elsevier ISBN: 0 444 10357 0

Publishers:

NORTH-HOLLAND PUBLISHING COMPANY – AMSTERDAM
NORTH-HOLLAND PUBLISHING COMPANY, LTD. – LONDON

Sole distributors for the USA and Canada:

AMERICAN ELSEVIER PUBLISHING COMPANY, INC.
52 VANDERBILT AVENUE
NEW YORK, N.Y. 10017

PRINTED IN GERMANY

PREFACE

This book is dedicated to the memory of Raphaël Salem: it contains most of his beautiful discoveries and the proof of his conjecture about the role played by Pisot numbers in the problem of spectral synthesis.

I wish to express my gratitude to Mrs Deschamps, S. Drury, J.-P. Kahane, Y. Katznelson, N. Lohoué, J.-P. Schreiber and N. Varopoulos for stimulating discussions.

It is a pleasure to thank Mme Dumas for her very careful preparation of the manuscript.

I have greatly appreciated the patience of the publisher, and the Department of Mathematics of the University of Nantes was a quiet place to write this book.

Institute of Mathematics, Nantes

CONTENTS

LIST OF SYMBOLS

INTRODUCTION

Harmonic Analysis was originally devoted to periodic phenomena. The analysis of a periodic function f is achieved by the computation of the Fourier coefficients and only the knowledge of f on an interval of one period is needed.

Almost periodic functions are the natural aperiodic generalization of continuous periodic functions. A representation of f by a Fourier series $\sum a_\lambda \exp 2\pi i\lambda t$ exists; the set Λ of all λ for which $a_\lambda \neq 0$ is the spectrum of f. But now for a computation of Fourier coefficients one needs to know the function up to the end of time.

This is very troublesome in applications and the modern approach is to try to get global estimates on all almost periodic functions with a given spectrum Λ by means of local knowledge.

Let Λ be a set of real numbers. We say that Λ is a coherent set of frequencies if there exist a $C > 0$ and a compact set K of real numbers such that $\sup_{\mathbf{R}} |P| \leqslant C \sup_{K} |P|$ for all trigonometric sums P whose frequencies belong to Λ.

A remarkable set of algebraic numbers, studied (among others) by Vijayaraghavan, Pisot and Salem, plays a fundamental role in the determination of coherent sets of frequencies (Chapter I, Section 6 and Chapter IV, Section 5).

The study of coherent sets of real (or p-adic) frequencies is the main subject of this book. But from one chapter to another the point of view and the methods vary considerably. (From algebraic number field theory to measure theory; from Diophantine approximations to Laplace transforms and complex methods.)

Chapter I contains old and new things about Diophantine approximations. The reader who does not like this subtle theory may go directly to Section 6 of Chapter I, where Pisot numbers are defined.

Now there are two possibilities. The reader may begin with Chapter V or Chapter VI or VIII, which are pairwise independent and independent of the other ones, or may follow a more logical course: Chapters II, III, IV and then VII.

In Chapter II, harmonious sets are studied in general groups. Defined by remarkable properties of Diophantine approximations, they give the most interesting examples of coherent sets of frequencies.

In Chapter III, we seek connections between the arithmetical structure of a compact set E of real numbers and the behavior at infinity of functions whose spectra lie in E. The famous problems of uniqueness and multiplicity receive here new treatments by the use of harmonious sets. Pisot numbers play a decisive role in this topic.

If Λ is a coherent set of frequencies, perturbed trigonometric sums $\sum_{\lambda=\Lambda} a_\lambda(t) \exp 2\pi i\lambda t$ whose coefficients have slow variations (corresponding to 'secular perturbations' in astronomy) have excellent local approximations by the uniformly bounded corresponding pure trigonometric sums $\sum_{\lambda=\Lambda} a_\lambda(t_0) \exp 2\pi i\lambda t$ (Chapter IV).

Complex methods are introduced in Chapter V and group theoretic methods in Chapter VI to get the best known results of the type: 'from local to global' for almost periodic functions whose spectra lie in a given set Λ of real numbers. Recent results about Sidon sets are given here.

Before giving the content of the remainder of the book, a mathematical situation which is fairly common in astronomy should be described. A function φ of the time is observed on the interval $0 \leqslant t \leqslant t_1$; φ seems to be periodic and — for the sake of simplicity — it will be assumed that $\varphi = a + b \cos \omega_1 t + c \sin \omega_1 t$ (the period T_1 of φ is $2\pi/\omega_1$).

Computations are made over many such intervals $[t, t + l_1]$, $t \leqslant l_2$ (l_2 is great compared with l_1); a, b and c are actually three functions of the time whose period T_2 is very great compared with T_1 (the 'secular perturbations' in astronomy). We write $a(t) = a_2 + a_2' \cos \omega_2 t + a_2'' \sin \omega_2 t$ and we get for b and c the corresponding relations. Now φ is an almost periodic function with nine coefficients and whose frequencies are of the form $(2\pi)^{-1}(\varepsilon_1\omega_1 + \varepsilon_2\omega_2)$, $\varepsilon_j = -1$, 0 or 1. More careful observations are made on longer and longer intervals and, at each stage, coefficients which seemed constants become periodic functions with greater and greater periods. However, φ remains bounded.

A mathematical model will be built in Chapter VIII; let E be the compact set of all convergent sums $\sum_1^\infty \varepsilon_k\omega_k$ ($\varepsilon_k = -1$, 0 or 1, and we assume that ω_k tends very rapidly to 0). The function φ described above is any complex valued bounded continuous function whose spectrum lies in

$(2\pi)^{-1} E$. The global behavior of φ may be very different from the expected one of an almost periodic function.

Spectral synthesis is the study of this local approximation of φ by uniformly bounded almost periodic functions; it works if ω_k/ω_{k+1} is a Pisot number or if ω_k tends to 0 very quickly but it fails in the general case (Chapters VII and VIII).

This book has been written with the idea that it can be read by a graduate student. However, it is not an introduction to harmonic analysis. Nevertheless important classical subjects have been exposed in full detail (Laplace transform, mean periodic functions, idempotent measures).

CHAPTER I

DIOPHANTINE APPROXIMATIONS TO REAL NUMBERS

A great number of problems in the theory of Diophantine approximations can be set in the following form. Let $\lambda_1, \ldots, \lambda_n$ be n real numbers and ε a positive real number. We seek a function $T(\varepsilon; \lambda_1, \ldots, \lambda_n) = T > 0$, such that each interval of real numbers of length T contains a solution t to the Diophantine inequalities $\|t\lambda_k\| \leqslant \varepsilon\ (1 \leqslant k \leqslant n)$. (For a real number x, $\|x\|$ is the distance from x to the nearest integer.)

For fixed $\lambda_1, \ldots, \lambda_n$, the dependence on ε of the function T has a deep connection with the 'badly approximable' real numbers (Sections 2 and 3). The dependence of this function $T(\varepsilon; \lambda_1, \ldots, \lambda_n)$ on $\lambda_1, \ldots, \lambda_n$ and on n leads to quite different problems: a sequence $(\lambda_k)_{k \geqslant 1}$ of real numbers is a harmonious set if, for each positive ε, there is a function $T = T(\varepsilon)$, such that each interval of length T contains at least one solution t of the infinite system of Diophantine inequalities $\|t\lambda_k\| < \varepsilon$, $1 \leqslant k$ (Section 5). The study of harmonious sets and their uses in harmonic analysis is one of the main subjects of this book. In Section 6 we recall classical results about Pisot and Salem numbers. In Section 7 we prove the following theorem; let θ be a real number greater than 1: then the set of all powers θ^k, $k \geqslant 0$, of θ is a harmonious set if and only if θ is a Pisot or a Salem number.

1. Some classical results in Diophantine approximations

In Section 1, we give, without proofs and for the entertainment of the reader some celebrated theorems which will not be used in the remainder of this book.

1.1. What is a problem of Diophantine approximation?

Roughly speaking, a problem of Diophantine approximation is always of the following type: there is a metrizable group G and a subgroup H of G, dense in G. An estimate of the 'complexity' of elements h in H is given by a sub-additive positive weight-function $\omega\colon H \to [0, +\infty]$; 'complicated' elements h in H are those for which $\omega(h)$ is great. Let d be a translation-invariant metric on G. The 'quality' of the approximation of an element g in G by elements h in H is defined by $d(h, g)$ and the 'cost' of this approximation by $\omega(h)$.

The problem is to specify the connection between 'cost' and 'quality' of the approximation.

1.2. Approximation of real numbers by rationals. Dirichlet's and Roth's theorem

Let, with the notations of Section 1.1, $G = \mathbf{R}$ be the field of real numbers and $H = \mathbf{Q}$ be the field of rational numbers. The complexity of a reduced fraction p/q is defined by $|q|$ (p and q are rational integers relatively prime and $q \neq 0$); $\omega(p/q) = \log |q|$ would be a subadditive weight function.

For a given irrational real number α, one would like to compute, for each positive ε, the least positive integer $q(\varepsilon)$ such that the system

$$|q\alpha - p| \leqslant \varepsilon, \quad 1 \leqslant q \leqslant q(\varepsilon) \tag{1.1}$$

has a solution.

In fact a precise estimate of $q(\varepsilon)$ is generally unavailable, and we seek two functions $q_1(\varepsilon)$ and $q_2(\varepsilon)$, not too different, such that, for each positive ε,

$$|q\alpha - p| \leqslant \varepsilon \quad \text{and} \quad q \geqslant 1 \quad \text{imply} \quad q \geqslant q_1(\varepsilon) \tag{1.2}$$

$$|q\alpha - p| \leqslant \varepsilon \quad \text{and} \quad 1 \leqslant q \leqslant q_2(\varepsilon) \quad \text{has a solution.} \tag{1.3}$$

The best case is when $q_2(\varepsilon) = 0\,(q_1(\varepsilon))$ but generally $q_1(\varepsilon) = o\,(q_2(\varepsilon))$ and there is a gap between $q_1(\varepsilon)$ and $q_2(\varepsilon)$. Dirichlet's theorem gives for any α an upper estimate of $q_2(\varepsilon)$.

THEOREM I. *For every real number* $\alpha \neq 0$ *and every* ε *in* $]0, 1[$ *we can solve the system*

$$|q\alpha - p| \leqslant \varepsilon \quad \textit{and} \quad 1 \leqslant q < \frac{1}{\varepsilon}.$$

On the other hand no 'universal' lower estimate for $q_1(\varepsilon)$ is possible. Liouville's numbers $\alpha = \sum_{k \geqslant 0} 10^{-n_k}$, constructed with a very rapidly growing sequence n_k of integers, give examples where q_1 is as small as desired. Such α are transcendental real numbers and if α is algebraic, we have the following celebrated result, known as Roth's theorem [82].

THEOREM II: *Let α be an algebraic number and δ a real number in* $]0,1[$ *arbitrarily close to* 1. *There exists a positive real number ε_0 depending on α and δ such that the following holds:* $0 < \varepsilon < \varepsilon_0$, $|q\alpha - p| \leqslant \varepsilon$ *and* $q \geqslant 1$ *imply* $q \geqslant \varepsilon^{-\delta}$.

If α is algebraic, $q_1(\varepsilon) \geqslant \varepsilon^{-\delta}$; there is a gap between the estimate of q_1 given by Roth's theorem and the estimate of q_2 given by Dirichlet's theorem.

1.3. Simultaneous approximation of n real numbers. Schmidt's theorems

Here $G = \mathbf{R}^n$, $n > 1$, and H is the dense subgroup of G consisting of all $\mathbf{x} = (p_1/q, \dots, p_n/q)$, where $q \geqslant 1$ and p_j are rational integers $(1 \leqslant j \leqslant n)$. If one of the fractions p_j/q is irreducible, q will be an estimate of the complexity of $\mathbf{x}$; such an $\mathbf{x}$ will be called a rational element in $\mathbf{R}^n$ and, if $q = 1$, $\mathbf{x}$ will be called an integer point in $\mathbf{R}^n$.

Let $\alpha_1, \dots, \alpha_n$ be n real numbers; the problem is to compute, for every positive ε the least integer $q(\varepsilon)$ such that the system

$$|q\alpha_1 - p_1| \leqslant \varepsilon, \dots, |q\alpha_n - p_n| \leqslant \varepsilon, \quad 1 \leqslant q \leqslant q(\varepsilon) \tag{1.4}$$

has a solution in rational integers $p_1, \dots, p_n$ and q.

As above, we can divide this problem into two parts:

(1) Find, for each positive ε, an estimate of a function $q_1(\varepsilon)$, such that

$$|q\alpha_1 - p_1| \leqslant \varepsilon, \dots, |q\alpha_n - p_n| \leqslant \varepsilon \quad \text{and} \quad q \geqslant 1 \quad \text{imply}$$
$$q \geqslant q_1(\varepsilon). \tag{1.5}$$

(2) Find, for each positive ε, an estimate of a function $q_2(\varepsilon)$, such that the system

$$|q\alpha_1 - p_1| \leqslant \varepsilon, \dots, |q\alpha_n - p_n| \leqslant \varepsilon \quad \text{and} \quad 1 \leqslant q \leqslant q_2(\varepsilon) \tag{1.6}$$

has a solution.

Here also, we have two theorems.

Dirichlet's theorem: *If* $q_2(\varepsilon) = \varepsilon^{-n}$ *and* $0 < \varepsilon < 1$, *then* (1.6) *has a solution in rational integers.*

Schmidt's theorem: *Let $\alpha_1, \dots, \alpha_n$ be n algebraic real numbers such that* $1, \alpha_1, \dots, \alpha_n$ *are linearly independent over* $\mathbf{Q}$. *Then, for each $\eta > 0$ there is an $\varepsilon_0 > 0$ (depending on $\alpha_1, \dots, \alpha_n$ and η) such that, for $0 < \varepsilon < \varepsilon_0$, every integral solution $q \geqslant 1$ of* $\|q\alpha_1\| \leqslant \varepsilon, \dots, \|q\alpha_n\| \leqslant \varepsilon$ is greater than $q_1(\varepsilon) = \varepsilon^{-n+\eta}$ ($\|x\|$ is the distance from the real number x to the nearest integer).

Schmidt has, in fact, proved what is a very striking improvement on the preceeding result: with the same hypothesis, for each $\eta > 0$ the following multiplicative inequality has only a finite number of solutions

$$\|q\alpha_1\| \cdots \|q\alpha_n\| \, q^{1+\eta} < 1. \tag{1.7}$$

This means that if, for instance, α_1 is well approximated by p_1/q, then the other α_j's cannot simultaneously be well approximated by any p_j/q ([84] or [85]).

2. Measure-theoretical methods in Diophantine approximations

We present here quite simple and powerful techniques, which will be very useful in the study of 'badly approximable' real numbers.

2.1. Minkowski's theorem

Let μ be Lebesgue measure on $\mathbf{R}^n$, and let C be the cube $0 \leqslant x_j \leqslant 1$, $1 \leqslant j \leqslant n$; μ is normalized by $\mu(C) = 1$. Let $\mathbf{Z}^n$ be the lattice of all integer points $\mathbf{x}$ in $\mathbf{R}^n$, let $\varrho : \mathbf{R}^n \to \mathbf{R}^n/\mathbf{Z}^n$ be the canonical homomorphism whose kernel is $\mathbf{Z}^n$, and ν be Lebesgue measure on $\mathbf{R}^n/\mathbf{Z}^n$ normalized by $\nu(\mathbf{R}^n/\mathbf{Z}^n) = 1$.

Minkowski's theorem runs as follows:

THEOREM III: *Let K be a compact subset of $\mathbf{R}^n$. Assume that $\mu(K) \geqslant 1$. There exist two distinct points $\mathbf{x}$ and $\mathbf{y}$ in K, such that $\mathbf{y} - \mathbf{x}$ belongs to $\mathbf{Z}^n$.*

An immediate consequence is the following result:

COROLLARY 1: *Let K be a compact convex subset of $\mathbf{R}^n$, symmetric about $\mathbf{0}$. If $\mu(K) \geqslant 2^n$, K contains an integer point $\mathbf{x}$ different from zero.*

To get Corollary 1, we just apply Theorem III to $\frac{1}{2}K$.

COROLLARY 2: *Let $l_1, \ldots, l_n$ be n linear forms over $\mathbf{R}^n$. Assume that $l_1, \ldots, l_n$ are linearly independent. Let D be the determinant of $l_1, \ldots, l_n$ and $c_1, \ldots, c_n$ n positive real numbers such that $c_1, \ldots, c_n \geqslant D$. There exists an integer point $\mathbf{x}$, different from zero, such that $|l_1(\mathbf{x})| \leqslant c_1, \ldots, |l_n(\mathbf{x})| \leqslant c_n$.* The same result holds for pairwise-conjugate complex linear forms.

Corollary 2 is a particular case of Corollary 1 and is a generalization of Dirichlet's theorem.

Now we shall prove Theorem III. It depends on the following lemma:

LEMMA 1: *Let E be any compact subset of $\mathbf{R}^n$ and let $F = \varrho(E)$ be the set of all residues modulo $\mathbf{Z}^n$. Then $\nu(F) \leqslant \mu(E)$. If two distinct points of E are never congruent modulo $\mathbf{Z}^n$, then $\nu(F) = \mu(E)$.*

Proof: We have an explicit formula to compute $\nu(F)$. In fact

$$\nu(F) = \lim_{N \to +\infty} N^{-n} \mu \Big(\bigcup_{1 \leqslant k_j \leqslant N} E + \mathbf{k} \Big) \quad \text{with} \quad \mathbf{k} = (k_1, \ldots, k_n). \tag{1.8}$$

The two parts of Lemma 1 follow from this equality; if two distinct points of E are never congruent modulo $\mathbf{Z}^n$, then, for distinct $\mathbf{k}$, $E + \mathbf{k}$ are pairwise disjoint.

To prove (1.8) we remark that the measure space $(\mathbf{R}^n/\mathbf{Z}^n, \nu)$ can be identified with (C, μ) where C is the unit cube $0 \leqslant x_j \leqslant 1$, $1 \leqslant j \leqslant n$ and μ the restriction to C of Lebesgue measure. In both cases the σ-field is the σ-field of Borel sets.

Let $\tilde{E}$ be the union of all $E + \mathbf{k}$, $\mathbf{k} \in \mathbf{Z}^n$. With this identification, F becomes the intersection of C and $\tilde{E}$.

We get $\nu(F) = \mu(\tilde{E} \cap C) = \lim_{N \to +\infty} N^{-n} \mu(\tilde{E} \cap NC)$ since $\tilde{E}$ is $\mathbf{Z}^n$-periodic. But, with only a finite number of exceptions whose number is $0(N^{n-1})$, the integer points $\mathbf{k}$ for which $\mathbf{k} + E$ intersects NC do satisfy $1 \leqslant k_j \leqslant N$ for $1 \leqslant j \leqslant n$. This ends the proof of (1.8).

2.2. *Inhomogeneous problems*

Minkowski's theorem allows to solve homogeneous linear Diophantine problems. Is there a similar method to solve inhomogeneous linear Diophantine systems? Consider, for instance, the system

$$|q\alpha_1 - p - x_1| \leqslant a_1, \quad |q\alpha_2 - p - x_2| \leqslant a_2; \tag{1.9}$$

we seek a condition on α_1, a_1, α_2 and a_2 implying the existence of an integer solution of (1.9) for every real number x_1 and x_2. In Section 3 we shall see that, if for a suitable $b = b(\alpha_1, \alpha_2)$ a sufficient condition is $a_1 a_2 \geqslant b(\alpha_1, \alpha_2)$, just as in Minkowski's theorem, then there is a positive ε such that if $a_1 a_2 \leqslant \varepsilon$, then $|q\alpha_1 - p| \leqslant a_1$ and $|q\alpha_2 - p| \leqslant a_2$ imply $p = q = 0$. Let K be defined by $|\alpha_1 y - x| \leqslant a_1$, $|\alpha_2 y - x| \leqslant a_2$. Our aim is to prove that, for each $\mathbf{x}$ in $\mathbf{R}^2$, $K + \mathbf{x}$ intersects $\mathbf{Z}^2$ and this is connected to the existence of integer points in λK for a small enough positive λ.

We give now a general result due to Kneser and Birch [5], which is fairly near to being the best possible.

THEOREM IV: *Let S be a compact subset of $\mathbf{R}^n$. Assume that the following two conditions are fulfilled:*

(a) *two distinct points of S are never congruent modulo $\mathbf{Z}^n$;*

(b) *the Lebesgue measure of S, $\mu(S)$, is positive.*

Let N be an integer defined by $N\mu(S) \geqslant 1$ and let K be the compact subset of $\mathbf{R}^n$ defined by $K = S + \cdots + S$ (N times). Then, for each $\mathbf{x}$ in $\mathbf{R}^n$, $K + \mathbf{x}$ intersects $\mathbf{Z}^n$.

We recall that $S + \cdots + S$ (N times) is the set of all sums $\mathbf{s}_1 + \cdots + \mathbf{s}_N$, where $\mathbf{s}_j$ belongs to S $(1 \leqslant j \leqslant N)$.

Before proving this theorem let us see, by means of an example, why it is the best possible. Let S be defined in $\mathbf{R}^2$, by $0 \leqslant x \leqslant m^{-1}$, $0 \leqslant y \leqslant l < 1$ ($m \geqslant 2$ is an integer). Then $\mu(S) = lm^{-1}$ and S fulfils (a). Here K is defined by $0 \leqslant x \leqslant Nm^{-1}$, $0 \leqslant y \leqslant lN$ and K fulfils the conclusion of Theorem IV if and only if $N \geqslant m$ or $N\mu(S) \geqslant l$. Since l can be chosen arbitrarily near to 1, N cannot, in general, be less than the integer defined by $N\mu(S) \geqslant 1$.

The part played by condition (a) of Theorem IV will be better understood in the study of badly approximable real numbers.

COROLLARY: *Let S be a convex compact subset of $\mathbf{R}^n$, symmetric about $\mathbf{0}$. Assume that the Lebesgue measure of S, $\mu(S)$, is positive and that the only integer point in S is $\mathbf{0}$. Let N be an integer defined by $N\mu(S) \geqslant 2^{n-1}$. Then, for each $\mathbf{x}$ in $\mathbf{R}^n$, $NS + \mathbf{x}$ intersects $\mathbf{Z}^n$.*

The proof of Theorem IV depends on the following lemma (proved in Appendix A).

LEMMA 2: *Let A and B be two compact subsets of $\mathbf{R}^n/\mathbf{Z}^n$. Then there are only two possibilities:*

(a) $\nu(A + B) \geqslant \nu(A) + \nu(B)$ (*ν is Lebesgue measure on $\mathbf{R}^n/\mathbf{Z}^n$*); *or*

(b) $A + B = \mathbf{R}^n/\mathbf{Z}^n$.

Now to the proof of Theorem IV. From Lemma 2, we deduce that if A is a compact subset of $\mathbf{R}^n/\mathbf{Z}^n$, and if N is an integer such that $N\nu(A) \geqslant 1$, then $A + \cdots + A$ (N times) is the whole $\mathbf{R}^n/\mathbf{Z}^n$. Applying this remark to $A = \varrho(S)$, we get Theorem IV, since $\nu(A) = \mu(S)$ by Lemma 1.

3. Badly approximable real numbers

3.1. More or less rapid ways of filling $\mathbf{R}/\mathbf{Z}$ *with residues modulo* 1 *of* $m\alpha$

Let α be a real number and $h: \mathbf{Z} \to \mathbf{R}/\mathbf{Z}$ the homomorphism defined by the condition that $h(m)$ is the residue modulo 1 of $m\alpha$. If α is an irrational number, $h(\mathbf{Z})$ is dense in $\mathbf{R}/\mathbf{Z}$ but there are more or less rapid ways of filling $\mathbf{R}/\mathbf{Z}$. To make this notion precise, a definition is needed.

DEFINITION 1: *For each positive* ε, *let* $T(\varepsilon)$ *be the least integer* Q *such that, for each real* x, *the system* $|q\alpha + p - x| \leqslant \varepsilon$, $0 \leqslant q \leqslant Q$ *can be solved. Such a* Q *can also be defined by the condition that each interval* $[j, j + Q]$, $j \in \mathbf{Z}$ *contains a* q *such that* $\|q\alpha\| \leqslant \varepsilon$.

There is a general relation between $T(\varepsilon)$ and ε.

PROPOSITION 1: *For each irrational real number* α, $T(\varepsilon) \geqslant 1/2\varepsilon$.

Proof: Let, for each integer Q, $\Omega(Q, \varepsilon)$ be the union of the $Q + 1$ 'intervals' of $\mathbf{R}/\mathbf{Z}$ defined by $\|q\alpha - x\| \leqslant \varepsilon, 0 \leqslant q \leqslant Q$ and ν be Lebesgue measure on $\mathbf{R}/\mathbf{Z}$ normalized by $\nu(\mathbf{R}/\mathbf{Z}) = 1$. We have $\nu(\Omega(Q, \varepsilon)) \leqslant (Q + 1)\,\nu(\Omega(0, \varepsilon)) = (2Q + 2)\,\varepsilon$. If $\Omega(Q, \varepsilon) = \mathbf{R}/\mathbf{Z}$ then $Q + 1 \geqslant 1/2\varepsilon$. If $Q + 1 = 1/2\varepsilon$ while $\Omega(Q, \varepsilon) = \mathbf{R}/\mathbf{Z}$, then $\nu(\Omega(Q, \varepsilon)) = (Q + 1)\,\nu(\Omega(0, \varepsilon))$. This implies that the intervals defining $\Omega(Q, \varepsilon)$ do not have any intersection of positive measure while their union is $\mathbf{R}/\mathbf{Z}$. This is only possible if α is rational.

This inequality leads to the following definition:

DEFINITION 2: *Let* α *be an irrational real number and* $h(j)$, *for each integer* j, *be the residue modulo* 1 *of* $j\alpha$. *We say that* $h(\mathbf{Z})$ *fills* $\mathbf{R}/\mathbf{Z}$ *rapidly if there is a positive number* c *such that, for each positive* ε, $T(\varepsilon) \leqslant c/\varepsilon$.

In other words, for each positive ε, $0 < \varepsilon < 1$, and each real number x the system $|q\alpha + p - x| \leqslant \varepsilon$, $0 \leqslant q \leqslant c/\varepsilon$ can be solved.

3.2. Badly approximable real numbers

We now connect the preceding notion with the quality of rational approximations of α.

THEOREM V: *Let* α *be an irrational real number and* h *the homomorphism from* $\mathbf{Z}$ *to* $\mathbf{R}/\mathbf{Z}$ *such that: for each* j *in* $\mathbf{Z}$, $h(j)$ *is the residue modulo* 1 *of* $j\alpha$. *Then the following three properties of* α *are equivalent:*

(a) $h(\mathbf{Z})$ *fills* $\mathbf{R}/\mathbf{Z}$ *rapidly;*

(b) *there is a positive* c *such that, for each integer* $q \geqslant 1$, $\|q\alpha\| \geqslant c/q$;

(c) *if* $\alpha = [a_0, a_1, \ldots, a_n \ldots]$ *is the expansion of* α *as a continued fraction, then the sequence of* a_n *is bounded.*

Before proving theorem *V*, some remarks should be made. If a real number α has one of the equivalent properties (a), (b) or (c) we shall say that α is a 'badly approximable' real number. As an example $\sqrt{2}$ is 'badly approximable' and each quadratic real number satisfies (b) or (c); in this case, the expansion into a continued fraction is, after some time, a periodic one. But property (c) shows that there are, apart from quadratic numbers, many badly approximable real numbers since the set of numbers defined by (c) has the same cardinality as **R**.

The proof of the equivalence between (b) and (c) is classical but uses the theory of continued fractions; for these two reasons we refer to [12].

We shall now prove that (a) implies (b).

If (b) is not fulfilled, α has good rational approximations and there is a sequence γ_n of real numbers tending to zero and a sequence q_n of positive integers, tending to infinity such that $|\alpha - p_n/q_n| \leqslant \gamma_n/q_n^2$; let m_n, $n \geqslant 1$, be a sequence of positive integers such that $m_n \to +\infty$ while $m_n\gamma_n \to 0$ $(n \to +\infty)$ and put $N_n = m_n q_n$, $\varepsilon_n = m_n\gamma_n$. For each positive ε, consider, with the notation of Proposition 1, the subset $\Omega(N_n, \varepsilon)$ of $\mathbf{R}/\mathbf{Z}$. Since, for $1 \leqslant j \leqslant m_n$, $\|jq_n\alpha\| \leqslant \varepsilon_n/q_n$, we get $\Omega(N_n, \varepsilon) \subset \Omega(q_n, \varepsilon + \varepsilon_n/q_n)$. Then $\Omega(N_n, \varepsilon) = \mathbf{R}/\mathbf{Z}$ implies $\Omega(q_n, \varepsilon + \varepsilon_n/q_n) = \mathbf{R}/\mathbf{Z}$ and, by Proposition 1, $2q_n(\varepsilon + \varepsilon_n/q_n) \geqslant 1$. If $\varepsilon = 1/3q_n$, for suitably large n $\Omega(N_m, \varepsilon)$ is not the whole $\mathbf{R}/\mathbf{Z}$, while $N_n\varepsilon$ tends to infinity with n; $h(\mathbf{Z})$ does not fill $\mathbf{R}/\mathbf{Z}$ rapidly.

Now it must be shown that (b) implies (a). Let $S(\varepsilon)$ be the compact subset of $\mathbf{R}^2$ defined by $|x + \alpha y| \leqslant \varepsilon$ and $|y| \leqslant c/6\varepsilon$ (c is defined by condition (b)). Then two distinct points of $S(\varepsilon)$ are never congruent modulo $\mathbf{Z}^2$ and $\mu S(\varepsilon) = c/6$. Let N be an integer defined by $cN \geqslant 6$ and K be the compact subset of $\mathbf{R}^2$ defined by $|x + \alpha y| \leqslant N\varepsilon$ and $|y| \leqslant cN/6\varepsilon$. Then Theorem IV shows that, for each $\mathbf{x}$ in $\mathbf{R}^2$, $K + \mathbf{x}$ intersects $\mathbf{Z}^2$. Changing ε into $N^{-1}\varepsilon$, this means that, for each real x and y and each positive ε, the Diophantine inequalities $|p + \alpha q - x| \leqslant \varepsilon$ and $|q - y| \leqslant cN^2/6\varepsilon$ can be solved.

THEOREM VI: *Let* α *and* β *be two distinct real numbers. The three following conditions are equivalent:*

(a) *there is a positive constant a such that for each* (p, q) *in* $\mathbf{Z}^2$ *distinct from* $(0, 0)$, $|q\alpha + p|\,|q\beta + p| \geqslant a$;

(b) *there is a constant b such that, if the product of the two positive real numbers* ε *and* η *exceeds b, then the Diophantine inequalities* $|q\alpha + p - x|$

$\leqslant \varepsilon$, $|q\beta + p - y| \leqslant \eta$ can be solved for each pair of real numbers x and y;

(c) *α and β are badly approximable.*

Proof: Let $c < a/4$ and, for each positive ε, let K_ε be the compact subset of $\mathbf{R}^2$ defined by $|y\alpha + x| \leqslant \varepsilon$, $|y\beta + x| \leqslant c\varepsilon^{-1}$. Then (a) implies that two distinct points of K_ε are never congruent modulo $\mathbf{Z}^2$. The Lebesgue measure of K_ε is $4c/(|\alpha - \beta|)$ and if an integer N is defined by $4Nc \geqslant |\alpha - \beta|$, Theorem V gives (b) with $b = N^2c$.

(b) $\Rightarrow$ (c). We shall prove that α is badly approximable. The same argument will be valid for β. If we apply (b) to $y = 0$ and $\varepsilon \leqslant 1$, the two inequalities $|q\alpha + p| \leqslant 1$ and $|q\beta + p| \leqslant k/\varepsilon$ imply, for a constant d, $|q| \leqslant d/\varepsilon$. For each ε in $(0, 1)$ and each real x, the inequalities $|q\alpha + p - x| \leqslant \varepsilon$, $|q| \leqslant d\varepsilon^{-1}$ can be solved; α is badly approximable.

(c) $\Rightarrow$ (a): Since α and β are distinct irrational real numbers, there is a positive constant γ such that $|q\alpha + p| \leqslant 1$ implies $|q\beta + p| \geqslant \gamma\, |q|$. If α is badly approximable and $|q\alpha + p| \leqslant 1$, (a) is a consequence of $|q|\, |q\alpha + p| \geqslant \delta > 0$. If $|q\beta + p| \leqslant 1$, we interchange the role played by α and β and if $|q\alpha + p| \geqslant 1$ and $|q\beta + p| \geqslant 1$, their product is greater than one. We have proved (a).

It is very surprising that quadratic numbers are the only naturally occurring examples of badly approximable real numbers, and hence it is all the more interesting that the set of badly approximable real numbers can be obtained constructively: by the continued fraction method. For the details we refer to [12].

3.3. *The Diophantine system $|q\alpha_j - \beta_j - p_j| \leqslant \varepsilon$, $1 \leqslant j \leqslant n$*

A natural generalization of the last problem is the following one: let $\omega_1, \ldots, \omega_n$ be n real numbers linearly independent over $\mathbf{Q}$ and $h: \mathbf{R} \to (\mathbf{R}/\mathbf{Z})^n$ be the continuous homomorphism defined by $h(t) = (\varrho\,(\omega_1 t), \ldots, \varrho\,(\omega_n t))$; it should be recalled that, for a real number t, $\varrho(t)$ is the residue modulo 1 of t; $\varrho(t)$ belongs to $\mathbf{R}/\mathbf{Z}$. Let H be the subgroup $h(\mathbf{R})$ of $(\mathbf{R}/\mathbf{Z})^n$. Then by Kronecker's theorem (Appendix A), H is dense in $(\mathbf{R}/\mathbf{Z})^n$ and $|t|$ is a natural estimate for the complexity of an element $h(t)$ of H. The distance between two elements $\varrho(\mathbf{x}) = (\varrho\,(x_1), \ldots, \varrho(x_n))$ and $\varrho(\mathbf{y}) = (\varrho(y_1), \ldots, \varrho(y_n))$ of $(\mathbf{R}/\mathbf{Z})^n$ is denoted $\|\varrho(x) - \varrho(y)\|$ and is defined by $\sup_{1 \leqslant j \leqslant n} \|x_j - y_j\|$ ($\mathbf{x} \in \mathbf{R}^n$; $x_1, \ldots, x_n \in \mathbf{R}$, $y \in \mathbf{R}^n$, $y_1, \ldots, y_n \in \mathbf{R}$ and for each real t, $\|t\|$ is the distance from t to the nearest integer).

Our problem of Diophantine approximation will be to study how $h(\mathbf{R})$ fills $(\mathbf{R}/\mathbf{Z})^n$.

Let, for each ε in $(0, \frac{1}{2})$, $V(\varepsilon)$ be the neighborhood of 0 in $(\mathbf{R}/\mathbf{Z})^n$ defined by $\|x_1\| \leqslant \varepsilon, \ldots, \|x_n\| \leqslant \varepsilon$ and $M(\varepsilon)$ be the set of all real t such that $h(t) \in V(\varepsilon)$. If $\omega_1 < \omega_2 < \cdots < \omega_n$, $t \in M(\varepsilon)$, $t' \in M(\varepsilon)$ and $|t - t'| \leqslant 1/2\omega_n$, then $|t - t'| \leqslant 2\varepsilon/\omega_n$ (if, for a real x, $|x| \leqslant \frac{1}{2}$, then $|x| = \|x\|$). This remark leads to the following definition:

DEFINITION 3: *With the preceding terminology, let $\tau(\varepsilon)$ be the first return-time into $V(\varepsilon)$: $\tau(\varepsilon)$ is the infimum of $t > 1/2\omega_n$ such that $h(t) \in V(\varepsilon)$. Let $t_1 = \tau(\varepsilon), t_2, \ldots, t_k, \ldots$ be the following return-times into $V(\varepsilon)$: t_{k+1} is the infimum of $t > t_k + 1/2\omega_n$, such that $h(t) \in V(\varepsilon)$. We set* $T(\varepsilon) = \sup_{k \geqslant 1} (t_{k+1} - t_k)$.

LEMMA 3: *If we accept for $T(\varepsilon)$ an error not exceeding $2\varepsilon/\omega_n$, $T(\varepsilon)$ can be defined by one of the following three equivalent properties:*

(a) *$T(\varepsilon)$ is the infimum of positive l such that each interval of positive real numbers of length l contains a t such that* $\sup_{1 \leqslant j \leqslant n} \|t\omega_j\| \leqslant \varepsilon$;

(b) *$T(\varepsilon)$ is the infimum of positive l such that each interval of real numbers of length l has the preceding property;*

(c) *$T(\varepsilon)$ is the infimum of positive l such that, for each finite set of real numbers $x_1, \ldots, x_n$ the Diophantine inequalities $0 \leqslant t \leqslant l$, $\|t\omega_j - x_j\| \leqslant \varepsilon$ $(1 \leqslant j \leqslant n)$ can be solved.*

Lemma 3 will be proved, in a more general context, in Section 5.2.

To present our fundamental definition, the following lemma will be needed:

LEMMA 4: *Let ε and T be two positive real numbers. Assume that, for each $\mathbf{r}$ in $(\mathbf{R}/\mathbf{Z})^n$, one can find a real number t in $[0, T]$ such that $\|h(t) - \mathbf{r}\| \leqslant \varepsilon$. Then $T \geqslant c\varepsilon^{1-n}$, where c depends only on $\omega_1, \ldots, \omega_n$.*

To prove Lemma 4, some terminology is needed. For each $\mathbf{x} = (x_1, \ldots, x_n)$ in $\mathbf{R}^n$, we put $\varrho(\mathbf{x}) = (\varrho(x_1), \ldots, \varrho(x_n))$, where $\varrho(x_1)$ is the residue modulo 1 of the real number x_1; $\varrho(x_1)$ belongs to $\mathbf{R}/\mathbf{Z}$. Let f and g be two positive functions defined on $]0, +\infty[$; we write $f \asymp g$ if there are two constants a and b such that, for each ε in $]0, +\infty[$, $af(\varepsilon) \leqslant g(\varepsilon) \leqslant bf(\varepsilon)$.

Let μ be Lebesgue measure on $\mathbf{R}^n$, ν Lebesgue measure on $\mathbf{R}^n/\mathbf{Z}^n$, and $C(T, \varepsilon)$ the set of all $(x_1, \ldots, x_n)$ in $\mathbf{R}^n$, such that $x_j = \omega_j t + r_j$, $0 \leqslant t \leqslant T$, $|r_j| \leqslant \varepsilon$.

The hypothesis of Lemma 4 implies that $\varrho(C(T, \varepsilon)) = \mathbf{R}^n/\mathbf{Z}^n$; using

Lemma 1, we get $1 = \nu(\varrho(C(T, \varepsilon)) \leqslant \mu C(T, \varepsilon) \asymp T\varepsilon^{n-1}$ $(T > 1, 0 < \varepsilon < 1)$. This ends the proof of Lemma 4.

We are led to the following definition, where $\omega_1, \dots, \omega_n$ and h are defined as above.

DEFINITION 4: *We say that $h(\mathbf{R})$ fills $(\mathbf{R}/\mathbf{Z})^n$ rapidly if there is a constant C (depending on h and n) such that, for each positive T and each $\mathbf{r}$ in $(\mathbf{R}/\mathbf{Z})^n$, the distance from $\mathbf{r}$ to $h([0, T])$ does not exceed $CT^{-(1/n-1)}$.*

The following result is a generalization of Theorem V; rougly speaking, Theorem VI asserts that $h(\mathbf{R})$ fills $\mathbf{T}^n$ rapidly if and only if $\omega_1, \dots, \omega_n$ are strongly independent over $\mathbf{Z}$ or have bad simultaneous approximations by rational numbers.

THEOREM VII: *Let $\omega_1, \dots, \omega_n$ be n real numbers linearly independent over $\mathbf{Q}$ and $h : \mathbf{R} \to (\mathbf{R}/\mathbf{Z})^n$ be the map defined by: $h(t)$ is the sequence of the residues modulo 1 of $\omega_1 t, \dots, \omega_n t$. The following three assertions are equivalent:*

(a) *$h(\mathbf{R})$ fills $(\mathbf{R}/\mathbf{Z})^n$ rapidly;*

(b) *there is a $\gamma > 0$ such that for each sequence $p_1, \dots, p_n$ of rational integers $|p_1\omega_1 + \cdots + p_n\omega_n| \geqslant \gamma(\sup_{1 \leqslant j \leqslant n} |p_j|)^{-n+1}$;*

(c) *there is a $\delta > 0$ such that for each real number $t \geqslant 1$, $\sup_{1 \leqslant j \leqslant n} \|t\omega_j\| \geqslant \delta t^{-(1/n-1)}$.*

Before this theorem can be proved, some remarks should be made. Condition (c) is equivalent to the following one: there is a positive a such that, for each $T \geqslant 1$ the distance between $h(0)$ and $h([1, T])$ exceeds $aT^{-(1/n-1)}$. In fact, by Dirichlet's theorem, here exists a sequence T_k tending to infinity in such a way that the distance between $h(0)$ and $h(T_k)$ is less than $T_k^{-(1/n-1)}$. The 'trajectory' reapproaches zero and, by Theorem VII, if the trajectory never reapproaches zero too closely, it fills $(\mathbf{R}/\mathbf{Z})^n$ rapidly.

To prove Theorem VII we replace (using arguments of Section 3.3) (c) with

(d) *for each $T \geqslant 1$ the distance from $h(0)$ to $h([1, T])$ exceeds $aT^{-(1/n-1)}$.*

(a) $\Rightarrow$ (d): Let us assume that (a) is true and (d) is not. Then there is a sequence ε_k of real numbers tending to zero and a sequence t_k of real numbers tending to infinity such that $\|h(0) - h(t_k)\| \leqslant \varepsilon_k t_k^{-(1/n-1)}$. Let m_k be a sequence of integers tending to infinity in such a way that $m_k\varepsilon_k = \alpha_k \to 0$; put $s_k = m_k t_k$. Let $\Omega(T, \alpha)$ be the set of all $\mathbf{r}$ in $(\mathbf{R}/\mathbf{Z})^n$ whose distance to $h([0, T])$ does not exceed α. Then $\Omega(s_k, \alpha) \subset \Omega(t_k, \alpha + m_k\varepsilon_k t_k^{-(1/n-1)})$. If $\Omega(s_k, \alpha) = (\mathbf{R}/\mathbf{Z})^n$, then, by Lemma 4,

$\alpha + m_k \varepsilon_k t_k^{-(1/n-1)} \geqslant a t_k^{-(1/n-1)}$ and

$$\alpha \geqslant (a - m_k \varepsilon_k)\, t_k^{-(1/n-1)} \sim a t_k^{-(1/n-1)}.$$

If (a) were true, α would be smaller than $c s_k^{-(1/n-1)}$. Since $s_k t_k^{-1}$ tends to infinity this is not the case.

(d) $\Rightarrow$ (a): Let L be the subset of $\mathbf{R}^n$ defined by $x_1 = \omega_1 t + r_1, \ldots, x_n = \omega_n t + r_n$, $1 \leqslant t \leqslant T$, $|r_j| \leqslant a/(6T^{1/n-1})$. Then, by (d), two distinct points of L cannot be congruent modulo $\mathbf{Z}^n$. The measure of L is greater than some fixed $b > 0$ as soon as $T \geqslant 2$. Let N be an integer greater than b^{-1}. Then, by Theorem IV, for each $\mathbf{x}$ in $\mathbf{R}^n$, $NL + \mathbf{x}$ intersects $\mathbf{Z}^n$. This proves (a).

As may easily be seen, the equivalence between (a) and (c) is based on a kind of duality in Diophantine approximations. Let us state a general result. To do so, some terminology is needed.

Let $\mathbf{x} \cdot \mathbf{y}$ be the bilinear form $x_1 y_1 + \cdots + x_n y_n$, on $\mathbf{R}^n \times \mathbf{R}^n$ and, for each compact subset K of $\mathbf{R}^n$, let K^* be the set of all $\mathbf{y}$ in R^n such that $\sup_{\mathbf{x} \in K} |\mathbf{x} \cdot \mathbf{y}| \leqslant 1$.

Lemma 5: *If* $K + \mathbf{Z}^n = \mathbf{R}^n$ *and* $0 < a < \frac{1}{2}$, *then* $(aK^*) \cap \mathbf{Z}^n = \{0\}$.

Proof: Let $\mathbf{y} \in (aK^*) \cap \mathbf{Z}^n$. If $\mathbf{y} \neq 0$, there is an $\mathbf{x}'$ in $\mathbf{R}^n$ defined by $\mathbf{y} \cdot \mathbf{x}' = \frac{1}{2}$. Since $K + \mathbf{Z}^n = \mathbf{R}^n$, we can find an $\mathbf{X}'$ in $\mathbf{Z}^n$ and $\mathbf{x}$ in K, such that $\mathbf{x}' = \mathbf{X}' + \mathbf{x}$. Then $\mathbf{y} \cdot \mathbf{x}' = \mathbf{y} \cdot \mathbf{X}' + \mathbf{y} \cdot \mathbf{x}$; $|\mathbf{y} \cdot \mathbf{x}| \leqslant a$ and $\mathbf{y} \cdot \mathbf{x}'$ would belong to $\mathbf{Z} + [-a, a]$. But $\mathbf{y} \cdot \mathbf{x}' = \frac{1}{2}$ does not belong to $\mathbf{Z} + [-a, a]$.

From now on c and d will denote fixed constants (not always the same) depending only on $\omega_1, \ldots, \omega_n$; whereas ε and η will be variable positive real numbers. Let D be the line of $\mathbf{R}^n$ defined by the parametric equations $x_1 = \omega_1 t, \ldots, x_n = \omega_n t$ for real t and let $C(c, \varepsilon)$ be the compact subset of $\mathbf{R}^n$ defined by the two following conditions: $(x_1, \ldots, x_n) \in C(c, \varepsilon)$ if $|\omega_1 x_1 + \cdots + \omega_n x_n| \leqslant \varepsilon$ and if the distance from $(x_1, \ldots, x_n)$ to this line D does not exceed $c\varepsilon^{-(1/n)}$. If $\eta = d/\varepsilon$, where d is small enough and $K = C(c, \varepsilon)$, then $C(d, \eta)$ is contained in K^*.

Using Lemma 5 and this remark, we shall finish the proof of the equivalence between (a) and (c). Hypothesis (a) means that, for a big enough ε, $C(c, \varepsilon) + \mathbf{Z}^n = \mathbf{R}^n$. Then, for small enough d and η, $C(d, \eta) \cap \mathbf{Z}^n = \{0\}$ and thus (b) is proved.

We now start from (b). For small enough d and η, two distinct points of $C(d, \eta)$ are never congruent modulo $\mathbf{Z}^n$; on the other hand, the Lebesgue measure of $C(d, \eta)$ exceeds a constant d_1 which is independent of η. Using Theorem IV, we can find an integer N, independent of η, such that $NC(d, \eta) + \mathbf{Z}^n = \mathbf{R}^n$. We now apply Lemma 5 to get (c).

DEFINITION 5: *Let $\omega_1, \ldots, \omega_n$ be a sequence of n real numbers. We say that this sequence is strongly independent over the integers if and only if one of the equivalent properties of Theorem VII is satisfied.*

If $\alpha_1, \ldots, \alpha_n$ is a sequence of n real numbers such that $1, \alpha_1, \ldots, \alpha_n$ is a strongly independent sequence, then the following equivalent properties are true:

(a) there is a constant C such that, for each positive ε and each sequence $\beta_1, \ldots, \beta_n$ of real numbers, the Diophantine inequalities $|q\alpha_1 + p_1 - \beta_1| \leqslant \varepsilon, \ldots, |q\alpha_n + p_n - \beta_n| \leqslant \varepsilon$, $|q| \leqslant C\varepsilon^{-n}$ can be solved.

(b) there is a constant c such that for each sequence $q, p_1, \ldots, p_n$ of rational integers we have $|p_1\alpha_1 + \cdots + p_n\alpha_n + q| \geqslant c(\sup_{1 \leqslant j \leqslant n} |p_j|)^{-n}$ (unless $p_1 = \cdots = p_n = 0$).

(c) there is a constant d such that, for each $q \geqslant 1$, $\sup_{1 \leqslant j \leqslant n} \|q\alpha_j\| \geqslant dq^{-(1/n)}$.

3.4. *Example. The algebraic case*

Let $\mathfrak{K}$ be a real algebraic number field, $n = d^0 [\mathfrak{K}; \mathbf{Q}]$, $\sigma_1, \ldots, \sigma_n$ be the n $\mathbf{Q}$-isomorphisms of $\mathfrak{K}$ into $\mathbf{C}$ indexed in such a way that $\sigma_1(\mathfrak{K})$ is contained in $\mathbf{R}$. Let $\omega_1, \ldots, \omega_n$ be any basis of $\mathfrak{K}$ over $\mathbf{Q}$.

PROPOSITION 2: *The sequence of real numbers $\sigma_1(\omega_1), \ldots, \sigma_1(\omega_n)$ is strongly independent over* $\mathbf{Z}$.

Since independence is not affected by changing $\sigma_1(\omega_1), \ldots, \sigma_1(\omega_n)$ into $\lambda\sigma_1(\omega_1), \ldots, \lambda\sigma_1(\omega_n)$ for a real λ, it may be assumed that $\omega_1, \ldots, \omega_n$ are algebraic integers of the field $\mathfrak{K}$. We shall prove condition (b). Let $a > 0$ be chosen in such a way that, for each algebraic integer $\lambda = p_1\omega_1 + \cdots + p_n\omega_n$ in $\mathfrak{K}$, $|\sigma_2(\lambda)| \leqslant a \sup_{1 \leqslant j \leqslant n} |p_j|, \ldots, |\sigma_n(\lambda)| \leqslant a \sup_{1 \leqslant j \leqslant n} |p_j|$.

We cannot have $|\sigma_1(\lambda)| < a^{-(1/n-1)}(\sup_{1 \leqslant j \leqslant n} |p_j|)^{-(1/n-1)}$ — we except the case when $p_1 = \cdots = p_n = 0$ — for then the rational integer $\sigma_1(\lambda) \cdots \sigma_n(\lambda) = m$ would have an absolute value less than 1 and this implies $m = 0$ and $\lambda = 0$.

The following result of Cassels [34] shows that, apart from algebraic numbers, there are other interesting examples of strongly independent sequences.

THEOREM VIII: *Let $\lambda_1, \ldots, \lambda_n$ be a sequence of real numbers. Assume that $1, \lambda_1, \ldots, \lambda_n$ are linearly independent over* **Q**. *The set of real t such that $\omega_1 = 1$, $\omega_2 = \lambda_1 + t, \ldots, \omega_{n+1} = \lambda_n + t$ is a strongly independent sequence has the same Cardinality as* **R**.

4. Diophantine approximations and prediction of the size of trigonometric sums

4.1. *What is prediction?*

Let Λ be a finite set of real numbers. We study the following trigonometric sums:

$$P(t) = \sum_{\lambda \in \Lambda} a_\lambda \exp 2\pi i \lambda t, \tag{1.10}$$

where a_λ are arbitrary complex numbers.

Our fundamental problem is the following: if we know a set Λ of frequencies and the values taken by P on a large interval $[-T, T]$, can we predict what P will be on the whole real line? What is the extent of the error?

To be more precise, we shall try to specify the connection between a positive T and an ε in $(0, 1)$ such that, for each P defined by (1.10),

$$\sup_{|t| \leqslant T} |P(t)| \geqslant (1 - \varepsilon) \sup_{t \in \mathbf{R}} |P(t)|. \tag{1.11}$$

This condition means that the relative error term on the size of P is less than ε. If we replace Λ by $\Lambda + \alpha$, $P(t)$ is replaced by $\exp 2\pi i t\alpha P(t)$ and (1.11) remains valid. Our problem is translation-invariant on Λ and in all the following we assume that 0 belongs to Λ.

The result stated below expresses that we cannot get excellent predictions. The quality of the prediction depends, when 0 belongs to Λ, on the dimension of the vector space over **Q** generated by Λ.

PROPOSITION 3: *Let n be the cardinality of Λ and m, $1 \leqslant m \leqslant n$, be the dimension of the vector space over* **Q** *generated by Λ. There is a constant A, depending only on Λ, such that* (1.11) *implies $T \geqslant A\varepsilon^{-(m-1)/2}$.*

The proof of Proposition 3 depends on the following lemma:

LEMMA 6: *Let $z_1, \ldots, z_n$ be n complex numbers of absolute value* 1. *Then for each ε in* $(0, 1)$, *the inequality $|z_1 + \cdots + z_n| \geqslant n(1 - \varepsilon)$ implies $|z_j - z_k| \leqslant 2\sqrt{2n\varepsilon}$ for $1 \leqslant j, k \leqslant n$.*

Proof: Multiplying the z_j's by a suitable complex number α of absolute value 1, we may assume that $(z_1 + \cdots + z_n)/n = s$ is real and positive.

Let us write $z_j = x_j + iy_j$. We then have $x_1 + \cdots + x_n = ns$ and since, by hypothesis, $s \geqslant 1 - \varepsilon$, we get $\sum_1^n (1 - x_j) \leqslant n\varepsilon$, which implies $1 - x_j \leqslant n\varepsilon$, $|z_j - 1| \leqslant \sqrt{2n\varepsilon}$ and $|z_j - z_k| \leqslant 2\sqrt{2n\varepsilon}$.

Let us return to Proposition 3. We define $P(t) = \sum_{\lambda \in \Lambda} \exp 2\pi i\lambda t$ and, for each real number s, let P_s be defined by: $P_s(t) = P(t - s)$. Then (1.11) and Lemma 1 imply that we can find in each interval of length $2T$ a real t such that $|\exp 2\pi i\lambda t - 1| \leqslant 2\sqrt{2n\varepsilon}$. We can write Λ as a sequence $\lambda_1, \ldots, \lambda_n$, such that $\lambda_1, \ldots, \lambda_m$ are linearly independent over $\mathbf{Q}$. Then, for $1 \leqslant j \leqslant m$, there is in each interval of length T a t such that $|\exp 2\pi i\lambda_j t - 1| \leqslant 2\sqrt{2n\varepsilon}$ $(1 \leqslant j \leqslant m)$. By Lemma 4 of Section 3.3 we get $T \geqslant A\varepsilon^{-(m-1)/2}$.

Proposition 3 leads to the following definition:

DEFINITION 6: *Let Λ be a finite set of real numbers. Assume that* 0 *belongs to Λ and let m be the dimension of the vector space over* $\mathbf{Q}$ *generated by Λ. We say that we can rapidly predict the size of trigonometric sums* (1.10) *if there is a constant $C > 0$ such that, for each ε in* $]0, 1[$, *the supremum of* $|P(t)|$ *for* $|t| \leqslant C\varepsilon^{-(m-1)/2}$ *is not less than* $(1 - \varepsilon) \sup_{\mathbf{R}} |P|$.

4.2. The main theorem concerning prediction

From now on, Λ will denote a finite set of real numbers containing 0; m will be the dimension of the vector space over $\mathbf{Q}$ generated by Λ; we can write Λ as a sequence $\{\lambda_1, \ldots, \lambda_n\}$ in such a way that $\lambda_1, \ldots, \lambda_m$ is a basis of this vector space.

As in Section 3, we define, for each real number x, the residue $\varrho(x)$ of x modulo 1; $\varrho(x)$ belongs to $\mathbf{R}/\mathbf{Z}$. Let $h : \mathbf{R} \to (\mathbf{R}/\mathbf{Z})^m$ be the homomorphism defined by: $h(t)$ is the sequence of the m residues $\varrho(\lambda_1 t), \ldots, \varrho(\lambda_m t)$.

With this terminology, we have the following theorem:

THEOREM IX: *The following two assertions are equivalent:*

(a) *we can rapidly predict the size of trigonometric sums whose frequencies belong to Λ;*

(b) $\lambda_1, \ldots, \lambda_m$ *are strongly independent over* $\mathbf{Z}$.

More precisely, we have that, for constants $a, b, c \ldots$ depending only on Λ but not on the real number ε in $]0, 1[$ the following four assertions are equivalent:

(a) *for each* $P(t) = \sum_{\lambda \in \Lambda} a_\lambda \exp 2\pi i\lambda t$, $\sup_{|t| \leqslant a\varepsilon^{-(m-1)/2}} |P(t)| \geqslant (1 - \varepsilon) \sup_{t \in \mathbf{R}} |P(t)|$;

(b) $\sup_{1 \leqslant j \leqslant m} \|t\lambda_j\| \geqslant b|t|^{-(1/m-1)}$ for each $t \geqslant 1$.

(c) *for a constant c, for every positive* ε *and every sequence* $x_1, \ldots, x_m$ *of real numbers, the Diophantine inequalities*

$$|t| \leqslant c\varepsilon^{-m+1}, \ \|t\lambda_1 - x_1\| \leqslant \varepsilon, \ldots, \|t\lambda_m - x_m\| \leqslant \varepsilon$$

can be solved;

(d) *for a constant d and every sequence* $p_1, \ldots, p_n$ *of rational integers we have either* $p_1\lambda_1 + \cdots + p_n\lambda_n = 0$ *or* $|p_1\lambda_1 + \cdots + p_n\lambda_n| \geqslant d(\sup_{1 \leqslant j \leqslant n} |p_j|)^{-m+1}$.

4.3. *Proof of the main theorem*

The proof of Theorem IX depends on the following lemmas:

LEMMA 7: *Let* $\lambda_1, \ldots, \lambda_n$ *be n real numbers and* $\omega_1, \ldots, \omega_m$ *be a basis for the* **Z**-*module generated by the* λ_j's, $1 \leqslant j \leqslant n$. *Let us assume that each interval of length 2T contains a t such that* $\|t\lambda_j\| \leqslant \varepsilon$ *for* $1 \leqslant j \leqslant n$. *Then there is a constant* $a > 0$ *such that, for each sequence* $x_1, \ldots, x_m$ *of real numbers, the Diophantine inequalities*

$$|t| \leqslant T, \ |t\omega_j - x_j - p_j| \leqslant a\varepsilon \quad \text{for} \quad 1 \leqslant j \leqslant m \tag{1.12}$$

can be solved.

Proof: We have, for rational integers $b_{j,k}$, $\omega_j = \sum_1^n b_{j,k}\lambda_k$. Hence $\|t\lambda_k\| \leqslant \varepsilon$ for $1 \leqslant k \leqslant n$ implies $\|t\omega_j\| \leqslant a\varepsilon$ for $1 \leqslant j \leqslant m$; each interval of length $2T$ contains such a real t. At this point, we use Lemma 3 in Section 3.3.

Remark 1: If $\lambda_1, \ldots, \lambda_m$ is a basis for the vector space over **Q** generated by Λ, then there is an integer p such that the inequalities $\|t\lambda_j\| \leqslant \varepsilon$ for $1 \leqslant j \leqslant m$ imply $\|pt\lambda_j\| \leqslant b\varepsilon$ for $1 \leqslant j \leqslant n$.

To prove this remark, let us write the λ_j's for which $m + 1 \leqslant j \leqslant n$ as linear combinations with rational coefficients of $\lambda_1, \ldots, \lambda_m$. Then, for a suitable integer p and integers $a_{j,k}$, $p\lambda_j = \sum_1^m a_{j,k}\lambda_k$ $(m + 1 \leqslant j \leqslant n)$. This relation implies our remark.

Remark 2: If each interval of length $2T$ contains a solution t of the Diophantine system $\|t\lambda_j\| \leqslant \varepsilon$, $1 \leqslant j \leqslant m$, then there is an integer p and a constant a sucht that, for each sequence $x_1, \ldots, x_m$ of real numbers, the system $|t| \leqslant pT$, $\|t\omega_j - x_j\| \leqslant a\varepsilon$ $(1 \leqslant j \leqslant m)$ can be solved.

In the remainder of the proof the following notation will be used: let E be a set and f a complex valued function defined on E. Then we put $\|f\|_\infty = \sup_{x \in E} |f(x)|$.

LEMMA 8: *Let* $f: \mathbf{R}^m \to \mathbf{C}$ *be an indefinitely differentiable function. Assume that* f *is bounded and that the partial derivatives* $\partial^2 f/(\partial u_j \partial u_k)$, $1 \leqslant j, k \leqslant m$ *are also bounded. Let* M *be* $\sup_{1 \leqslant j,k \leqslant m} \|\partial^2 f/(\partial u_j \partial u_k)\|_\infty$. *For* $\mathbf{u} = (u_1, \ldots, u_m) \in \mathbf{R}^m$ *we put* $|\mathbf{u}|^2 = u_1^2 + \cdots + u_m^2$. *Then, if we can find a* $\mathbf{v}$ *in* $\mathbf{R}^m$ *such that* $|f(\mathbf{v})| = \|f\|_\infty$, *we have, for every* $\mathbf{u}$ *in* $\mathbf{R}^m$,

$$|f(\mathbf{u})| \geqslant \|f\|_\infty - |\mathbf{u} - \mathbf{v}|^2 \, mM/2. \tag{1.12}$$

Proof: The Taylor formula gives, for a linear form l over $\mathbf{R}^m$ and a $\mathbf{u}'$ in $\mathbf{R}^m$, $f(\mathbf{u}) = f(\mathbf{v}) + l(\mathbf{u} - \mathbf{v}) + \frac{1}{2}\sum_{j,k}(u_j - v_j)(u_k - v_k)\, \partial^2 f/(\partial u_j \partial u_k)(\mathbf{u}')$. This implies $|f(\mathbf{u})| \geqslant |f(\mathbf{v}) + l(\mathbf{u} - \mathbf{v})| - M/2 \sum_{j,k} |u_j - v_j|\,|u_k - v_k| \geqslant |f(\mathbf{v}) + l(\mathbf{u} - \mathbf{v})| - mM/2\, |\mathbf{u} - \mathbf{v}|^2$.

For each $\mathbf{u}$ in $\mathbf{R}^m$ $|f(\mathbf{v}) + l(\mathbf{u} - \mathbf{v})| \geqslant |f(\mathbf{v})|$. Otherwise we would find a positive ε and points $\mathbf{v}$ arbitrarily near to $\mathbf{u}$, such that $|f(\mathbf{v}) + l(\mathbf{u} - \mathbf{v})| \geqslant |f(\mathbf{v})| + \varepsilon\, |\mathbf{u} - \mathbf{v}|$; $f(\mathbf{u})$ would be greater than $f(\mathbf{v})$. This ends the proof of Lemma 8.

LEMMA 9: *Let* $\Lambda = \{\lambda_1, \ldots, \lambda_n\}$ *be a finite subset of* $\mathbf{R}$ *and* P *be a finite trigonometric sum whose frequencies belong to* Λ.

Assume that each interval of length $2T$ *contains a solution* t *of the system* $\|t\lambda_j\| \leqslant \alpha$ *for* $1 \leqslant j \leqslant n$.

Then, for a constant a, *depending only on* Λ, *we have*

$$\sup_{|t| \leqslant T} |P(t)| \geqslant (1 - a\alpha^2)\, \|P\|_\infty.$$

Proof: Let $\omega_1, \ldots, \omega_m$ be a basis of the $\mathbf{Z}$-module generated by Λ. Then $\lambda_j = \sum_1^m p_{j,k}\omega_k$, where $p_{j,k} \in \mathbf{Z}$. For complex numbers $z_1, \ldots, z_m$ of absolute value 1, we put $Q_j(z_1, \ldots, z_m) = \prod_1^m z_k^{p_{j,k}}$; then

$$P(t) = \sum_1^n a_j Q_j(\exp 2\pi i\omega_1 t, \ldots, \exp 2\pi i\omega_m t).$$

Let $f: \mathbf{R}^m \to \mathbf{C}$ be the $\mathbf{Z}^m$-periodic function defined by $f(u_1, \ldots, u_m) = \sum_1^n a_j Q_j(\exp 2\pi i u_1, \ldots, \exp 2\pi i u_m)$; f is indefinitely differentiable and there exists $\mathbf{v}$ in $\mathbf{R}^m$ such that $|f(\mathbf{v})| = \|f\|_\infty$.

By Lemma 7, we can find a real number t and integers p_j, $1 \leqslant j \leqslant m$, such that $|t| \leqslant T$ and $|t\omega_j - v_j - p_j| \leqslant a\alpha$. Let $\mathbf{u} \in \mathbf{R}^m$ be defined by $u_j = t\omega_j - p_j$. By Lemma 8,

$$|f(\mathbf{u})| \geqslant \|f\|_\infty - b\alpha^2 M. \tag{1.13}$$

Let $\mathscr{E}$ be the vector space over $\mathbf{C}$ of all trigonometric sums $\sum_{\lambda \in \Lambda} a_\lambda \exp 2\pi i\lambda t$. The dimension of $\mathscr{E}$ is n and all norms on $\mathscr{E}$ are equivalent. In particular there exists a constant c, such that, for each P in $\mathscr{E}$, $\sum_{\lambda \in \Lambda} |a_\lambda|$

$\leqslant c\|P\|_\infty$. This implies immediately that for a constant C depending only on Λ, $M \leqslant C\|P\|_\infty$.

Kronecker's theorem (Appendix B) shows that, for each $\mathbf{v}$ in $\mathbf{R}^m$, we can find a sequence t_p, $p \geqslant 1$, of real numbers such that $\exp 2\pi i\omega_j t_p \to \exp 2\pi i v_j$ for $1 \leqslant j \leqslant m$ as p tends to infinity. It follows that $\|f\|_\infty = \|P\|_\infty$. Then (1.13) can be rewritten

$$|P(t)| \geqslant \|P\|_\infty - bC\alpha^2 \|P\|_\infty, \quad |t| \leqslant T.$$

Now let us return to Theorem IX.

(a) $\Rightarrow$ (b): This follows immediately by the same proof as that used for Proposition 3.

(b) $\Rightarrow$ (a): We already know (Theorem VI) that (b) implies the following: for a constant A, each interval of length $A\varepsilon^{-m+1}$ contains a solution t of the Diophantine inequalities $\|t\lambda_j\| \leqslant \varepsilon\, (1 \leqslant j \leqslant m)$. By the first remark following Lemma 7, each interval of length $A'\varepsilon^{-m+1}$ contains a solution t of the enlarged system $\|t\lambda_j\| \leqslant \varepsilon$ $(1 \leqslant j \leqslant n)$. Now using Lemma 9, we get (a).

(b) $\Leftrightarrow$ (c): This is an immediate consequence of Theorem VII.

5. Kronecker's theorem and problems with a large number of equations

5.1. We shall state the usual forms of Kronecker's theorem and another equivalent formulation (a proof of Kronecker's theorem is given in Appendix B).

5.1.1. Let $\lambda_1, \ldots, \lambda_n$ be n real numbers linearly independent over $\mathbf{Q}$. Then for each positive ε and each sequence $x_1, \ldots, x_n$ of real numbers, the Diophantine inequalities $\|t\lambda_j - x_j\| \leqslant \varepsilon$, $1 \leqslant j \leqslant n$ have a solution. If $1, \lambda_1, \ldots, \lambda_n$ are linearly independent over $\mathbf{Q}$, we can find an integral solution q of the Diophantine inequalities

$$\|q\lambda_j - x_j\| \leqslant \varepsilon\, (1 \leqslant j \leqslant n). \tag{1.14}$$

Proof: (1.14) follows from 5.1.1 applied to the sequence $x_0 = 1$, $x_1, \ldots, x_n$ of real numbers; then, for a suitable integer q, $|t - q| \leqslant \varepsilon$ and replacing t in 5.1.1 by q, we get (1.14).

Now if, $\lambda_1, \ldots, \lambda_n$ are not $\mathbf{Q}$-linearly independent, we are led to another formulation.

5.1.2. Let $\lambda_1, \ldots, \lambda_n$ and $x_1, \ldots, x_n$ be $2n$ real numbers. The following two conditions are equivalent:

(a) for each positive ε, there exists a real number t such that $\|t\lambda_j - x_j\| \leqslant \varepsilon$ for $1 \leqslant j \leqslant n$;

(b) for each sequence $p_1, \ldots, p_n$ of rational integers, $p_1\lambda_1 + \cdots + p_n\lambda_n = 0$ implies $p_1x_1 + \cdots + p_nx_n \equiv 0 \pmod 1$.

Condition (b) can be stated in the following form: let H be the subgroup of $\mathbf{R}$ generated by Λ; then there is a homomorphism χ from H to $\mathbf{R}/\mathbf{Z}$, such that $\chi(\lambda_j) \equiv x_j \pmod 1$ for $1 \leqslant j \leqslant n$.

Proof of 5.1.3:

(a) $\Rightarrow$ (b): If $p_1\lambda_1 + \cdots + p_n\lambda_n = 0$ then, for every real t, $p_1t\lambda_1 + \cdots + p_nt\lambda_n = 0$ and using (a) we get, for each positive ε, $\|p_1x_1 + \cdots + p_nx_n\| \leqslant n\varepsilon$, which implies (b).

(b) $\Rightarrow$ (a): Using the structure theorems about submodules of free modules on principal rings, we can find m real numbers $\omega_1, \ldots, \omega_m$ linearly independent over $\mathbf{Q}$, such that $H = \mathbf{Z}\omega_1 + \cdots + \mathbf{Z}\omega_m$. By 5.1.1, for each positive ε, there exists a real number t such that $\|t\omega_j - \chi(\omega_j)\| \leqslant \varepsilon$ for $1 \leqslant j \leqslant m$. Since $\lambda_j \in \mathbf{Z}\omega_1 + \cdots + \mathbf{Z}\omega_m$, $1 \leqslant j \leqslant n$, there is a constant C, independent of ε, such that $\|t\lambda_j - \chi(\lambda_j)\| \leqslant C\varepsilon$ for $1 \leqslant j \leqslant n$.

5.2. *A function $T(\varepsilon_1, \lambda, \ldots, \lambda_n)$*

DEFINITION 7: *The following two properties of a positive number T are equivalent*

(a) *for each sequence $x_1, \ldots, x_n$ of real numbers satisfying condition* (b) *of* 5.1.3, *there is a solution t of $\|t\lambda_j - x_j\| \leqslant \varepsilon$, $|t| \leqslant T$;*

(b) *each interval of real numbers of length $2T$ contains a real number t such that $\|t\lambda_j\| \leqslant \varepsilon$ for $1 \leqslant j \leqslant n$.*

Proof:

(a) $\Rightarrow$ (b): Apply (a) to $x_j = sx_j$, where s is the center of the given interval of length $2T$.

(b) $\Rightarrow$ (a): Using 5.1.3. (b) let, $s_p, p \geqslant 1$, be a sequence of real numbers such that $\|s_p\lambda_j - x_j\| \to 0\ (p \to +\infty)$. The interval of center s_p and length $2T$ contains a real number x_p such that $\|x_p\lambda_j\| \leqslant \varepsilon$, $1 \leqslant j \leqslant n$. Let $t_p = s_p - x_p$; then $|t_p| \leqslant T$ and we can extract from the t_p's a subsequence tending to t. We have $|t| \leqslant T$ and $\|t\lambda_j - x_j\| \leqslant \varepsilon$, $1 \leqslant j \leqslant n$.

5.3. *Open problems*

Our aim is to find such a function $T(\varepsilon, \lambda_1, \ldots, \lambda_n) = T$. If we are only interested in the dependence of T on ε for fixed $\lambda_1, \ldots, \lambda_n$, the problem can be reduced to that studied in Section 3, but instead of the λ_j's, we use a basis $\omega_1, \ldots, \omega_m$ of the $\mathbf{Z}$-module generated by Λ.

If we are also interested in the dependence on n and on $\lambda_1, \ldots, \lambda_n$, it is quite a different matter. The reduction to a problem dealing with $\omega_1, \ldots, \omega_m$ drastically alters the estimates. We need new methods which will be introduced in Section 5.4. Nevertheless many problems remain unsolved. Let us give an example. Let $\mathfrak{K}$ be a real algebraic number field, n be the degree of $\mathfrak{K}$ over $\mathbf{Q}$ and $\sigma_1, \ldots, \sigma_n$ the n $\mathbf{Q}$-isomorphisms from $\mathfrak{K}$ into $\mathbf{C}$ indexed in such a way that σ_1 is real ($\sigma_1(\mathfrak{K}) \subset \mathbf{R}$). For each positive ε, let $T(\varepsilon, \mathfrak{K}) = T$ be the least positive real number satisfying the following condition: for every real number t there exists an algebraic integer λ in $\mathfrak{K}$, such that $|t - \sigma_1(\lambda)| \leqslant T$ while $|\sigma_2(\lambda)| \leqslant \varepsilon^{-1}, \ldots, |\sigma_n(\lambda)| \leqslant \varepsilon^{-1}$. We do not hope to get an explicit value for T but only to find a function $f(\varepsilon, \mathfrak{K})$ and two absolute constants a and b (independent of ε and $\mathfrak{K}$) such that $af(\varepsilon, \mathfrak{K}) \leqslant T \leqslant bf(\varepsilon, \mathfrak{K})$. Using the preceding methods, all we can show is that, for constants A and B depending on $\mathfrak{K}$, we have $A\varepsilon^{n-1} \leqslant T \leqslant B\varepsilon^{n-1}$.

5.4. *Linear Diophantine problems with an infinite number of equations*

Let λ_k, $k \geqslant 1$, be a given sequence of real numbers. We study the sequences β_k, $k \geqslant 1$, of real numbers such that, for each positive ε, there exists a solution t of the system $\|t\lambda_k - \beta_k\| \leqslant \varepsilon$. As in Section 5, we find a necessary condition on the β_k's: for every integer n and rational integers $p_1, \ldots, p_n$, we have the following implication:

$$\sum_1^n p_k\lambda_k = 0 \Rightarrow \sum_1^n p_k\beta_k \equiv 0 \pmod 1). \tag{1.15}$$

DEFINITION 8: *Let Λ be a set of real numbers, H the subgroup of* $\mathbf{R}$ *generated by Λ and χ a map from Λ to* $\mathbf{R}/\mathbf{Z}$. *We write* $\chi \in \operatorname{Hom}(\Lambda, \mathbf{R}/\mathbf{Z})$ if χ *is the restriction to Λ of a homomorphism from H to* $\mathbf{R}/\mathbf{Z}$.

With this definition, (1.15) can be written: $\chi \in \operatorname{Hom}(\Lambda, \mathbf{R}/\mathbf{Z})$, where χ is defined by $\chi(\lambda_k) \equiv \beta_k \pmod 1$.

DEFINITION 9: *The set Λ of the λ_k's is harmonious if for each positive ε and each sequence β_k, $k \geqslant 1$, of real numbers satisfying the necessary condition* (1.15) *there exists a real number t such that, for each $k \geqslant 1$,* $\|t\lambda_k - \beta_k\| \leqslant \varepsilon$.

In other words, for each $\chi \in \operatorname{Hom}(\Lambda, \mathbf{R}/\mathbf{Z})$ and each positive ε, we can solve the infinite sequence of Diophantine inequalities $\|t\lambda_k - \chi(\lambda_k)\| \leqslant \varepsilon$, $k \geqslant 1$.

Roughly speaking we try to guess whether a given sequence β_k, $k \geqslant 1$, real numbers, can be, modulo 1, a sequence $t\lambda_k$, $k \geqslant 1$, for a suitable t. A set Λ is harmonious if each sequence β_k, $k \geqslant 1$, of real numbers which is not excluded for evident algebraic reasons is, with an arbitrarily small error (not depending on k), such a sequence of residues for a suitable t.

THEOREM X: *Let $(\lambda_k)_{k \geqslant 1}$ be a sequence of real numbers and Λ be the set of λ_k. The following four conditions are equivalent:*

(a) *Λ is harmonious;*

(b) *for each positive ε there is a positive $T(\varepsilon)$ such that, for each $\chi \in$ Hom$(\Lambda, \mathbf{R}/\mathbf{Z})$, the Diophantine inequalities $\|t\lambda_k - \chi(\lambda_k)\| \leqslant \varepsilon, k \geqslant 1, |t| \leqslant T(\varepsilon)$ can be solved;*

(c) *for each positive ε, there is a positive $T(\varepsilon)$, such that there exists, in each interval of real numbers of length $2T(\varepsilon)$, a solution t of the Diophantine inequalities $\|t\lambda_k\| \leqslant \varepsilon$, $k \geqslant 1$;*

(d) *for each positive ε, there is a positive $T(\varepsilon)$, such that for each integer $n \geqslant 1$, every interval of real numbers of length $2T(\varepsilon)$ contains a solution t of the finite sequence of Diophantine inequalities $\|t\lambda_k\| \leqslant \varepsilon$, $1 \leqslant k \leqslant n$.*

Condition (d) of Theorem X shows that, using the terminology of Section 5.3, we are interested into getting bounds independent of n for the function $T(\varepsilon, \lambda_1, \ldots, \lambda_n)$.

Proof of Theorem X: The equivalences between (b), (c) and (d) are easy consequences of Section 5.2. For example, let us show why (c) implies (b). For each $n \geqslant 1$, there exists a t_n such that $|t_n| \leqslant T(\varepsilon)$ and $\|t_n\lambda_k - \chi(\lambda_k)\| \leqslant \varepsilon$ $(1 \leqslant k \leqslant n)$; this was proved in Section 5.2. Now we extract from the t_n's a subsequence tending to t; t is a solution of the inequalities $\|t\lambda_k - \chi(\lambda_k)\| \leqslant \varepsilon$ and $|t| \leqslant T(\varepsilon)$.

We have finally to prove that (a) implies (b). Let $\mathfrak{G}$ be the subgroup of $(\mathbf{R}/\mathbf{Z})^{\mathbf{N}}$ consisting of all sequences $(\beta_k)_{k \geqslant 1}$ of real numbers modulo 1 satisfying (1.15). If convergence in $(\mathbf{R}/\mathbf{Z})^{\mathbf{N}}$ means that each coordinate converges to a limit, the well-known Cantor diagonal process shows that $(\mathbf{R}/\mathbf{Z})^{\mathbf{N}}$ is a compact group and $\mathfrak{G}$ is a closed subgroup of $(\mathbf{R}/\mathbf{Z})^{N}$. Hence $\mathfrak{G}$ is a compact metrisable group and Baire's theorem is valid for $\mathfrak{G}$.

For each real number x, $\varrho(x)$ is the residue modulo 1 of x; $\varrho(x)$ belongs to $\mathbf{R}/\mathbf{Z}$. Let $h: \mathbf{R} \to \mathfrak{G}$ be the continuous map defined by $h(t) = (\varrho(\lambda_1 t), \ldots, \varrho(\lambda_n t), \ldots)$. By Kronecker's theorem (Section 5.1) $h(\mathbf{R})$ is a dense subgroup of $\mathfrak{G}$. Let, for each positive ε, $v(\varepsilon)$ be the closed subset of $\mathfrak{G}$ defined by $\|\beta_k\| \leqslant \varepsilon$ for all $k \geqslant 1$.

With this terminology, Λ is harmonious if for each positive ε, $h(\mathbf{R}) + v(\varepsilon)$

$= \mathfrak{G}$. For each integer m, let K_m be the compact subset of $\mathfrak{G}$ defined by $K_m = h([-m, m)] + v(\varepsilon)$. If Λ is harmonious, the union of all K_m, $m \geqslant 1$, is $\mathfrak{G}$. Baire's theorem shows that, for a suitable m, K_m contains an open subset Ω of $\mathfrak{G}$. Since $h(\mathbf{R})$ is dense in $\mathfrak{G}$, $h(\mathbf{R}) + \Omega = \mathfrak{G}$; hence $\bigcup_{t \in \mathbf{R}} h(t) + \Omega = \mathfrak{G}$ and again using the compactness of $\mathfrak{G}$, we can find a finite subset F of $\mathbf{R}$ such that $h(F) + \Omega$ is $\mathfrak{G}$. Define an integer l by $F + [-m, m] \subset [-l, l]$. Then $\mathfrak{G} = h(F) + \Omega \subset h(F) + K_m = h(F) + h([-m, m]) + v(\varepsilon) \subset h([-l, l]) + v(\varepsilon) = K_l$; this is (b).

6. Pisot and Salem numbers

6.1. Definition of Pisot numbers

Throughout this section, the following definitions and terminology will be used. The ring of polynomials in one variable with integral coefficients will be denoted by $\mathbf{Z}[X]$. We say that P in $\mathbf{Z}[X]$ is distinguished if the leading coefficient of P is one; this means that $P(X) = X^n + c_1X^{n-1} + \cdots + c_n$, where $c_j \in \mathbf{Z}$, $1 \leqslant j \leqslant n$. We say that $P(X) = c_0X^n + c_1X^{n-1} + \cdots + c_n$ is reduced if $c_0, c_1, \ldots, c_n$ are relatively prime and if $c_0 \geqslant 1$.

A real algebraic integer is any real root of a distinguished P in $\mathbf{Z}[X]$. An algebraic number is any root (real or complex) of a P in $\mathbf{Z}[X]$. The minimal polynomial of an algebraic number θ is the reduced element Q of $\mathbf{Z}[X]$ of the least degree such that $Q(\theta) = 0$. Gauss has proved that the minimal polynomial of an algebraic integer is distinguished.

If θ is an algebraic number, the roots of its minimal polynomial are simple. The degree of an algebraic number is the degree of its minimal polynomial. In what follows, where θ is an algebraic number, n will be the degree of θ, $P(X)$ the minimal polynomial of θ and $\theta_2, \ldots, \theta_n$ the $n - 1$ other distinct (real or complex) roots of $P(X)$. We write $\theta_2, \ldots, \theta_n$ for conjugates of θ different from θ.

DEFINITION 10: *A Pisot number is a real algebraic integer θ greater than* 1 *whose other conjugates $\theta_2, \ldots, \theta_n$ satisfy the inequalities* $|\theta_2| < 1, \ldots, |\theta_n| < 1$.

The set of all Pisot numbers is denoted by S.

6.2. Characterization of Pisot numbers among algebraic numbers

THEOREM XI: *Let $\theta > 1$ be a real algebraic number. The following two properties of θ are equivalent:*

(a) *θ is a Pisot number;*

(b) *there is a $\lambda > 0$ such that $\|\lambda\theta^k\| \to 0$ $(k \to +\infty)$.*

(We recall that, for each real x, $\|x\|$ is the distance from x to the nearest integer.)

It is not known whether transcendental numbers θ can have property (b).

(a) $\Rightarrow$ (b): If θ is an algebraic integer, so is θ^k for each $k \geqslant 0$. The conjugates of θ^k are $\theta_2^k, \ldots, \theta_n^k$ and $\theta^k + \theta_2^k + \cdots + \theta_n^k$ is a rational integer for each $k \geqslant 0$. Since $|\theta_2| < 1, \ldots, |\theta_n| < 1$ we get, for a r in $]0, 1[$, $\|\theta^k\| = 0\,(r^k)$, which is an improvement on (b).

(b) $\Rightarrow$ (a): The proof depends on a well-known theorem due to Fatou. We say that a formal series $\sum_{k\geqslant 0} c_k X^k$ with real coefficients represents a rational fraction if there exist two polynomials $P(X)$ and $Q(X)$ with real coefficients such that $P(X) = Q(X) \sum_{k\geqslant 0} c_k X^k$. We write $\sum_{k\geqslant 0} c_k X^k = P(X)/Q(X)$.

Fatou's lemma: *Assume that the formal series $\sum_{k\geqslant 0} c_k X^k$ has the following two properties:*

(a) *$c_k \in \mathbf{Z}$ for all $k \geqslant 0$;*

(b) *$\sum_{k\geqslant 0} c_k X^k$ represents a rational fraction.*

Then there exist two elements $A(X)$ and $B(X)$ in $\mathbf{Z}[X]$, such that

(a) *$A(X)$ and $B(X)$ are relatively prime;*

(b) *$B(0) = 1$;*

(c) *$\sum_{k\geqslant 0} c_k X^k = A(X)/B(X)$.*

The proof of this result is given in Section 6.3. We can now conclude the proof of Theorem XI. If θ is a real algebraic number let n be the degree of θ and $P(X)$ the minimal polynomial of θ; $P(X) = c_0 X^n + \cdots + c_n$.

By hypothesis $\lambda\theta^k = p_k + r_k$, where r_k tends to 0. Put $s_k = c_0 p_{k+n} + \cdots + c_n p_k = -c_0 r_{k+n} - \cdots - c_n r_k$; s_k is a rational integer which tends to zero and if k is large enough, $s_k = 0$. For $k \geqslant k_0$, r_k satisfies the recurrence relations $c_0 r_{k+n} + \cdots + c_n r_k = 0$; since the roots of P are simple, we have $r_k = \lambda_1\theta^k + \lambda_2\theta_2^k + \cdots + \lambda_n\theta_n^k$. We now use the following lemma (proved in Section 6.3).

LEMMA 10: *Let $u_1, \ldots, u_n$ and $v_1, \ldots, v_n$ be $2n$ complex numbers. Assume that $f(k) = u_1 v_1^k + \cdots + u_n v_n^k$ tends to 0 $(k \to +\infty)$. Then, for each j in*

$[1, n]$, $|v_j| \geqslant 1$ *implies* $u_j = 0$. *If* $f(k)$ *is bounded for* $k \geqslant 1$, *then* $|v_j| > 1$ *implies* $u_j = 0$.

Returning to r_k we have $r_k = \sum_2^n \lambda_j \theta_j^k$, where $\lambda_j = 0$ if $|\theta_j| \geqslant 1$. The formal series $\sum_{k \geqslant k_0} p_k X^k$ represents the rational fraction

$$\frac{\lambda \theta^{k_0} X^{k_0}}{1 - \theta X} - \sum_2^n \frac{\lambda_j \theta_j^{k_0} X^{k_0}}{1 - \theta_j X}.$$

Omitting terms for which $\lambda_j = 0$, the last fraction can be rewritten $U(X)/V(X)$, where $V(X) = (1 - \theta X) \prod'_{2 \leqslant j \leqslant n} (1 - \theta_j X)$. The dash means that the product is restricted to those j for which $\lambda_j \neq 0$; U and V are then relatively prime.

At this point we use Fatou's lemma. We get $U(X)/V(X) = \sum_{k \geqslant k_0} p_k X^k = A(X)/B(X)$; $A(X)$ and $B(X)$ are relatively prime, as are $U(X)$ and $V(X)$, and $B(0) = V(0) = 1$. This implies $B(X) = V(X)$. Let m be the degree of $B(X)$ and $C(X) = X^m B(1/X)$ be the reciprocal polynomial of $B(X)$; then $C(X)$ is distinguished, θ is a root of $C(X)$ and the θ_j for which $\lambda_j \neq 0$ are the other roots of $C(X)$. Hence θ is a Pisot number.

As a complement to Theorem XI it is interesting to study the set of λ such that $\|\lambda \theta^k\|$ tends to 0. We need the following definition:

DEFINITION 11: *Let K be an algebraic number field and $\omega_1, \ldots, \omega_n$ a basis of K over* $\mathbf{Q}$. *Then a dual basis $\omega_1^*, \ldots, \omega_n^*$ of K over $\mathbf{Q}$ is defined by the properties* $\mathrm{Tr}(\omega_i \omega_j^*) = 1$ *if* $i = j$, $\mathrm{Tr}(\omega_i \omega_j^*) = 0$ *if* $i \neq j$.

We recall that, for x in K, $\mathrm{Tr}(x) = \sigma_1(x) + \cdots + \sigma_n(x)$ is a rational number. (We refer to [3] for the proof.)

PROPOSITION 4: *Let θ be a Pisot number, K be the field of θ, $K = \mathbf{Q}(\theta)$ and $\omega_1, \ldots, \omega_n$ be the dual basis of* $1, \theta, \ldots, \theta^{n-1}$. *Then* $\|\lambda \theta^k\| \to 0$ $(k \to +\infty)$ *if and only if for an integer l, $l \geqslant 0$, and n rational integers $p_1, \ldots, p_n$, we have* $\lambda = \theta^{-l}(p_1 \omega_1 + \cdots + p_n \omega_n)$.

This last condition is sufficient. Since we are interested in the products $\lambda \theta^k$ for $k \geqslant l$, we may assume that $l = 0$ and $\lambda = p_1 \omega_1 + \cdots + p_n \omega_n$. Let $\lambda_2, \ldots, \lambda_n$ be the conjugates of λ; then, for each $k \geqslant 0$, $\lambda \theta^k + \lambda_2 \theta_2^k + \cdots + \lambda_n \theta_n^k$ is a rational integer (for $0 \leqslant k \leqslant n - 1$ it is a consequence of the definition of $\omega_1, \ldots, \omega_n$ and for $k \geqslant n$ we use the recurrence relations $\theta^k + c_1 \theta^{k-1} + \cdots + c_n \theta^{k-n} = 0$).

The latter condition is necessary. Returning to the proof of Theorem XI, we see that for $k \geqslant k_0$, $\lambda \theta^k - \lambda_2 \theta_2^k - \cdots - \lambda_n \theta_n^k \in \mathbf{Z}$. Putting $\mu = \lambda \theta^{k_0}$, $\mu_2 = -\lambda_2 \theta_2^{k_0}, \ldots, \mu_n = -\lambda_n \theta_n^{k_0}$, we get for each $k \geqslant 0$

$$\mu \theta^k + \mu_2 \theta_2^k + \cdots + \mu_n \theta_n^k = q_k \in \mathbf{Z}. \tag{1.16}$$

Let A be the matrix whose rows are θ^k, $0 \leqslant k \leqslant n-1$ and θ_j^k, $0 \leqslant k \leqslant n-1$ and B the matrix whose rows, indexed by j, are $\sigma_k(\omega_j)$, $1 \leqslant k \leqslant n$. Then $AB = I$ and (1.16) gives $\mu = q_0\omega_1 + \cdots + q_{n-1}\omega_n$.

6.3. *Proof of Fatou's lemma*

The proof uses the following two lemmas:

LEMMA 11: *Let* $\sum_{k\geqslant 0} c_k X^k = S(X)$ *be a formal series with rational coefficients. Assume that two polynomials* P *and* Q *can be found in* $\mathbf{C}[X]$ *such that* $S(X) = P(X)/Q(X)$. *Then, there are two irreducible polynomials* $A(X)$ *and* $B(X)$ *in* $\mathbf{Z}[X]$ *such that* $B(0) \neq 0$ *and* $S(X) = A(X)/B(X)$.

Proof of Lemma 11: We can assume that P and Q are relatively prime. Let us write $Q(X) = q_0 + q_1X + \cdots + q_nX^n$; two elements Q_1 and Q_2 of $\mathbf{C}[X]$ are called equivalent if, for some complex $\lambda \neq 0$, $Q_2 = \lambda Q_1$. Up to an equivalence, Q is determined by the relations $c_kq_n + c_{k+1}q_{n-1} + \cdots + c_{k+n}q_0 = 0$, $k \geqslant k_0$. Since these linear equations have rational coefficients and a well-defined solution, this solution can be chosen to be rational. Hence P and Q are relatively prime elements of $\mathbf{Z}[X]$. If $Q(0) = 0$ while $P(0) \neq 0$, $P(X)/Q(X)$ cannot be a formal series with positive powers.

Using Lemma 11, we can find two relatively prime polynomials P and Q with integral coefficients such that $S(X) = P(X)/Q(X)$ and $Q(X) = q_0 \neq 0$. We may assume that the coefficients of P and Q do not have a common integral divisor. If $S(X)$ has integral coefficients, the coefficients of $Q(X)$ are relatively prime; otherwise since $P(X) = S(X)\,Q(X)$, a divisor d of these coefficients would be a divisor of all coefficients of $P(X)$.

By Bezout's theorem, we can find two polynomials U and V with integral coefficients such that $PU + QV = d$. Then $d/Q = \sum_{k\geqslant 0} d_kX^k$ is a formal series with integral coefficients. At this point we use the following lemma:

LEMMA 12: *If the* $n+1$ *integers* $q_0, \ldots, q_n$ *are relatively prime and if the series*

$$d/(q_0 + \cdots + q_nX^n) = \sum_{k\geqslant 0} d_kX^k \tag{1.17}$$

has integral coefficients, then $q_0 = \pm 1$.

It can be assumed that $d \geqslant 1$. If $|q_0| > 1$, we shall show that (1.17) implies a relation of the same type $d'/(q_0 + \cdots + q_nX^n) = \sum_{k\geqslant 0} d_k'X^k$ with $1 \leqslant d' < d$. If we repeat this descent argument d times, we get a contradiction. Assume that $|q_0| > 1$ and let p be a prime divisor of q_0; since $q_0, \ldots,$

q_n are relatively prime, there is an integer h defined by the following property: p divides $q_0, \ldots, q_h$ but not q_{h+1}. We have $d = (q_0 + \cdots + q_n X^n) \sum_{k \geqslant 0} d_k X^k$, whence

$$d_{h+1} q_0 + \cdots + d_1 q_h + d_0 q_{h+1} = 0 \tag{1.18}$$

$$d_{h+2} q_0 + \cdots + d_2 q_h + d_1 q_{h+1} + d_0 q_{h+2} = 0. \tag{1.19}$$

(1.18) implies that p divides d_0; (1.19) that p divides d_1 and so on; p also divides $d = q_0 d_0$. Writing $d = pd'$ and $d_k = pd'_k$ we get the 'prime relation'.

Proof of Lemma 10: Reordering $v_1, \ldots, v_n$ if necessary, we can define an integer $m \leqslant n$ by the condition $|v_1| = \cdots = |v_m| > |v_{m+1}| \geqslant \cdots \geqslant |v_n|$. Let $R = |v_1|$, $v_j = \mathrm{R}e^{i\omega_j}$, $1 \leqslant j \leqslant m$ and let $g(k) = u_1 e^{ik\omega_1} + \cdots + u_m e^{ik\omega_m}$; then $R^{-k} f(k) - g(k)$ tends to 0 ($k \to \infty$). If $R < 1$ there is nothing to prove. If $R \geqslant 1$ and $f(k)$ tends to 0, so does $g(k)$. But an easy computation gives $\lim\limits_{K \to +\infty} 1/K \sum_1^K |g(k)|^2 = |u_1|^2 + \cdots + |u_m|^2$; if $\lim\limits_{k \to +\infty} g(k) = 0$, this square mean is 0 and the lemma is proved.

6.4. *Characterization of Pisot numbers among real numbers*

Let θ be a real number greater than 1 and $\lambda > 0$. It is not known whether or not the assumption $\|\lambda\theta^k\| \to 0$ ($k \to +\infty$) implies that θ is a Pisot number. Nevertheless we have the following theorem due to Pisot.

THEOREM XII: *Let θ be a real number greater than* 1 *and $\lambda > 0$. Assume that $\sum_1^\infty \|\lambda\theta^k\|^2 < +\infty$. Then θ is a Pisot number and λ belongs to the field of θ.*

All that has to be proved is that θ is algebraic. Let us write $\lambda\theta^k = a_k + r_k$, where $a_k \in \mathbf{Z}$ and $|r_k| \leqslant \frac{1}{2}$. To prove that θ is algebraic, we seek an $n \geqslant 1$ and integers $p_0, \ldots, p_n$, such that $p_0 + p_1\theta + \cdots + p_n\theta^n = 0$. Let $\sigma_k = p_0 a_k + p_1 a_{k+1} + \cdots p_n a_{k+n}$; if there exist n and $p_0, \ldots, p_n$ such that, for $k \geqslant k_0$, $\sigma_k = 0$, then θ is an algebraic integer. In fact, $\lambda\theta^k (p_0 + p_1\theta + \cdots + p_n\theta^n) = \sigma_k + p_0 r_k + \cdots + p_n r_{k+n}$ and the first term tends to zero as k tends to infinity. Since $\theta > 1$, this implies $p_0 + p_1\theta + \cdots + p_n\theta^n = 0$.

It therefore remains to find such an n and such $p_0, \ldots, p_n$. Let $\eta_k = \theta r_k - r_{k+1}$ ($k \geqslant 0$). We have $\sigma_{k+1} - \theta\sigma_k = p_0\eta_k + \cdots + p_n\eta_{k+n}$. Our aim is now to solve all the Diophantine inequalities $|\sigma_0| \leqslant \frac{1}{2}$, $|\sigma_1 - \theta\sigma_0| \leqslant \frac{1}{2}, \ldots, |\sigma_{k+1} - \theta\sigma_k| \leqslant \frac{1}{2}, \ldots$ Since σ_0 is a rational integer, the first inequality will imply $\sigma_0 = 0$; substituting this value in the second one we get $\sigma_1 = 0$, and so on.

To solve such a system of Diophantine inequalities, we consider Minkowski's theorem. But here a technical difficulty arises: we need to compute the measure of the subset of $\mathbf{R}^{n+1}$ defined by $|x_0a_0 + \cdots + x_na_n| \leqslant \frac{1}{2}$, $|x_0\eta_0 + \cdots + x_n\eta_n| \leqslant \frac{1}{2}$, $|x_0\eta_1 + \cdots + x_n\eta_{n+1}| \leqslant \frac{1}{2}, \ldots$ This computation is not feasible and for this reason we are led to restrict our attention to the first $n + 1$ equations $|x_0a_0 + \cdots + x_na_n| \leqslant \frac{1}{2}$, $|x_0\eta_0 + \cdots + x_n\eta_n| \leqslant \frac{1}{2}, \ldots, |x_0\eta_{n-1} + \cdots + x_n\eta_{2n-1}| \leqslant \frac{1}{2}$. Let D_n be the determinant of these $n + 1$ linear forms. If $2^{n+1}D_n \leqslant 1$ a non trivial integral solution $p_0, \ldots, p_n$ of this finite system can be found and we obtain $\sigma_0 = 0, \sigma_1 = 0, \ldots, \sigma_n = 0$. New, although we appear to be a long way from our goal, the following lemma enables us to conclude the proof in a few lines:

LEMMA 13: *Let $a_k, k \geqslant 0$, be a sequence of rational integers. Assume that, for each sufficiently large value of n there exists a non-trivial integral solution of $a_0x_0 + \cdots + a_nx_n = 0, \ldots, a_nx_0 + \cdots + a_{2n}x_n = 0$; then there is an integer n and a non-trivial sequence of integers $y_0, \ldots, y_n$ such that for each $k \geqslant 0$, $a_ky_0 + \cdots + a_{k+n}y_n = 0$.*

This lemma will be proved in Section 6.5. To finish the proof of the theorem, we have therefore to prove that for sufficiently large n, $2^{n+1}D_n \leqslant 1$. In fact, if $\sum_0^\infty \eta^2 < +\infty$, $\lim_{n\to+\infty} 2^{n+1}D_n = 0$ as the following inequality of Hadamard shows.

LEMMA 14: *For each $X = (x_1, \ldots, x_n)$ in $\mathbf{R}^n$,* let $|X|^2 = x_1^2 + \cdots + x_n^2$. *Then the determinant of the n vectors $X_1, \ldots, X_n$ in $\mathbf{R}^n$ is bounded by $|X_1| \cdots |X_n|$.*

In our case we put $\varrho_k = (\eta_k^2 + \cdots + \eta_{k+n}^2)^{1/2}$ and, for a constant C depending only on θ and not on n, we get $|D_n| \leqslant C\theta^n\varrho_0, \ldots, \varrho_{n-1}$. Since $2\theta\varrho_k$ tends to 0, the infinite product $\prod_0^\infty (2\theta\varrho_k)$ is 0 and 2^nD_n tends to 0.

6.5. *Proof of the main lemma*

For each integer $m \geqslant 1$, let Δ_m be the determinant

$$\begin{vmatrix} a_0 & \cdots & a_m \\ a_m & \cdots & a_{2m} \end{vmatrix}.$$

The hypothesis implies that, if m is large enough $\Delta_m = 0$. We define n as the least integer m for which $\Delta_m = 0$. Then $\Delta_{n-1} \neq 0$. We define $(y_0, \ldots,$

y_n) to be the non-trivial integral solution $(x_0, \ldots, x_n)$. Let $\sigma_k = a_k x_0 + \cdots + a_{k+n} x_n$. We shall prove, by induction on k, that $\sigma_k = 0$.

We already know that $\sigma_0 = \cdots = \sigma_n = 0$. Writing Δ_{n+1} and combining each row with the n preceding rows and with coefficients $x_0, \ldots, x_k$, we get

$$\Delta_{n+1} = \begin{bmatrix} a_0 & \cdots a_{n-1} & 0 & 0 \\ a_1 & \cdots a_n & 0 & 0 \\ a_{n-1} & \cdots a_{2n-2} & 0 & 0 \\ \cdots & \cdots & \cdots & \cdots \\ a_n & \cdots a_{2n-1} & 0 & \sigma_{n+1} \\ a_{n+1} & \cdots a_{2n} & \sigma_{n+1} & \sigma_{n+2} \end{bmatrix} = -\Delta_{n-1}\sigma_{n+1}.$$

Since $\Delta_{n+1} = 0$, while $\Delta_{n-1} \neq 0$, we get $\sigma_{n+1} = 0$. Using this new fact we compute $\Delta_{n+2} = \Delta_{n-1}\sigma_{n+2}^3$ to get $\sigma_{n+2} = 0$, and so on.

6.6. *The set S of Pisot numbers is closed*

This remarkable fact was proved by Salem [17].

The proof depends on the following unexpected theorem:

THEOREM XIII: *For each Pisot number θ, there exist a real number λ, $1 \leqslant \lambda \leqslant \theta$ and a sequence c_k, $k \geqslant 0$, of rational integers such that $\sum_0^\infty (\lambda\theta^k - c_k)^2 \leqslant 0$.*

Before proving theorem XIII, let us see why it implies Salem's theorem. Let $\|t\|$ be the distance from t to the nearest integer. The conclusion of Theorem XIII can be written $\sum_0^\infty \|\lambda\theta^k\|^2 \leqslant 9$.

Let θ_m be a sequence of Pisot numbers and assume that $\theta_m \to \theta$ $(m \to +\infty)$. For each θ_m let λ_m be the corresponding real number given by Theorem XIII. Then the λ_m are bounded and there is a subsequence of the λ_m tending to $\lambda \geqslant 1$. By Fatou's inequality $\sum_0^\infty \|\lambda\theta^k\|^2 \leqslant 9$. Pisot's theorem implies that θ is a Pisot number.

Now to Theorem XIII! The proof depends on the following three observations:

6.6.1. Let D be the closed disc $|z| \leqslant 1$, g any holomorphic function on a neighbourhood of D, and $g(z) = \sum_0^\infty r_k z^k$ the Taylor expansion of g. Then, by Parseval's equation, $\sum_0^\infty |r_k|^2 = 1/2\pi \int_0^{2\pi} |g(e^{i\theta})|^2 \, d\theta$.

6.6.2. Let $P(z)$ be any polynomial in z with real coefficients. Then if $|z| = 1$, $|P(z)| = |P(1/z)|$.

6.6.3. Let θ be a Pisot number. Assume that a conjugate θ_j of θ is $1/\theta$. Then we can find an integer $p \geqslant 3$ such that $\theta^2 - p\theta + 1 = 0$.

To prove the third observation, write $\theta \prod_2^n \theta_k = \theta\theta_j \prod_{k \neq j}^{k \geqslant 2} \theta_k$. If θ is not a quadratic number and if $\theta\theta_j = 1$, this product has an absolute value less than 1. But, for any algebraic integer $\theta \prod_2^n \theta_k$ is a non-zero rational integer.

Now to the proof! We first deal with the case when θ is not a quadratic unit. Let $P(z) = z^n + a_1 z^{n-1} + \cdots + a_n$ the minimal polynomial of θ and $Q(z) = 1 + a_1 z + \cdots + a_n z^n$ the reciprocal polynomial. Then $f(z) = P(z)/Q(z)$ has the following properties:

(a) $f(z)$ is meromorphic in the complex plane;

(b) the poles of f are $1/\theta, 1/\theta_2, \ldots, 1/\theta_n$ (by 6.6.3);

(c) the Taylor expansion of f in $|z| < 1/\theta$ is $f(z) = \sum_0^\infty c_k z^k$, where $c_k \in \mathbf{Z}$.

Since $1/\theta$ is the unique pole z of f, such that $|z| \leqslant 1$, and since this pole is simple, there exists a complex number $\mu \neq 0$ such that $g(z) = f(z) - \mu/(1 - \theta z)$ is holomorphic in a neighbourhood of D.

To estimate μ, let us write

$$P(z) = (z - \theta)\, P_1(z), \quad Q(z) = (1 - \theta z)\, Q_1(z).$$

Then P_1 and Q_1 are two reciprocal polynomials of degree $n - 1$ with real coefficients. We have $\mu = (1/\theta - \theta)\, P_1\,(1/\theta)/Q_1\,(1/\theta)$; therefore μ is real. On $|z| = 1$, 6.6.2 gives $|P_1(z)/Q_1(z)| = 1$, and since P_1/Q_1 is holomorphic in a neighbourhood of D, the maximum modulus principle gives

$$\left| \frac{P_1\,(1/\theta)}{Q_1\,(1/\theta)} \right| \leqslant 1;$$

therefore $|\mu| \leqslant \theta - (1/\theta) \leqslant \theta$.

On $|z| = 1$ we have $|g(z)| \leqslant 1 + |\mu|/(\theta - 1)$; using the bound $|\mu| \leqslant \theta - (1/\theta)$, we get $|\mu| \leqslant 2\theta - 2$ and $|g(z)| \leqslant 3$.

The Taylor expansion of g in $|z| < 1/\theta$ is $\sum_0^\infty (c_k - \mu\theta^k)\, z^k$. Since g is holomorphic on a neighbourhood of D, this expansion is valid on a neighbourhood of D, and 6.6.1 implies $\sum_0^\infty (c_k - \mu\theta^k)^2 \leqslant 9$. We have $|\mu| \leqslant \theta$ and $\mu \neq 0$. If $\mu < 0$ we replace μ with $-\mu$ and c_k with $-c_k$. If $0 < \mu < 1$ let h be the first k such that $\mu\theta^k \geqslant 1$ and replace μ with $\mu\theta^h = \lambda$. We have $9 \geqslant \sum_0^\infty (c_{h+j} - \mu\theta^{h+j})^2 = \sum_0^\infty (d_j - \lambda\theta^j)^2$.

If θ is a quadratic unit such that $\theta^2 - p\theta + 1 = 0$ for some $p \geqslant 3$, then $\theta \geqslant (3 + \sqrt{5})/2$ and since $\theta^k + \theta^{-k}$ is a natural integer, we get

$$\sum_0^\infty \|\theta^k\|^2 \leqslant \sum_1^\infty \left(\frac{3 - \sqrt{5}}{2} \right)^{2k} \leqslant 9.$$

6.7. Construction of Pisot numbers in a given real algebraic number field

Let K be a real algebraic field, $\mathscr{A} \subset K$ be the ring of algebraic integers in K and $\omega_1, \ldots, \omega_n$ a basis of the free $\mathbf{Z}$-module $\mathscr{A}$. Let $\sigma_1, \ldots, \sigma_n$ be the n $\mathbf{Q}$-isomorphisms from K into $\mathbf{Z}$ indexed in such a way that $\sigma_1 = \mathrm{Id}$ (the identity map from K to $\mathbf{R}$). With this terminology we have the following result:

THEOREM XIV: *Each positive real number* $\lambda = p_1\omega_1 + \cdots + p_n\omega_n$ $(p_j \in \mathbf{Z})$ *such that* $|p_1\sigma_j(\omega_1) + \cdots + p_n\sigma_j(\omega_n)| < 1$ *for* $2 \leqslant j \leqslant n$ *is a Pisot number of degree n in K. Each such Pisot number can be written in this way.*

Proof: If $\lambda = p_1\omega_1 + \cdots + p_n\omega_n$, λ is an algebraic integer. Assume that $\lambda \neq 0$. Then $\lambda > 1$, for otherwise the product $\lambda\sigma_2(\lambda) \cdots \sigma_n(\lambda)$ would have an absolute value less than 1. Let m be the degree of λ. If $m < n$, then $n = mp$ and the complex numbers $\lambda = \sigma_1(\lambda), \sigma_2(\lambda), \ldots, \sigma_n(\lambda)$ are in fact m distinct complex numbers, each of them written p times. Hence $\sigma_1(\lambda) = \lambda$ is written p times; this is incompatible with $\lambda > 1$ and $|\sigma_j(\lambda) < 1$ for $2 \leqslant j \leqslant n$.

We have proved that the degree of λ is n. In this case $\sigma_2(\lambda), \ldots, \sigma_n(\lambda)$ are the conjugates of λ different from λ and λ is a Pisot number. The converse implication is obvious.

COROLLARY: *The set of Pisot numbers of degree n lying in K is closed with respect to multiplication.*

Proof: Since the σ_j, $2 \leqslant j \leqslant n$, are multiplicative, the set of positive λ in Λ such that $|\sigma_j(\lambda)| < 1$ for $2 \leqslant j \leqslant n$ is closed with respect to multiplication.

Let r be the number of j such that $\sigma_j(K) \subset \mathbf{R}$ and put $n = r + 2s$, $1 \leqslant r \leqslant n$, $s \geqslant 0$. Assume that σ_j and σ_{j+s} are pairwise-conjugate for $r + 1 \leqslant j \leqslant r + s$. Let d be the determinant whose elements are $\sigma_j(\omega_k)$, $1 \leqslant j \leqslant r$, $\mathscr{R}e\sigma_j(\omega_k)$ and $\mathrm{Im}\,\sigma_j(\omega_k)$ for $r + 1 \leqslant j \leqslant r + s$, $1 \leqslant k \leqslant n$. Theorem VIII of Chapter V gives in our case:

THEOREM XV: *With the preceding terminology, each interval of real numbers of length T contains* $2^{r-1}\pi^s|d|^{-1}T + o(T)$ *Pisot numbers of degree n belonging to the real algebraic number field K; the little o is uniform with respect to the position of the interval I.*

6.8. *The Salem class T*

DEFINITION 12: *Let θ be a real number greater than 1. We say that θ is a Salem number and we write $\theta \in T$ if θ is an algebraic integer, of degree say n, whose other conjugates $\theta_2, \ldots, \theta_n$ satisfy the inequalities $|\theta_j| \leqslant 1$, $2 \leqslant j \leqslant n$, with equality for at least one j.*

If the latter condition was not satisfied, a Salem number would be a Pisot number.

Let $P(X) = X^n + a_1 X^{n-1} + \cdots + a_n \in \mathbf{Z}[X]$ be the minimal polynomial of θ and, reordering if necessary, let θ_2 be a root of P such that $|\theta_2| = 1$.

The following lemma will be useful in constructing Salem numbers:

LEMMA 15: *Let P in $\mathbf{Z}[X]$ be irreducible and let n be the degree of P. If one root ζ of $P(z) = 0$ satisfies $|\zeta| = 1$, then $P(X) = X^n P(1/X)$.*

Proof: Let $Q(X) = X^n P(1/X)$. Then $Q(X) \in \mathbf{Z}[X]$ and $Q(\zeta) = 0$ (since $P(\bar{\zeta}) = \overline{P(\zeta)} = 0$). Hence P and Q are not relatively prime and since P is irreducible over $\mathbf{Z}[X]$, this implies $P = cQ$, where c belongs to $\mathbf{Q}$. Using the fact that P is distinguished, we get $c = 1$ ($c = -1$ would lead to P being divisible by $X - 1$).

This lemma at once gives the following result:

THEOREM XVI: *Let θ be a Salem number. The degree n of θ is an even integer, say $n = 2m$, and the conjugates of θ are $1/\theta$ and $n - 2$ pairwise-conjugate complex numbers of absolute value 1, $(\zeta_1, \bar{\zeta}_1)$, $(\zeta_2, \bar{\zeta}_2)$, …, $(\zeta_{m-1}, \bar{\zeta}_{m-1})$.*

Construction of Salem numbers. We say that an algebraic number field K is totally real if the n $\mathbf{Q}$-isomorphisms from K to $\mathbf{C}$ are real.

THEOREM XVII: *Let K be a totally real field, $\sigma_1, \ldots, \sigma_n$ the n $\mathbf{Q}$-isomorphisms from K into $\mathbf{R}$ and Λ the set of all real numbers t greater than 2 of the form $t = \sigma_1(\lambda)$, where the algebraic integer λ satisfies the inequalities $|\sigma_2(\lambda)| < 2, \ldots, |\sigma_n(\lambda)| < 2$. Let θ be defined by $\theta + \theta^{-1} = t$. Then θ is a Salem number and each Salem number can be constructed in this way.*

Proof: Let θ be a Salem number and let $P(X)$ be the minimal polynomial of θ. Using the transformation $X + X^{-1} = Y$, we find a distinguished polynomial Q of degree m such that $P(X) = X^m Q(X + X^{-1})$. The roots of Q are exactly the m distinct real numbers $\theta + \theta^{-1}, 2\,\mathrm{Re}\,\zeta_1, \ldots, 2\,\mathrm{Re}\,\zeta_{m-1}$. On the other hand, if K is totally real, let λ be an algebraic integer in K such that $\sigma_1(\lambda) > 2$, $-2 < \sigma_j(\lambda) < 2$ for $2 \leqslant j \leqslant m$. Let $\theta > 1$ be defined by $\theta + \theta^{-1} = \sigma_1(\lambda)$; then θ is an algebraic integer and the conjugates of θ are θ^{-1} and the $2m - 2$ complex numpers of absolute value 1, $(\zeta_1, \bar{\zeta}_1)$, $\ldots$, $(\zeta_{m-1}, \bar{\zeta}_{m-1})$ defined by $2 \operatorname{Re} \zeta_j = \sigma_j(\lambda)$ $(2 \leqslant j \leqslant m)$.

Let d^2 be the discriminant of the field K. It can be proved (Chapter V) that each interval of real numbers of length T contains $4^{n-1}T\,|d|^{-1} + \circ(T)$ distinct such Salem numbers θ such that $\theta + \theta^{-1}$ belongs to K; the little $\circ$ is uniform with respect to the position of the given interval but not uniform with respect to K.

6.9. *Characterization of Pisot and Salem numbers.*

Characterization of Pisot and Salem numbers among algebraic numbers

THEOREM XVIII: *Let θ be an algebraic real number greater than* 1; *the following two properties are equivalent:*

(a) *θ is a Pisot or a Salem number;*

(b) *for each positive ε, there exists a positive λ such that, for each $k \geqslant 0$, $\|\lambda\theta^k\| \leqslant \varepsilon$.*

Proof: If θ is a Pisot number, we put $\lambda = \theta^m$ and (b) is fulfilled provided m is sufficiently large. If θ is a Salem number, (b) will be proved in Section 7.2. The implication (b) $\Rightarrow$ (a) can be proved exactly as in Theorem XI.

Charakterization of Pisot and Salem numbers among real numbers

THEOREM XIX: *Let θ be a real greater than* 1. *The two following properties of θ are equivalent:*

(a) *θ is a Pisot or a Salem number;*

(b) *for each positive ε, there exists a positive T such that any interval of real numbers of length T contains a solution λ to all the Diophantine inequalities $\|\lambda\theta^k\| \leqslant \varepsilon$ $(k \geqslant 0)$.*

This theorem will be proved in full detail in Section 7.2.

7. Pisot numbers, Salem numbers and harmonious sets of real numbers

Let λ_k, $k \geqslant 0$, be a sequence of real numbers. Let Λ be the set of all λ_k and H the subgroup of $\mathbf{R}$ generated by Λ (H is the set of all finite sums $\sum_{k\geqslant 0} p_k\lambda_k$, $p_k \in \mathbf{Z}$. Some definitions should be recalled.

DEFINITION 13: *Let $\chi: \Lambda \to \mathbf{R}/\mathbf{Z}$ be a mapping defined by $\chi(\lambda_k) \equiv \alpha_k$ (mod* 1), *where α_k is a given sequence of real numbers.*

We write $\chi \in \mathrm{Hom}(\Lambda, \mathbf{R}/\mathbf{Z})$ *if one of the following equivalen conditions holds:*

(a) χ *is the restriction to* Λ *of a homomorphism from* H *to* $\mathbf{R}/\mathbf{Z}$;

(b) *for each finite sequence of rational integers* p_k, $k \geqslant 0$, *thefollowing implication is true* $\sum_{k\geqslant 0} p_k\lambda_k = 0 \Rightarrow \sum_{k\geqslant 0} p_k\alpha_k \equiv 0 \pmod 1$.

Here is an example. Let θ be an algebraic integer, n the degree of θ and $\lambda_k = \theta^k$, $k \geqslant 0$. With this terminology we have:

PROPOSITION 5: *Let* $\alpha_0, \ldots, \alpha_{n-1}$ *be any sequence of n real numbers. There is one and exactly one* χ *in* $\mathrm{Hom}(\Lambda, \mathbf{R}/\mathbf{Z})$, *such that* $\chi(\theta^k) \equiv \alpha_k \pmod 1$ *for* $0 \leqslant k \leqslant n-1$.

(a) Proof of the uniqueness of χ: Let $P(X) = X^n + a_1X^{n-1} + \cdots + a_n$ be the minimal polynomial of θ. We have, for each positive k, $\theta^{k+n} + a_1\theta^{k+n-1} + \cdots + a_n\theta^k = 0$, whence $\chi(\theta^{k+n}) + a_1\chi(\theta^{k+n-1}) + \cdots + a_n\chi(\theta^k) \equiv 0 \pmod 1$ and by induction on k, all the $\chi(\theta^k)$ can be found from those for $0 \leqslant k \leqslant n-1$.

(b) Proof of the existence of χ. Let $H = \mathbf{Z} + \cdots + \mathbf{Z}\theta^{n-1}$; the independence over $\mathbf{Q}$ of $1, \theta, \ldots, \theta^{n-1}$ implies that H is a free $\mathbf{Z}$-module and there is a homomorphism χ' from H to $\mathbf{R}/\mathbf{Z}$, defined by $\chi'(\theta^k) \equiv \alpha_k \pmod 1$ for $0 \leqslant k \leqslant n-1$. The restriction of χ' to Λ is χ.

Assume now that the α_k, $0 \leqslant k \leqslant n-1$ are rational real numbers. Then we have:

PROPOSITION 6: *With the preceding terminology the mapping* $k \to \chi(\theta^k)$ *is periodic for* $k \geqslant k_0$.

Let p be an integer such that $p\alpha_k \equiv 0 \pmod 1$, $0 \leqslant k \leqslant n-1$ and let Γ_p be the finite subgroup of $\mathbf{R}/\mathbf{Z}$ consisting of residues modulo 1 of k/p, $k \in \mathbf{Z}$. Let E be $(\Gamma_p)^n$ and let x_k be the element $\chi(\theta^k), \ldots, \chi(\theta^{k+n-1})$ of E. The proof of the uniqueness of χ shows that a mapping φ from E to E exists such that $x_{k+1} = \varphi(x_k)$ for all $k \geqslant 0$. A standard argument then shows that $k \to x_k$ is, eventually, a periodic mapping.

7.1. *Weak and strong elements of* Hom $(\Lambda, \mathbf{R}/\mathbf{Z})$

Among the $\chi \in \mathrm{Hom}(\Lambda, \mathbf{R}/\mathbf{Z})$, we can distinguish weak and strong elements: let t be a real number and let α_k be the residue modulo 1 of $t\lambda_k$; then $\chi(\lambda_k) = \alpha_k$ defines an element of $\mathrm{Hom}(\Lambda, \mathbf{R}/\mathbf{Z})$ which will be called strong. On the other hand an element of $\mathrm{Hom}(\Lambda, \mathbf{R}/\mathbf{Z})$ which is not a strong homomorphism will be called a weak homomorphism. The periodic element of $\mathrm{Hom}(\Lambda, \mathbf{R}/\mathbf{Z})$ defined above is a weak homomorphism.

If, however, the weak elements of Hom $(\Lambda, \mathbf{R}/\mathbf{Z})$ do not give any substantial information about the possible sequences of residues of $t\lambda_k$, $k \geqslant 0$, they are of no interest. We recall the definition of harmonious sets.

DEFINITION 14: *A set* $\Lambda = \{\lambda_0, \lambda_1, \ldots, \lambda_k, \ldots\}$ *of real numbers is harmonious if, for each* χ *in* Hom $(\Lambda, \mathbf{R}/\mathbf{Z})$ *and each positive* ε *there exists a real number* t *such that,* $\|\chi(\lambda_k) - t\lambda_k\| \leqslant \varepsilon$ *for each positive* k.

7.2. The connection with Pisot and Salem numbers

THEOREM XX: *Let* $\theta > 1$ *be a real number and* $\Lambda = \{1, \theta, \theta^2, \ldots\}$ *be the set of all powers of* θ. *Then* Λ *is harmonious if and only if* θ *is a Pisot or a Salem number.*

Before proving Theorem XX, some remarks should be made.

(a) Combined with Theorem X it gives Theorem XIX.

(b) Let $\theta = (1 + \sqrt{5})/2$, $\chi \in$ Hom $(\Lambda, \mathbf{R}/\mathbf{Z})$ be defined by the following properties: $\chi(1) = 0$, $\chi(\theta) = \frac{1}{2}$, $\chi(\theta^2) = \frac{1}{2}$ and $k \to \chi(\theta^k)$ is periodic of period 3. For each positive ε, let I be the interval $[-\varepsilon, \varepsilon]$ and J the interval $[\frac{1}{2} - \varepsilon, \frac{1}{2} + \varepsilon]$. Then a real number t can be found such that for $k = 0, 1, 2, 3, 4, 5, 6, \ldots$

$$t\theta^k \in I, J, J, I, J, J, I, \ldots \text{ modulo } 1.$$

This kind of periodic repetition modulo 1 can be found for all Pisot or Salem numbers.

Now to the proof of Theorem XX!

(A) θ is a Pisot or a Salem number implies Λ is harmonious.

The proof depends on the following lemma:

LEMMA 16: *Let* θ *be an algebraic integer,* n *the degree of* θ, $P(X) = X^n + a_1X^{n-1} + \cdots + a_n$ *the minimal polynomial of* θ *and* ε *any positive real number. Let* Λ *be the set* $\{1, \theta, \ldots, \theta^k, \ldots\}$. *For each* $\chi \in$ Hom $(\Lambda, \mathbf{R}/\mathbf{Z})$ *there exist a real number* t *and* $n - 1$ *complex numbers* $t_2, \ldots, t_n$ *with the following properties:*

(a) *for each* $k \geqslant 0$, $\chi(\theta^k) \equiv t\theta^k + \sum_2^n t_j\theta_j^k \pmod 1$;

(b) *if* $\theta_m = \bar{\theta}_l$, *then* $t_m = \bar{t}_l$ *for any* $2 \leqslant l \leqslant m \leqslant n$;

(c) $|t_j| \leqslant \varepsilon$ *for* $2 \leqslant j \leqslant n$ *and* $|t| \leqslant C(\theta)\, \varepsilon^{-n+1}$ *where* $C(\theta)$ *depends only on* θ.

Proof of the lemma: Let $\chi'(\theta^k)$ be the second term in (a) with an unspecified sequence $(t_1, t_2, \ldots, t_n)$ satisfying (b) and (c); then $\chi' \in \mathrm{Hom}(\Lambda, \mathbf{R}/\mathbf{Z})$ and to prove (a) for all $k \geqslant 0$, it suffices to prove it for $0 \leqslant k \leqslant n-1$. Let $\chi(\theta^k) \equiv \alpha_k \pmod 1$. We are led to solve the n congruences $\alpha_k \equiv t\theta^k + \sum_2^n t_j \theta^k \pmod 1$, $0 \leqslant k \leqslant n - 1$. Let C_ε be the set of all $(x_1, \ldots, x_n)$ in $\mathbf{R}^n$, where $x_k = t\theta^k + \sum_2^n t_j \theta_j^k$ for real t and complex t_j satisfying (b) and $|t_j| \leqslant \varepsilon$ $(2 \leqslant j \leqslant n)$. Since the vectors $(1, \theta, \ldots, \theta^{n-1}), (1, \theta_2, \ldots, \theta_2^{n-1}), \ldots, (1, \theta_n, \ldots, \theta_n^{n-1})$ are independent over $\mathbf{C}$, C_ε contains an infinite open cylinder whose axis is the straight line D defined by $x_k = t\theta^k$, $0 \leqslant k \leqslant n - 1$, $t \in \mathbf{R}$. But $1, \theta, \ldots, \theta^{n-1}$ are $\mathbf{Z}$-linearly independent. Let $\varrho : \mathbf{R}^n \to \mathbf{R}^n/\mathbf{Z}^n$ be the canonical homomorphism; then $\varrho(D)$ is dense in $\mathbf{R}^n/\mathbf{Z}^n$ and $\varrho(C_\varepsilon)$ is the whole $\mathbf{R}^n/\mathbf{Z}^n$. Estimate $|t| \leqslant C(\theta)\, \varepsilon^{-n+1}$ follows from Theorem VI.

COROLLARY 1: *Let θ be a Pisot or a Salem number; let n be the degree of θ. There exists a constant $C(\theta)$ depending only on θ, such that for each $\chi \in \mathrm{Hom}(\Lambda, \mathbf{R}/\mathbf{Z})$ and each positive ε, the Diophantine inequalities $\|t\theta^k - \chi(\theta^k)\| \leqslant \varepsilon$, $|t| \leqslant \varepsilon^{-n+1} C(\theta)$ can be solved.*

COROLLARY 2: *With the same terminology, each interval of real numbers of length $\varepsilon^{-n+1}C(\theta)$ contains a solution t of the Diophantine inequalities $\|t\theta^k\| \leqslant \varepsilon$ $(k \geqslant 0)$.*

The two corollaries follow from the lemma and the inequalities $|\theta_j| \leqslant 1$, $2 \leqslant j \leqslant n$; instead of (a), we get $\|\chi(\theta^k) - t\theta^k\| \leqslant (n - 1)\, \varepsilon$, which is the required approximation.

(B) The set $\Lambda = \{1, \theta, \ldots, \theta^k, \ldots\}$ is harmonious implies θ is a Pisot or a Salem number.

We distinguish between two cases:

(a) θ is an algebraic number. By Theorem X, each interval of length $T(\varepsilon)$ contains a solution of the Diophantine inequalities $\|t\theta^k\| \leqslant \varepsilon$ $(k \geqslant 0)$ and by Theorem XVIII, θ is a Pisot or a Salem number.

(b) θ is a transcendental number. Then $1, \theta, \ldots, \theta^k, \ldots$ are linearly independent over $\mathbf{Q}$. Let X be the set of all mappings from Λ to the subset $\{0, \frac{1}{2}\}$ of $\mathbf{R}/\mathbf{Z}$. Each χ in X belongs to $\mathrm{Hom}(\Lambda, \mathbf{R}/\mathbf{Z})$ and $\mathrm{Card}(X) = \mathrm{Card}(\mathbf{R})$. The following lemma is now needed.

LEMMA 17: *Let t_k, $k \geqslant 0$, be an increasing sequence of real numbers tending to infinity in such a way that $t_0 = 1$ and $t_{k+1} \leqslant \theta t_k$. Then, for $\varepsilon < (1 + \theta)^{-1}$, the inequalities $|x| \leqslant \varepsilon$, and $\|x\theta^k\| \leqslant \varepsilon$ for all positive k imply $x = 0$.*

Proof: We write $xt_k = p_k + r_k$ and assume that $x \neq 0$ Since $t_0 = 1$ and t_k tends to infinity there is a least integer $h \geqslant 1$ for which $p_h \neq 0$. Then $|xt_h| \geqslant 1 - \varepsilon$, while $|xt_{h-1}| \leqslant \varepsilon$ and this implies $t_h/t_{h-1} \geqslant (1 - \varepsilon)/\varepsilon$, which contradicts $t_{k+1} \leqslant \theta t_k$ if $\varepsilon < (1 + \theta)^{-1}$.

For $\varepsilon > 0$, let T_ε be the set of all real numbers t such that, for some χ in X and each $k \geqslant 0$, $\|\chi(\theta^k) - t\theta^k\| \leqslant \varepsilon$. If $\varepsilon < \frac{1}{2}$ and χ_1 and χ_2 are different elements of X, then $t_1 \neq t_2$. Hence Card T_ε = Card X if Λ is harmonious. If we can prove that T_ε is countable for sufficiently small ε, we get a contradiction.

Since $\{0, \frac{1}{2}\}$ is a finite subgroup of $\mathbf{R}/\mathbf{Z}$, X is a group; hence $T_\varepsilon - T_\varepsilon \subset T_{2\varepsilon}$. If $t \in T_\varepsilon$, then $\|2t\theta^k\| \leqslant 2\varepsilon$ for all $k \geqslant 0$. Using Lemma 1, we get that, if $2\varepsilon < (1 + \theta)^{-1}$, then either $t = 0$ or $|t| \geqslant \varepsilon$. Since $T_\varepsilon - T_\varepsilon \subset T_{2\varepsilon}$, the distance between two distinct points of T_ε exceeds 2ε if $4\varepsilon < (1 + \theta)^{-1}$ and so the set T_ε is countable.

8. Notes

1. A proof of Roth's theorem can be found in [82] or in [4]. Schmidt's theorem is proved in [84] or in [85].

2. A more general version of Theorem III is given in [22]. For a very complete discussion of inhomogeneous problems, the reader is referred to Chapter XI of [5].

3. In [34] or [83], thick sets of badly approximable sequences are constructed.

4. Theorem IX is an unpublished result of the author.

5. A first version of harmonious sets (where independence over $\mathbf{Q}$ is required) can be found in [44].

6. Recent work on Pisot numbers has led in four directions.

(A) Pisot's conjecture is "$\theta > 1$, $\lambda \geqslant 1$ and $\lim \|\lambda\theta^n\| = 0$ imply that θ is a Pisot number". Some partial results can be found in [33] or [74].

(B) Generalizations of Pisot numbers can be defined in p-adic fields or in adeles rings ([28], [29], [35] and [36]).

(C) Rational or analytie transformations connected with the set of Pisot numbers have been studied in [77] and [78].

(D) Explicit processes to find all Pisot numbers are investigated in [25].

Other rings for which Fatou's lemma is valid are studied in [27].

Another proof of Salem's theorem (Section 6.6) can be found in [76].

CHAPTER II

DIOPHANTINE APPROXIMATIONS AND ADDITIVE PROBLEMS IN LOCALLY COMPACT ABELIAN GROUPS

In many problems of approximation of functions or distributions defined on the real line, discretization is a very useful tool: **R** is replaced by discrete subgroups $2^{-k}\mathbf{Z}$ for larger and larger k and the given distribution becomes a discrete measure carried by $2^{-k}\mathbf{Z}$.

The p-adic field $\mathbf{Q}_p$ does not contain discrete subgroups. How can distributions on $\mathbf{Q}_p$ be discretized? By using harmonious sets instead of discrete subgroups: this fact will be proved in Chapter V, Section 8, and is the reason why we try to construct harmonious sets in locally compact abelian (l.c.a.) groups.

Harmonious sets are defined in Section 2 by certain remarkable properties of Diophantine approximations, and their basic properties can be found in Section 3. For our purpose, harmonious sets Λ such that $\Lambda + K = G$, for a suitable compact set K, are of special interest; they are characterized by Theorem IV.

At this point we know how to construct the most interesting examples of harmonious sets; they can be defined by remarkable additive properties.

Algebraic number field theory plays a decisive role in the determination of harmonious sets closed under multiplication.

1. Preliminaries on l.c.a. groups

1.1. Let G be a locally compact abelian group (we write an l.c.a. group), G_d be the same group with the discrete topology and H another l.c.a. group; then Hom (G, H) will denote the topological group of all continuous homomorphisms from G to H, the topology being that of uniform convergence on compact subsets of G. In this context, Hom (G_d, H) is the group of all homomorphisms from G to H and the topology of Hom (G_d, H) is that of pointwise convergence.

Let $\mathbf{T}$ be the group of complex numbers of absolute value 1 with the usual topology. If $H = \mathbf{T}$, Hom $(G, \mathbf{T})$ is called the dual group of G; we write Hom $(G, \mathbf{T}) = G^\wedge$ and the elements of $G^\wedge$ are called the characters of G. We shall use the following notation: for x in G and γ in $G^\wedge$, the value at x of the character γ is written $\chi_\gamma(x)$; if $G = \mathbf{R}$, $G^\wedge = \mathbf{R}$ and, for real numbers x and y, $\chi_y(x) = \exp 2\pi ixy$.

Let K run over all compact subsets of G and ε over all positive real numbers; the sets $V(K, \varepsilon)$ of all y in $G^\wedge$ such that $\sup_{x \in K} |\chi_y(x) - 1)| \leqslant \varepsilon$ are then a fundamental family of neighborhoods of 0 for the topology of $G^\wedge$.

Now each x in G defines a character on $G^\wedge$ by $\gamma \to \chi\ (x)$ and this mapping from G to Hom $(G^\wedge, \mathbf{T})$ is an isomorphism ([15] p. 27); thus G is the dual group of $G^\wedge$. This is Pontryagin's duality theorem.

If χ belongs to Hom $(G, \mathbf{T})$ and if $\sup_{x \in G} |\chi(x) - 1| \leqslant \frac{1}{2}$, the subgroup $\chi(G)$ of $\mathbf{T}$ reduces to $\{1\}$ and $\chi = 1$. Hence, if G is compact, $G^\wedge$ is discrete; if G is discrete, $G^\wedge$ is compact as either the duality theorem or an easy direct argument shows.

The following lemma shows that the equation of closed subgroups H of G can be written with characters belonging to the dual group $G^\wedge$ of G.

LEMMA 1: *Let G be a l.c.a. group and H a closed subgroup of G. Each element of* Hom $(H, \mathbf{T})$ *is the restriction to H of an element of* Hom $(G, \mathbf{T})$. *If x in G does not belong to H, there exists a χ in* Hom $(G, \mathbf{T})$ *such that χ is* 1 *everywhere on H but* $\chi(x) \neq 1$.

This remarkable fact is proved in ([15] Section 2.1).

Let G and H be two l.c.a. group and $h \in$ Hom (G, H). We define a dual homomorphism $h^\wedge \in$ Hom $(H^\wedge, G^\wedge)$ by the following rule: if v belongs to $H^\wedge$, the value of $h^\wedge(v)$ at a point x of G is $\chi_v\ (h(x))$.

LEMMA 2: *If h is injective, $h^\wedge$ has a dense range and if h has a dense range, $h^\wedge$ is injective.*

Assume that h is injective. Let N be the closure of $h^\wedge(H^\wedge)$ in $G^\wedge$. If $N \neq G^\wedge$ an $x \neq 0$ can be found in G such that, for all y in N, $\chi_x(y) = 1$. But if $y = h^\wedge(v)$, $\chi_x(y)$ equals $\chi_v\ (h(x))$ and then $\chi_v\ (h(x))$ equals 1 for all v in $H^\wedge$; we have $h(x) = 0$ and hence $x = 0$ since h is injective.

1.2. A special case of Lemma 2 appears in the following situation. Let G be a l.c.a. group and G_d the same group with the discrete topology; let $i : G_d \to G$ be the identity map. Put $\Gamma = G^\wedge$ and $\tilde{\Gamma} = (G_d)^\wedge$. Then both the maps i and $i^\wedge : \Gamma \to \tilde{\Gamma}$ are injective and have dense range. The compact group $\tilde{\Gamma}$ is called the Bohr compactification of Γ. Since the topology of $\tilde{\Gamma}$ is that of uniform convergence on finite subsets of G, we can rewrite the last statement in the following way:

Kronecker's theorem: *Let $F = \{\lambda_1 \ldots, \lambda_n\}$ be a finite subset of G and $\chi : F \to \mathbf{T}$ a mapping such that, for each sequence $p_1, \ldots, p_n$ of rational integers, each relation $\sum_1^n p_j\lambda_j = 0$ implies $\prod_1^n (\chi(\lambda_j))^{p_j} = 1$. Then for each positive ε, there exists a y in $G^\wedge$ such that $|\chi(\lambda_1) - \chi_y(\lambda_1)| \leqslant \varepsilon, \ldots, |\chi(\lambda_n) - \chi_y(\lambda_n)| \leqslant \varepsilon$.*

Proof: Let H be the subgroup of G_d generated by F. Then the assumption on χ means that χ is the restriction to F of an element of Hom $(H_d, \mathbf{T})$ which, by Lemma 1, can be lifted to an element of Hom $(G_d, \mathbf{T}) = \tilde{\Gamma}$. At this stage, we use the density of $i^\wedge(G^\wedge)$ in $\tilde{\Gamma}$.

1.3. Annihilators

DEFINITION 1: *Let G be a l.c.a. group and H a closed subgroup of G. The annihilator of H is the closed subgroup H' of $G^\wedge$ consisting of all continuous characters on G which are 1 everywhere on H.*

LEMMA 3: *The annihilator of H' is H.*

Obviously H is contained in the annihilator of H'. If $x \notin H$ there exists, by Lemma 1, a y in H' such that χ_y is 1 everywhere on H, but $\chi_y(x) \neq 1$ and so x does not belong to the annihilator of H'.

LEMMA 4: *The dual group of G/H is H'.*

The proof is obvious.

2. Harmonious sets

Let Λ be a subset of G and χ a map from Λ to $\mathbf{T}$.

DEFINITION 2: *We write $\chi \in$ Hom $(\Lambda_d, \mathbf{T})$ if for each integer $n \geqslant 1$, each sequence $p_1, \ldots, p_n$ of rational integers and each sequence $\lambda_1, \ldots, \lambda_n$ of elements of Λ, each relation $\sum_1^n p_j\lambda_j = 0$ implies $\prod_1^n (\chi(\lambda_j))^{p_j} = 1$.*

If H is the subgroup of G consisting of all sums $\sum_1^n p_j\lambda_j$, $p_j \in \mathbf{Z}$, $\lambda_j \in \Lambda$, equipped with the discrete topology, χ is the restriction to Λ of a χ' in Hom $(H, \mathbf{T})$.

PROPOSITION 1: *Let χ be a mapping from Λ to $\mathbf{T}$. The following three properties of Λ are equivalent:*

(1) *χ belongs to* Hom $(\Lambda_d, \mathbf{T})$;

(2) *χ is the restriction to Λ of a χ' in* Hom $(G_d, \mathbf{T})$;

(3) *for each positive* ε *and each finite subset F of* Λ, *there exists a* γ *in* $G^\wedge$ *such that* $\sup_{\lambda \in F} |\chi(\lambda) - \chi_\gamma(\lambda)| \leqslant \varepsilon$.

The proof of Kronecker's theorem applies here.

In Chapter I we were interested in a continuous homomorphism ϱ from $\mathbf{R}$ to $\mathbf{R}/\mathbf{Z}$ whose kernel was $\mathbf{Z}$; for a real number t, $\varrho(t)$ was the residue modulo 1 of t. The group $\mathbf{T}$ is isomorphic to $\mathbf{R}/\mathbf{Z}$ and, using this isomorphism, we can replace ϱ by the homomorphism h from $\mathbf{R}$ to $\mathbf{T}$ defined by $h(t) = \exp 2\pi i t$. In a general context, $\mathbf{R}$ becomes a l.c.a. group G and h a character χ of G. The generalization of the residue modulo 1 $\varrho(t)$ of the real number t is now the element $\chi(x)$ of $\mathbf{T}$ for a given x in G (χ being fixed in $G^\wedge$).

Let Λ be a subset of G and let $(z_\lambda)_{\lambda \in \Lambda}$ a family of complex numbers of absolute value 1. We try to guess whether $(z_\lambda)_{\lambda \in \Lambda}$ is the family of $(\chi(\lambda))_{\lambda \in \Lambda}$ for a suitable χ in Hom $(G, \mathbf{T})$ or, at least, whether for each positive ε, there exists such a χ with $\sup_{\Lambda} |\chi(\lambda) - z_\lambda| \leqslant \varepsilon$. A necessary condition is: the mapping $\lambda \to z_\lambda$ belongs to Hom $(\Lambda_d, \mathbf{T})$. We are led to the following definition:

DEFINITION 3: *Let* Λ *be a subset of a l.c.a. group G. We say that* Λ *is harmonious if, for each* χ *in* Hom $(\Lambda_d, \mathbf{T})$ *and each positive* ε *there is a* ξ *in* Hom $(G, \mathbf{T})$ *such that* $\sup_{\Lambda} |\chi(\lambda) - \xi(\lambda)| \leqslant \varepsilon$.

Kronecker's theorem shows that a finite set is harmonious.

By Proposition 1, each subset of an harmonious set is still harmonious.

3. Basic properties of harmonious sets

3.1. In all the following we shall restrict our attention to l.c.a. groups G with the two extra properties (a) and (b):

(a) G is a metrizable l.c.a. group: the topology of G is defined by a translation-invariant metric;

(b) G contains a countable dense subset, in which case G is said to be separable. Then the dual group $G^\wedge$ of G also satisfies (a) and (b). With these additional assumptions on G, we have the following result:

THEOREM I: *Let G be a metrizable and separable l.c.a. group,* Λ *a subset of G, and for each positive* ε, *let* M_ε *be the set of all* γ *in* $G^\wedge$ *(the dual group of G) such that* $\sup_{\Lambda} |\chi_\gamma(\lambda) - 1| \leqslant \varepsilon$. *Then the following three porperties of* Λ *are equivalent:*

(a) Λ *is harmonious in G;*

(b) *for each positive* ε, *there exists a compact subset* K_ε *of* $G^\wedge$ *such that for every* χ *in* $\mathrm{Hom}(\Lambda_d, \mathbf{T})$ *there is a* γ *in* K_ε *satisfying the Diophantine inequalities* $\sup_\Lambda |\chi(\lambda) - \chi_\gamma(\lambda)| \leqslant \varepsilon$;

(c) *for each positive* ε, *there exists a compact subset* K_ε *of* $G^\wedge$ *such that* $M_\varepsilon + K_\varepsilon = G^\wedge$.

Proof: Let $\Lambda = G^\wedge$ be the dual group of G and $\tilde{\Gamma}$ the Bohr compactification of Γ; Γ can be regarded as a dense subgroup of $\tilde{\Gamma}$. For each χ in $\tilde{\Gamma}$ and x in G, $\chi(x)$ will denote the value at x of the character $\chi \in \mathrm{Hom}(G_d, \mathbf{T})$.

Additive notation for the group operation in $\tilde{\Gamma}$ will be used.

(a) $\Rightarrow$ (b): Let $V(\varepsilon)$ be the set of all χ in $\tilde{\Gamma}$ such that $\sup_\Lambda |\chi(\lambda) - 1| \leqslant \varepsilon$; $V(\varepsilon)$ is a closed subset of $\tilde{\Gamma}$ and hence, equipped with the induced topology, $V(\varepsilon)$ is compact. Let $(K_n)_{n\geqslant 1}$ be an increasing sequence of compact subsets of Γ such that $\bigcup_{n\geqslant 1} K_n = \Gamma$; such a sequence exists since Γ is separable. Let $L_n = K_n + V(\varepsilon)$; L_n is a compact subset of $\tilde{\Gamma}$. Condition (a) means that the union of all L_n, $n \geqslant 1$, is the whole of $\tilde{\Gamma}$. Baire's theorem, applied to the compact set $\tilde{\Gamma}$, shows that, if n is large enough, the interior $\mathring{L}_n$ of L_n is not empty. For such an n, consider the translates $L_n + \gamma$, $\gamma \in \Gamma$. Since Γ is a dense subgroup of $\tilde{\Gamma}$, $\bigcup_{\gamma\in\Gamma} L_n + \gamma = L_n + \Gamma = \tilde{\Gamma}$. Again using the compactness of $\tilde{\Gamma}$, a finite subset F of Γ can be found such that $L_n + F = \tilde{\Gamma}$. Property (b) then holds with $K_\varepsilon = K_n + F$.

(b) $\Rightarrow$ (c): We apply (b) to $\chi = \chi_\gamma$ for some γ in Γ: we get a γ' in K_ε such that $\sup_\Lambda |\chi_\gamma(\lambda) - \chi_{\gamma'}(\lambda)| \leqslant \varepsilon$. Then $\gamma - \gamma'$ belongs to M_ε and $M_\varepsilon + K_\varepsilon = \Gamma$.

(c) $\Rightarrow$ (a): Let $V(\varepsilon)$ be defined as at the beginning of the proof. Then $V(\varepsilon)$ contains $M(\varepsilon)$ and the compact subset $V(\varepsilon) + K_\varepsilon$ of $\tilde{\Gamma}$ contains $M(\varepsilon) + K_\varepsilon = \Gamma$. But Γ is dense in $\tilde{\Gamma}$. Hence $K_\varepsilon + V(\varepsilon) = \tilde{\Gamma}$ and Λ is harmonious.

Remark: Following Besicovitch, a subset M *of* Γ *is called relatively dense if there is a compact subset* K *of* Γ *such that* $M + K = \Gamma$. *Condition (c) of Theorem I can be rewritten:* $M(\varepsilon)$ *is relatively dense for each positive* ε.

3.2. *Union with a finite set*

The topology of G will be defined by a translation invariant metric $d(x, x')$.

PROPOSITION 2: *Let Λ be an harmonious subset of G. There exists a positive δ such that the distance between any two distinct elements λ in Λ exceeds δ.*

Before giving the proof we observe that, if Λ is harmonious, so are $\Lambda + \Lambda$ and $\Lambda - \Lambda$. To get our result, it is sufficient to prove that, if Λ is harmonious, 0 is isolated in Λ. Let us assume that a sequence $\lambda_k \neq 0$, $k \geqslant 1$, tending to 0 can be found in Λ. Let M_ε and K_ε be as in Theorem I. Since convergence in G means uniform convergence on compact subsets of Γ, we have large $\sup_{\gamma \in K_\varepsilon} |\chi_\gamma(\lambda_k) - 1| \leqslant \varepsilon$ for sufficiently large k; the inequality $\sup_{\gamma \in M_\varepsilon} |\chi_\gamma(\lambda_k) - 1| \leqslant \varepsilon$ then gives $\sup_{\gamma \in \Gamma} |\chi_\gamma(\lambda_k) - 1| \leqslant 2\varepsilon$. But for a given k, the set of all $\chi_\gamma(\lambda_k)$, as γ runs through Γ, is a subgroup of $\mathbf{T}$, not equal to $\{1\}$, which, provided ε is chosen small enough, cannot be contained in $|z - 1| \leqslant 2\varepsilon$.

COROLLARY 1: *If G is a compact abelian group, each harmonious set is finite and conversely.*

COROLLARY 2: *If G is a metrizable and separable l.c.a. group, each harmonious set can be written as a sequence $(\lambda_k)_{k \geqslant 1}$.*

If $G = \mathbf{R}, \Lambda_1 = \mathbf{Z}$ and $\Lambda_2 = \mathbf{Z}\sqrt{2}$ are harmonious sets; in this case each χ in Hom $(\Lambda, \mathbf{T})$ is the restriction to Λ of a χ' in Hom $(\mathbf{R}\ \mathbf{T})$. But $\Lambda_1 \cup \Lambda_2$ is not harmonious, and the conclusion of Proposition 2 is not satisfied. Nevertheless the following result is valid:

THEOREM II: *Let Λ be an harmonious set in G and F a finite subset of G. Then the sets $\Lambda \cup F$ and $\Lambda + F$ are harmonious.*

It is sufficient to prove that $\Lambda \cup F$ is harmonious.

Let ε be a positive real number and M_ε the set of all γ in Γ such that $\sup_{\Lambda} |\chi_\gamma(\lambda) - 1| \leqslant \varepsilon$. We have $M_{\varepsilon/2} - M_{\varepsilon/2} \subset M_\varepsilon$ and Theorem II is now an easy corollay of the following lemma.

LEMMA 5: *Let Γ be a l.c.a. group, M a relatively dense subset of Γ and Ω an open subset of the Bohr compactification $\tilde{\Gamma}$ of Γ. Then $(M - M) \cap \Omega$ is relatively dense.*

To get Theorem II from Lemma 5, let Ω be the set of all χ in $\tilde{\Gamma}$ ($\chi \in$ Hom $(G_d, \mathbf{T})$) such that $\sup_{t \in F} |\chi(t) - 1| < \varepsilon$ and let $M = M(\varepsilon/2)$. For each γ in $(M - M) \cap \Omega$, we have $\sup_{x \in \Lambda \cup F} |\chi(x) - 1| \leqslant \varepsilon$ and condition (c) of Theorem I gives Theorem II.

To prove Lemma 5, let $\overline{M}$ be the closure of M in $\tilde{\Gamma}$, $N = M - M$ and $\bar{N} = \overline{M} - \overline{M}$ be the closure of N in $\tilde{\Gamma}$. For each subset Ω of $\tilde{\Gamma}$ and each γ in M we have $(\gamma + \Omega) \cap \overline{M} \subset \gamma + \Omega \cap \bar{N}$. If Ω is open, the union $\bigcup_{\gamma \in M} (\gamma + \Omega) = M + \Omega$ contains $\overline{M}$. But $\overline{M}$ is compact and so there is a finite subset $A \subset M$ such that $\overline{M} \subset A + \Omega$. It follows that $\overline{M} = (A + \Omega) \cap \overline{M}$ is contained in $A + \Omega \cap \bar{N}$. Let L be a compact subset of Γ such that $L + \overline{M} = \tilde{\Gamma}$; this implies $L + A + \Omega \cap \bar{N} = \tilde{\Gamma}$ and Lemma 5 is proved (with 2ε instead of ε).

4. Construction of harmonious sets in l.c.a. groups (the lacunary case)

4.1. When G is a compact abelian group, harmonious sets Λ in G are finite; when G is a discrete abelian group, each subset Λ of G is harmonious. The construction of harmonious sets in G is interesting when G is a l.c.a. group which is neither compact nor discrete.

To begin with, we shall see that many l.c.a. groups G contain infinite harmonious sets; in general, the sets will be lacunary sequences and the reason for this fact will appear in the study of relatively dense harmonious sets.

Some lemmas are needed.

LEMMA 6: *Let Λ be a subset of a l.c.a. group G. Assume that, for each positive ε, there exists a relatively dense subset M_ε of $G^\wedge$ and a finite subset F_ε of Λ such that $\lambda \in \Lambda$, $\lambda \notin F_\varepsilon$ and $\gamma \in M_\varepsilon$ imply $|\chi_\gamma(\lambda) - 1| \leqslant \varepsilon$. Then Λ is harmonious.*

Proof: In the terminology of Lemma 5, let $M = M(\varepsilon/2)$ and let Ω be the open subset of $\tilde{\Gamma}$ ($\Gamma = G^\wedge$), defined by $|\chi_\gamma(t) - 1| \leqslant \varepsilon$ for all t in $F_{\varepsilon/2}$.

Then M is relatively dense and, by Lemma 5, so is $\Omega \cap (M - M) = N$. But each $\gamma \in N$ satisfies $|\chi_\gamma(\lambda) - 1| \leqslant \varepsilon$ for all λ in Λ and Theorem I now gives the result.

COROLLARY 2: *Let $G =$ **R** and $(t_k)_{k \geqslant 1}$ be an increasing sequence of real numbers such that* $\lim t_{k+1}/t_k = +\infty$. *Then $\Lambda = \{t_1, \ldots, t_k, \ldots\}$ is an harmonious set of real numbers.*

Proof: For each positive ε, let the integer n be defined by the condition: $k \geqslant n$ implies $(1 + 2\varepsilon)/t_{k+1} \leqslant 2\varepsilon/t_k$. Then the inequalities $\|xt_k\| \leqslant \varepsilon$ are satisfied by real numbers x belonging to intervals of length $2\varepsilon/t_k$ whose centres belong to $\mathbf{Z}/t_k$; if $k \geqslant n$ each such interval contains a complete interval of the same type but with $k + 1$ replacing k; hence each interval of this type with $k = n$ contains an element of M_ε where M_ε is relatively dense and F_ε is the complement of $t_1, \ldots, t_{n-1}$ in Λ. At this stage we apply Lemma 6.

In a l.c.a. group Γ, the order of an element γ is the smallest integer $n \geqslant 1$ such that $n\gamma = 0$.

THEOREM III: *Let G be a l.c.a. group. Assume that G is not compact and that each neighborhood W of 0 in the dual group $G^\wedge$ contains elements of arbitrarily high order. Let K be a compact subset of G. Then we can find a sequence $(\lambda_k)_{k\geqslant 1}$ of elements of G which is a K-stable harmonious set: this means that if $(\lambda_k')_{k\geqslant 1}$ is another sequence such that $\lambda_k' - \lambda_k$ belongs to K for all $k \geqslant 1, \Lambda = \{\lambda_1', \ldots, \lambda_k', \ldots\}$ is also harmonious.*

4.2. This theorem establishes the existence of stable harmonious sets in many l.c.a. groups. These sets cannot enjoy a rigid arithmetical structure in contrast with relatively dense harmonious sets, which do (Section 5).

The proof of Theorem III depends on the following lemmas. The l.c.a. group G satisfies the conditions of Theorem III.

LEMMA 7: *We say that a subset E of* **T** *is ε-dense, for some positive ε, if each interval of* **T** *of length ε contains a z in E. Let W be a neighborhood of 0 in $G^\wedge$ and let ε be a positive real number. Then we can find a γ in W such that $\chi_\gamma(\Gamma^\wedge)$ is ε-dense in* **T**.

Proof: If the lemma were false, the group $\chi_\gamma(\Gamma^\wedge)$ would be of order $\leqslant n$ for all χ in W and for some n independent of γ. Hence each γ in W would be of order $\leqslant n$. This contradicts the assumptions of Theorem III.

LEMMA 8: *Let K be any compact subset of G and V any neighborhood of 0 in $\Gamma = G^\wedge$. Then for each positive ε, there exists an x in G such that $\bigcup_{\gamma\in V} \chi_\gamma(y)$ is ε-dense in* **T** *for every y in K + x.*

Proof: Let $n \geqslant 16/\varepsilon$ be a positive integer, and let W be a neighbourhood of 0 sufficiently small to satisfy $W + \cdots + W$ (n times) $\subset V$ and $\sup_{\gamma\in W, x\in K} |\chi_\gamma(x) - 1| \leqslant 1/n$. Using Lemma 7, we find a $\gamma \neq 0$ in W such that $\chi_\gamma(G)$ is $1/n$-dense in **T** and hence there is an x in G such that $\chi_\gamma(x)$ lies in the arc of **T** defined by the parametric condition $z = \exp 2\pi i t$, $2/n \leqslant t \leqslant 3/n$. For all y in $K + x$, $\chi_\gamma(y)$ lies in the arc defined by $z = \exp 2\pi i t$, $1/n \leqslant t \leqslant 4/n$. But $\gamma, 2\gamma, \ldots, n\gamma$ are in V and $z, z^2, \ldots, z^n$ are in $\bigcup_{\gamma\in V} \chi_\gamma(y)$. Hence this latter set is ε-dense in **T**.

4.3. Now we return to the proof of Theorem III. We inductively define a sequence $\lambda_1, \ldots, \lambda_n, \ldots$ of elements of G and a sequence $V_1, \ldots, V_n, \ldots$ of neighborhoods of 0 in $\Gamma = G^\wedge$ such that, if $\lambda'_n - \lambda_n \in K$, each translate of V_{n-1} contains a translate of V_n on which $|\chi_\gamma(\lambda'_n) - 1| \leqslant 1/n$. Once this is done, Lemma 6 shows that the set Λ of all λ'_n, $n \geqslant 1$, is harmonious.

Assume that $\lambda_1, \ldots, \lambda_n$, $V_1, \ldots, V_n$ are defined and let us construct λ_{n+1} and V_{n+1}. There is a neighborhood W_{n+1} of 0 in $\Gamma = G^\wedge$ such that $W_{n+1} + W_{n+1} \subset V_n$ and, by Lemma 8, there is a λ_{n+1} in G such that $\bigcup_{\gamma \in W_{n+1}} \chi_\gamma(\lambda')$ is $1/(2(n+1))$-dense in $\mathbf{T}$ for each λ' in $\lambda_{n+1} + K$. Let $V_{n+1} \subset W_{n+1}$ be the set of all γ such that $\sup_{\lambda' \in \lambda_{n+1} + K} |\chi_\gamma(\lambda') - 1| \leqslant (2n+2)^{-1}$ and consider a translate $\gamma_1 + V_n$ of V_n; taking $z = \chi_{\gamma_1}(\lambda'_{n+1})$, we can find a γ_2 in $\gamma_1 + W_{n+1}$ such that $|\chi_{\gamma_2}(\lambda'_{n+1}) - 1| \leqslant 1/(2(n+1))$. Thus $\gamma_2 + V_{n+1} \subset \gamma_1 + V_n$ and we have $|\chi_\gamma(\lambda'_{n+1}) - 1| \leqslant 1/(n+1)$ for all γ running in $\gamma_2 + V_{n+1}$.

5. Construction of relatively dense harmonious sets

5.1. The previous construction provides examples of lacunary stable harmonious sets. On the other hand, relatively dense harmonious sets have a fairly rich arithmetical structure that we shall now define. We characterize relatively dense harmonious sets in a l.c.a. group G in terms of models. But what is a model?

DEFINITION 4 *(The models): Let G be a l.c.a. group, $n \geqslant 0$ an integer, D a closed, discrete and relatively dense subgroup of $G \times \mathbf{R}^n$, and $p_1 : G \times \mathbf{R}^n \to G$, $p_2 : G \times \mathbf{R}^n \to \mathbf{R}^n$ the canonical projections.*

Assume that the intersection of D and $\{0\} \times \mathbf{R}^n$ is equal to $\{0, 0\}$ and that $p_2(D)$ is dense in $\mathbf{R}^n$.

Let Ω be an open subset of $\mathbf{R}^n$ with compact closure. The model Λ of G defined by n, p_1, p_2, D and Ω is the set of all $x = p_1(d)$ in G for all d in D such that $p_2(d)$ lies in Ω.

A model Λ in G is a set which is so defined for suitable n, p_1, p_2, D and Ω.

Before giving the main theorem, some remarks should be made. If $n = 0$ we define $\mathbf{R}^0$ as $\{0\}$. A subgroup D of $G \times \mathbf{R}^n$ is discrete if, for a suitable neighborhood $\mathscr{V}$ of 0 in $G \times \mathbf{R}^n$, $\mathscr{V} \cap D = \{0, 0\}$.

THEOREM IV: *Each model in G is harmonious and relatively dense. For each harmonious and relatively dense subset Λ of G there exists a model Λ' in G and a finite subset F of G such that Λ is contained in $\Lambda' + F$.*

Since the sum $\Lambda' + F$ is harmonious when Λ' is, and since each part of an harmonious set is harmonious, this theorem is a characterization of relatively dense harmonious sets in G in terms of models.

The proof is divided into several stages. Firstly we enlarge the definition of models by replacing $\mathbf{R}^n$ by a general l.c.a. group; we obtain a result similar to Theorem IV. Secondly using structure theorems for l.c.a. group, we reduce this general l.c.a. group to $\mathbf{R}^n$ and we introduce the finite set F.

5.2. *More general definition of models and the first part of Theorem IV*

DEFINITION 5: *Let G and H be two l.c.a. groups and $p_1 : G \times H \to G$, $p_2 : G \times H \to H$ the canonical projections.*

We say that G produces H if there exists a closed subgroup D of $G \times H$ with the following properties:

(a) *D is discrete in $G \times H$;*

(b) *D is relatively dense in $G \times H$;*

(c) *if, for some y in H, $(0, y) \in D$, then $y = 0$;*

(d) *$p_2(D)$ is dense in H.*

We shall see later that such H can be defined in terms of G and relatively dense harmonious subsets of G, whence the name G 'produces' H.

Let $\mathfrak{G}$ be the compact group $(G \times H)/D$ and π the canonical homomorphism from $G \times H$ to $\mathfrak{G}$ whose kernel is D. In all of what follows we shall write G instead of $G \times \{0\}$ and H instead of $\{0\} \times H$. Since $D \cap H = \{0\}$, the restriction of π to H is injective; since $p_2(D)$ is dense in H, $D + G$ is dense in $G \times H$ and $\pi(G)$ is dense in $\mathfrak{G}$.

Let $G^\wedge$, $H^\wedge$ be the dual groups of G and H and Δ the annihilator of D in $G^\wedge \times H^\wedge$ ($G^\wedge \times H^\wedge$ is the dual group of $G \times H$ and Δ is the subgroup of all characters in $G^\wedge \times H^\wedge$ which are 1 everywhere on D). Then Δ is isomorphic to $\mathfrak{G}^\wedge$; we shall write $\Delta = \mathfrak{G}^\wedge$ (Lemma 4, Section 1).

Let $D^\wedge$ be the dual group of D; $D^\wedge$ is compact and isomorphic to $(G^\wedge \times H^\wedge)/\Delta$ and we shall write $D^\wedge = (G^\wedge \times H^\wedge)/\Delta$. Hence Δ is relatively dense in $G^\wedge \times H^\wedge$. Since D is relatively dense, Δ is discrete (the proof is exactly as in Proposition 2).

Let q_1 and q_2 be the two canonical projections $q_1 : G^\wedge \times H^\wedge \to G^\wedge$ and $q_2 : G^\wedge \times H^\wedge \to H^\wedge$ and $Q : G^\wedge \times H^\wedge \to D^\wedge$ the canonical homomorphism whose kernel is Δ. The dual homomorphism of $\pi : H \to \mathfrak{G}$ is $q_2 : \Delta \to H^\wedge$; the first one is injective and hence the second has a dense range. The dual homomorphism of $p_2 : D \to H$ is $Q : H^\wedge \to D^\wedge$; the first one has a dense range and hence the second is injective. We have proved.

LEMMA 9: *If G produces H, then $G^\wedge$ produces $H^\wedge$.*

Before giving an interesting result, another lemma is needed:

LEMMA 10: *If G produces H, then for each open subset Ω of H, there exists a compact subset C of G such that $G \times H = D + (C \times \Omega)$.*

Proof: Let C_1 and C_2 be two compact subsets of G and H such that $D + (C_1 \times C_2) = G \times H$. Since $p_2(D)$ is dense in H, we can find a finite subset F of D such that the compact set C_2 is contained in $p_2(F) + \Omega$. Let $C = C_1 - p_1(F)$. We have $G \times H = D + C_1 \times (p_2(F) + \Omega) = D + (C_1 - p_1(F)) \times \Omega = D + C \times \Omega$.

5.3. The first half of Theorem IV can now be proved.

PROPOSITION 3: *If G produces H, then for each open subset Ω of H with compact closure, the set Λ of all $p_1(d)$, $d \in D$, such that $p_2(d) \in \Omega$ is harmonious and relatively dense.*

Such a Λ will be called a model in the wide sense.

Proof: (a) Λ is relatively dense. We apply Lemma 10 to $-\Omega$ instead of Ω; we find a compact subset C of G such that each $(x, 0)$ can be written $(x, 0) = d + (t, \omega)$ where $d \in D$, $t \in C$ and $\omega \in -\Omega$. Hence $x = p_1(d) + t$, $0 = p_2(d) + \omega$. Then $\lambda = p_1(d) \in \Lambda$ and $G = \Lambda + C$.

(b) Λ is harmonious. Let χ be in Hom $(\Lambda_d, \mathbf{T})$; we can find ξ in Hom $(D, \mathbf{T})$ such that $\xi = \chi \circ p_1$ on the subset B of D defined by $p_2(d) \in \Omega$. This ξ belongs to the dual group of D, which is $(G^\wedge \times H^\wedge)/\Delta$ and can be lifted to yield an element X of $G^\wedge \times H^\wedge$. For each d in D, we get $\chi_\xi(d) = \chi_X(d)$ because Δ is the annihilator of D.

Let $V \subset H^\wedge$ be an open neighbourhood of 0 sufficiently small to satisfy $\sup_{\theta \in V, y \in \Omega} |\chi_\theta(y) - 1| \leqslant \varepsilon$; such a V can be found because Ω has compact closure.

Using Lemma 10, we get a compact subset K of $G^\wedge$ such that $G^\wedge \times H^\wedge = \Delta + K \times V$. Hence for each X in $G^\wedge \times H^\wedge$, there is a δ in Δ, an η in K and a θ in V such that $X = \delta + (\eta, \theta)$. By restriction of these characters to D, we have, for all d in D, $\chi_X(d) = \chi_\xi(d) = \chi_\eta(p_1(d))\, \chi_\theta(p_2(d))$. Assume that $p_2(d) \in \Omega$ (if $p_1(d) \in \Lambda$) and $\theta \in V$; we obtain $|\chi_\theta(p_2(d)) - 1| \leqslant \varepsilon$, which implies $|\chi_\xi(d) - \chi_\eta(p_1(d))| \leqslant \varepsilon$ or $|\chi(\lambda) - \chi_\eta(\lambda)| \leqslant \varepsilon$ for all λ in Λ.

5.4. *The first step towards the second part of Theorem IV*

The main result here is

PROPOSITION 4: *Let Λ be a subset of a l.c.a. group G. If Λ is harmonious and relatively dense, then Λ is contained in a model in the wide sense.*

To prove Proposition 4, the main problem is to find H 'produced' by Λ and G. Consider the group $\mathrm{Hom}(\Lambda_d, \mathbf{T})$ with the topology of pointwise convergence and let V_1 be the compact subset of $\mathrm{Hom}(\Lambda_d, \mathbf{T})$ defined by $\sup_{\Lambda} |\chi(\lambda) - 1| \leqslant \frac{1}{2}$. Then, the dual group $H^\wedge$ of H will be the subgroup of $\mathrm{Hom}(\Lambda_d, \mathbf{T})$ generated by V_1; the topology of $H^\wedge$ and of $\mathrm{Hom}(\Lambda_d, \mathbf{T})$ coincide on V_1 and H will be defined as the dual group of $H^\wedge$.

Some new terminology is needed before giving details of the proof.

5.5. *Some terminology*

Λ is the given harmonious set; R the subgroup of G_d generated by Λ (R is not the real line); $R^\wedge$ the compact dual group; $j: R \to G$ is the identical injection and $h: G^\wedge \to R^\wedge$ is the dual homomorphism; $V(\varepsilon)$ is the set of all χ in $R^\wedge$ such that $\sup_{\lambda \in \Lambda} |\chi(\lambda) - 1| \leqslant \varepsilon$; $M(\varepsilon)$ is the set of all χ in $G^\wedge$ such that $\sup_{\lambda \in \Lambda} |\chi(\lambda) - 1| \leqslant \varepsilon$; E is the group (without any topology) generated by $V(\frac{1}{2})$ in $R^\wedge$; i.e. E is the set of all sums $p_1 v_1 + \cdots + p_n v_n$ for all $n \geqslant 1$, $p_j \in \mathbf{Z}$ and v_j in $V(\frac{1}{2})$.

5.6. *Some lemmas*

LEMMA 11: *$h(G^\wedge)$ is dense in $R^\wedge$.*

Proof: j is injective.

LEMMA 12: $E + h(G^\wedge) = R^\wedge$.

Proof: Λ is harmonious.

LEMMA 13: *There is a compact neighbourhood $\mathscr{U}$ of* 0 *in $G^\wedge$, symmetric about* 0 *such that $\mathscr{U} \cap M(\frac{1}{2}) = \{0\}$ and $h(\mathscr{U}) \cap V(\frac{1}{2}) = \{0\}$.*

Proof: There is a compact subset T in G such that $\Lambda + T = G$.

Let $\mathscr{U}$ be defined by $\sup_{u \in \mathscr{U}, x \in T} |\chi_u(x) - 1| \leqslant \varepsilon$. For $v \in V(\frac{1}{2}) \cap h(\mathscr{U})$ we have $\sup_{x \in G} |\chi_v(x) - 1| \leqslant \frac{1}{2} + \varepsilon$; if $\frac{1}{2} + \varepsilon < \sqrt{3}$ this implies $v = 0$. The same proof holds for $\mathscr{U} \cap M(\frac{1}{2})$.

5.7. Let $V_n = V(2^{-n})$. Then $V_n = -V_n$ and $V_{n+1} + V_{n+1} \subset V_n$. $(V_n)_{n\geqslant 1}$ is a sequence of neighborhoods of 0 in E defining topological group structure on E. Let $\mathscr{T}$ be the corresponding topology on E and $\mathscr{S}$ the topology of the compact group $R^\wedge$.

LEMMA 14: *For each $p \geqslant 1$, $\mathscr{S}$ and $\mathscr{T}$ induce the same topology on V_p.*

Proof: We write M_n for $M(2^{-n})$. Since Λ is harmonious, M_n is relatively dense and there exists a finite subset F_n of $G^\wedge$ such that $\mathscr{U} + F_n + M_n = G^\wedge$. We have $h(M_n) \subset V_n$ and, since $\mathscr{U}$ is compact, so is $V_n + h(\mathscr{U}) + h(F_n)$ which contains $h(G^\wedge)$ and is thus the whole of $R^\wedge$.

From Baire's theorem or elementary arguments, we deduce that $V_n + h(\mathscr{U})$ has a non-empty interior. But $V_{n+1} - V_{n+1} \subset V_n$ and we can replace $\mathscr{U}$ by a $\mathscr{V}$ such that $\mathscr{V} - \mathscr{V} \subset \mathscr{U}$; hence for a suitable $\mathscr{V}$ not depending on n $V_n + h(\mathscr{V})$ is a neighbourhood of 0 in $R^\wedge$ for each $n \geqslant 2$.

Lemma 13 shows that, if $n \geqslant p$, V_n is the intersection of V_p and $V_n + h(\mathscr{V})$; hence the topology $\mathscr{T}$ defined in V_p by the V_n is weaker than the topology induced in V_p by $\mathscr{S}$. But $\bigcap_{n\geqslant 1} V_n = \{0\}$; V_p is thus a Hausdorff space under $\mathscr{T}$ and a compact Hausdorff space under $\mathscr{S}$. Hence $\mathscr{T} = \mathscr{S}$ on V_p.

5.8. We have proved that E is a locally compact abelian group. We shall attempt to construct in $G^\wedge \times E$ the discrete relatively dense subgroup Δ. Let $\mathscr{D}$ be the intersection of E and $h(G^\wedge)$; we have

LEMMA 15: *$\mathscr{D}$ is dense in E.*

Proof: Since Λ is harmonious, we have, for each $n \geqslant 1$, $h(G^\wedge) + V_n \supset E$, which implies $\mathscr{D} + V_n = E$.

Let Δ be the subgroup of $G^\wedge \times E$ defined as the set of all pairs (γ, d) where $d \in \mathscr{D}$ and $h(\gamma) + d = 0$.

LEMMA 16: *Δ is a discrete relatively dense subgroup of $G^\wedge \times E$; the intersection of Δ and $\{0\} \times E$ is $\{0, 0\}$ and $q_2(\Delta)$ is dense in E.*

(We define $q_1 : G^\wedge \times E \to G^\wedge$ and $q_2 : G^\wedge \times E \to E$ as in Section 6.)

Proof: Let $\mathscr{U}$ be as given by Lemma 13; we shall prove that the intersection of $\mathscr{U} \times V_1$ with Δ is equal to $\{0, 0\}$. If $(\gamma, d) \in \Delta \cap (\mathscr{U} \times V_1)$, then $d \in V_1 \cap h(\mathscr{U}) = \{0\}$ and $\gamma \in M_1 \cap \mathscr{U} = \{0\}$

Let K be a compact set in $G^\wedge$ such that $K + M_1 = G^\wedge$. Then $K \times V_1 + \Delta = G^\wedge \times E$ and Δ is relatively dense. To prove this, we need, for each x in $G^\wedge$, y in E, to solve $x = d + r$, $y = -h(d) + s$ with $d \in \mathscr{D}$, $r \in K$ and

$s \in V_1$. This is equivalent to solving $x = d + r$ and $y + h(x) = s + h(r)$. But $V_1 + h(K) = R^\wedge$ and the second pair of equations has a solution (s, r) in $V_1 \times K$. We put $d = x - r$ and we have to show that $d \in \mathscr{D}$; but $h(d) = s - y = h(x - r) \in E \cap h(G^\wedge)$.

Finally if (γ, d) belongs to Δ and to $\{0\} \times E$, we have $\gamma = 0$ and $d = -h(\gamma) = 0$.

5.9. Let H be the dual group of E and D the annihilator of Δ in $G \times H$; $p_1 : G \times H \to G$ is the first projection and $p_2 : G \times H \to H$ the second one. Then $\Delta \cap \{0\} \times H = \{0, 0\}$ and $p_2(D)$ is dense in H (these two properties are not changed under duality).

LEMMA 17: *There is a compact subset L of H such that Λ is contained in the set of all $p_1(d)$, $d \in D$, for which $p_2(d) \in L$.*

Proof: Let λ be an element of Λ. Then λ defines a character on $R^\wedge$ since λ belongs to R. By restriction to E, λ defines a character χ_η on $E = H^\wedge$ where η belongs to H.

For each $d = h(\gamma)$ in $\mathscr{D}$, we have $\chi_\lambda(\gamma) = \chi_\eta(d)$. The pair (λ, η) then belongs to D, since for every (γ, d) in Δ, we have $\chi_\lambda(\gamma)\, \chi_\eta(d) = \chi_\eta(d + h(\gamma)) = 1$.

Let Ω be the set of all η for λ in Λ. Then $\sup_{v \in V_n} |\chi_\eta(v) - 1| \leqslant 2^{-n}$. The following lemma will be used to prove that Ω is relatively compact in H.

LEMMA 18: *Let H be a l.c.a. group and Ω a subset of H. Assume that, for each positive ε there exists a neighbourhood $V(\varepsilon)$ of 0 in the dual group $H^\wedge$ such that for all u in Ω and v in $V(\varepsilon)$, $|\chi_v(u) - 1| \leqslant \varepsilon$. Then Ω is relatively compact in H.*

Proof: Let U be a neighbourhood of 0 in H. We have to find a finite subset F of H such that $\Omega \subset U + F$. We can assume that, for some compact subset K of $H^\wedge$ and some positive ε, U is the set of all u in H such that $|\chi_v(u) - 1| \leqslant \varepsilon$ uniformly over v in K. Let N be a finite subset of $H^\wedge$ such that $V(\varepsilon/3) + N$ contains K. Consider the restriction to N of the functions $v \to \chi_v(u)$, where u belongs to Ω. There exists a finite number $u_1, \ldots, u_n$ of such u with the following property: for each u in Ω there is a j, $1 \leqslant j \leqslant n$, such that $\sup_{v \in N} |\chi_v(u) - \chi_v(u_j)| \leqslant \varepsilon/3$. Let F be $\{u_1, \ldots, u_n\}$; then $U + F$ contains Ω. In fact if u belongs to Ω, we have, for a suitable j, $|\chi_v(u) - \chi_v(u_j)| \leqslant \varepsilon/3$ and $|\chi_w(u) - 1| \leqslant \varepsilon/3$ for each v in N and each w in $V(\varepsilon/3)$, and $|\chi_w(u_j) - 1| \leqslant \varepsilon/3$. Hence $|\chi_{v+w}(u) - \chi_{v+w}(u_j)| \leqslant \varepsilon$ for all v in N and w in $V(\varepsilon/3)$ and $u - u_j$ belongs to U.

5.10. Reduction of H to $\mathbf{R}^n$

Let G be a given l.c.a. group, H a l.c.a. group generated by G, Ω a relatively compact subset of H and Ω' a relatively compact neighbourhood of 0 in H. Let $\Lambda(\Omega)$ and $\Lambda(\Omega')$ be the models (in the wide sense) defined by Ω and Ω' (the relatively dense discrete subgroup D of $G \times H$ is the same in each of these two cases).

There exists a finite subset F of G such that $\Lambda(\Omega)$ is contained in $\Lambda(\Omega')$ $+ F$: since $p_2(D)$ is dense in H, there is a finite subset A of D such that $p_2(A) + \Omega' \supset \Omega$; we put $F = p_1(A)$.

If we wish to know all harmonious and relatively dense subsets of G, it suffices to determine the sets $\Lambda(\Omega')$ for small Ω' since the family of harmonious and relatively dense subsets of G is closed under the operation of taking the sum with a finite set.

At this stage, we use the structure theorem for l.c.a. groups ([15], Section 2.4).

LEMMA 19: *Let H be a l.c.a. group. Then H contains a closed and open subgroup H_1 which is the product of* $\mathbf{R}^n$, $n \geqslant 0$ *with a compact group* H_2.

For effecting the reduction to $\mathbf{R}^n$, the following two lemmas will be useful:

LEMMA 20: *Let G be a l.c.a. group, H a l.c.a. group produced by G, and H_1 an open subgroup of H. Then H_1 is produced by G.*

Proof: Define D_1 to be the set of all d in D such that $p_2(d) \in H_1$.

LEMMA 21: *Let G be an l.c.a. group, H_2 a compact group and* $n \geqslant 0$. *Assume that* $\mathbf{R}^n \times H_2$ *is produced by G. Then* $\mathbf{R}^n$ *is produced by G.*

Proof: Define D_2 to be the set of all d_2 in $G \times \mathbf{R}^n$ such that for an h in H_2 and a d in D, $d_2 = (x, y)$, while $d = (x, y, h)$ where $x \in G$, $y \in \mathbf{R}^n$ and $h \in H_2$.

To get the reduced model, let Ω_1 be $\Omega \cap H_1$ and Λ_1 be the model defined by H_1 and Ω_1. Then $\Lambda \subset \Lambda_1 + F$, where F is finite. Let Ω_2 be the relatively compact subset of $\mathbf{R}^n$ which is the projection on $\mathbf{R}^n$ of Ω_1 $\subset \mathbf{R}^n \times H_2$ and Λ_2 the model defined by $\mathbf{R}^n$ and Ω_2. Then $\Lambda_1 \subset \Lambda_2$ and $\Lambda \subset \Lambda_2 + F$. Theorem IV is thus proved.

6. Relatively dense harmonious sets of real numbers

6.1. Let G be the group $\mathbf{R}$ of real numbers. A closed and relatively dense discrete subgroup D of $\mathbf{R} \times \mathbf{R}^{n-1}$ $(n \geqslant 1)$ is isomorphic to $\mathbf{Z}^n$ and there exist n independent linear forms over $\mathbf{R}^n$, $l_1, \ldots, l_n$, such that each d in D is a sequence $(l_1(\mathbf{p}), \ldots, l_n(\mathbf{p}))$ for some $\mathbf{p} \in \mathbf{Z}^n$.

The models of **R** are given by an integer $n \geqslant 1$, n such linear forms $l_1, \dots, l_n$ and a relatively compact neighbourhood Ω of 0 in $\mathbf{R}^{n-1}$: Λ is the set of all $\lambda = l_1(\mathbf{p})$ such that $l_2(\mathbf{p}), \dots, l_n(\mathbf{p}) \in \Omega$. The conditions on $l_1, \dots, l_n$ are

(a) $l_1, \dots, l_n$ are linearly independent over the field **R** (D is relatively dense in $\mathbf{R}^n$);

(b) $l_1(x_1, \dots, x_n) = \omega_1 x_1 + \cdots + \omega_n x_n$, where $\omega_1, \dots, \omega_n$ are **Z**-linearly independent (the intersection of D and $\{0\} \times \mathbf{R}^{n-1}$ is equal to $\{0, 0\}$);

(c) if $l(x_1, \dots, x_n) = b_1 x_1 + \cdots + b_n x_n$, where $b_1, \dots, b_n \in \mathbf{Z}$, then $l, l_2, \dots, l_n$ are **R**-linearly independent (the second projection of D is dense in $\mathbf{R}^{n-1}$).

In this case, p_1 and p_2 restricted to D are both injective.

Each harmonious and relatively dense set of real numbers is contained in the sum $\Lambda + F$ of a model Λ and a finite set F of real numbers. In fact we can prove more:

6.2. PROPOSITION 5: *Let t_k, $k \geqslant 1$, be an increasing sequence of real numbers such that $\overline{\lim}_{k \to +\infty} t_{k+1}/t_k < +\infty$. If the set $T = \{t_1, \dots, t_k, \dots\}$ is harmonious, then T is contained in the sum $\Lambda + F$ of a model Λ and a finite set F.*

For the proof we use Lemma 8 of Chapter I, Section 7; then with the notations of Lemma 13, we have $\mathscr{U} \cap M(\varepsilon) = \{0\}$ and $h(\mathscr{U}) \cap V(\varepsilon) = \{0\}$ for ε and $\mathscr{U}$ sufficiently small. The proof now runs as in Section 5.4.

This result is accurate since if $\lim t_{k+1}/t_k = +\infty$, T is harmonious (Section 4.1).

7. Quantitative problems

7.1. Let Λ be a model of real numbers. Can any information be given about the two positive numbers $\tau(\varepsilon)$ and $T(\varepsilon)$ defined in the following way: if $0 = t_0 < t_1 \cdots < t_k < \cdots$ is the increasing sequence of all positive t such that $\sup_{\lambda \in \Lambda} \|t\lambda\| \leqslant \varepsilon$, then $t_1 = \tau(\varepsilon)$ and $T(\varepsilon) = \sup_{k \geqslant 0} (t_{k+1} - t_k)$?

To formulate results, we need the following definition:

DEFINITION 6: *Let Λ be a model of real numbers. The degree of Λ is the integer used in the definition of Λ; it is the dimension of the vector space over **Q** generated by Λ.*

To substantiate this remark, we use the obvious lemma:

LEMMA 22: *Let Γ be a dense subgroup of $\mathbf{R}^n$ and Ω a neighbourhood of 0 in $\mathbf{R}^n$. Then Γ is generated by $\Omega \cap \Gamma$.*

With the notation of Theorem IV, Γ is here equal to $p_2(D)$, which is dense in $\mathbf{R}^{n-1}$. Since p_1 and p_2 restricted to D are injective mappings, $p_1(D)$ is generated by Λ. The two definitions of n are thus equivalent.

The constants, not always the same, de noted by $a, b, \varepsilon_0, \ldots$ depend on Λ but not on $\varepsilon > 0$.

THEOREM V: *Let Λ be a model of real numbers, n the degree of Λ, $\omega_1, \ldots, \omega_n$ a basis of the* **Z***-module generated by Λ and $h : \mathbf{R} \to \mathbf{T}^n$ the continuous homomorphism defined by $h(t) = (\exp 2\pi i\omega_1 t, \ldots, \exp 2\pi i\omega_n t)$ for $t \in \mathbf{R}$. Then the following five statements are equivalent:*

(a) *for each ε in $]0, \varepsilon_0[$, $\tau(\varepsilon) \geqslant a\varepsilon^{-n+1}$;*

(b) *for each ε in $]0, \varepsilon_0[$, $T(\varepsilon) \leqslant b\varepsilon^{-n+1}$;*

(c) *$h(\mathbf{R})$ fills $\mathbf{T}^n$ rapidly* (Definition 4 of Chapter I);

(d) *for all rational integers $m_1, \ldots, m_n$, $|m_1\omega_1 + \cdots + m_n\omega_n| \geqslant d(\sup\limits_{1 \leqslant j \leqslant n} |m_j|)^{-n+1}$ (we exclude the case where $m_1 = 0, \ldots, m_n = 0$);*

(e) *for all rational integers $m_1, \ldots, m_n$ and all elements $\lambda_1, \ldots, \lambda_n$ of Λ we have*

$$m_1\lambda_1 + \cdots + m_n\lambda_n = 0 \quad \text{or} \quad |m_1\lambda_1 + \cdots + m_n\lambda_n| \geqslant d(\sup_{1 \leqslant j \leqslant n} |m_j|)^{-n+1}.$$

7.2. We shall use bold letters $\mathbf{x}$ to denote elements of $\mathbf{R}^n$ and put $x' = p_1(\mathbf{x})$, $\mathbf{x}'' = p_2(\mathbf{x})$; $|\mathbf{x}''|$ is the Euclidean norm of $\mathbf{x}''$ in $\mathbf{R}^{n-1}$. To get Theorem V, we use the following lemma; the notation is that of Theorem IV.

LEMMA 23: *Let $\mathbf{y} = (y', \mathbf{y}'') \in \mathbf{R}^n$ and let B be the set of all $\mathbf{d}$ in D, $\mathbf{d} = (d', \mathbf{d}'')$ such that $\mathbf{d}'' \in \Omega$. Then for constants ε_0, a and b, we have (1) $\Rightarrow$ (2) $\Rightarrow$ (3) for each ε in $]0, \varepsilon_0[$, where (1), (2), and (3) are the following statements:*

(1) *there is a $(\delta', \boldsymbol{\delta}'')$ in Δ (the annihilator of D) such that $y' = \delta'$, $|\mathbf{y}'' - \boldsymbol{\delta}''| \leqslant a\varepsilon$;*

(2) $\sup\limits_{\mathbf{d} \in B} \|d'y' + \mathbf{d}''.\mathbf{y}''\| \leqslant \varepsilon$ ($\|t\|$ is the distance from t to the nearest integer);

(3) *there is a $(\delta', \boldsymbol{\delta}'')$ in Δ such that $y' = \delta'$, $|\mathbf{y}'' - \boldsymbol{\delta}''| \leqslant b\varepsilon$.*

Proof of Lemma 23:

(1) $\Rightarrow$ (2): Obvious since Ω is relatively compact.

(2) $\Rightarrow$ (3): Let F be a finite subset of B such that D is generated by F. Then $\sup\limits_{\mathbf{x} \in F} \|x'y' + \mathbf{x}''.\mathbf{y}''\| \leqslant \varepsilon$ implies the existence of a $\delta = (\delta', \boldsymbol{\delta}'')$ in Δ such that $y' = \delta' + r'$, $\mathbf{y}'' = \boldsymbol{\delta}'' + \mathbf{r}''$ and $|r'| \leqslant b\varepsilon$, $|\mathbf{r}''| \leqslant b\varepsilon$. It must be proved that $r' = 0$. Replacing $\mathbf{y}'$ and $\mathbf{y}''$ in (2), we get $\sup\limits_{\mathbf{d} \in B} \|d'r' + \mathbf{d}''.\mathbf{r}''\|$

$\leqslant \varepsilon$ which implies, since Ω is bounded, that $\sup_{\lambda \in \Lambda} \|r'\lambda\| \leqslant c\varepsilon$. We have $|r'| \leqslant b\varepsilon$ and Lemma 13 now gives $r' = 0$, for ε sufficiently small. (Λ is relatively dense.)

7.3. We now turn to Theorem V. If Λ were replaced by $\{\omega_1, \ldots, \omega_n\}$, Theorem V would become a restatement of Theorem VII of Chapter I. But in Theorem V, $\tau(\varepsilon)$ and $T(\varepsilon)$ are defined by an infinite set of Diophantine inequalities. Nevertheless, Theorem V is a corollary of Theorem VII of Chapter I. We shall give the proofs of the implications (a) $\Rightarrow$ (b) $\Rightarrow$ (e) in full. The other implications can be obtained by similar methods.

Let Δ be the annihilator of D; Δ is defined by the parametric representation $y_1 = m_1(\mathbf{p}), \ldots, y_n = m_n(\mathbf{p})$ for n linearly independent linear forms $m_1, \ldots, m_n$ on $\mathbf{R}^n$ and $\mathbf{p}$ running through $\mathbf{Z}^n$. For each positive η, let $\Lambda'(\eta)$ be the set of all $y_1 = m_1(\mathbf{p})$ for all $\mathbf{p}$ in $\mathbf{Z}^n$ such that $|m_2(\mathbf{p})| \leqslant \eta, \ldots, |m_n(\mathbf{p})| \leqslant \eta$ and for each positive ε, let $M(\varepsilon)$ be the set of all real t such that $\sup_{\lambda \in \Lambda} \|t\lambda\| \leqslant \varepsilon$.

Lemma 23 shows that for each positive ε, $\Lambda'(a\varepsilon) \subset M(\varepsilon) \subset \Lambda'(b\varepsilon)$. For each positive ε and η, let $S(\varepsilon, \eta)$ be the subset of $\mathbf{R}^n$ defined by $|y'| \leqslant \eta\varepsilon^{-n+1}$ and $|\mathbf{y}''| \leqslant \varepsilon$. If (a) is satisfied, if η is small enough and if ε lies in $]0, \varepsilon_1[$, $S(\varepsilon, \eta) \cap \Delta = \{0\}$ (we here use $\Lambda'(a\varepsilon) \subset M(\varepsilon)$. Theorem IV of Chapter I (Section 2.2) shows that, for large enough η and each ε in $]0, \varepsilon_2[$, $S(\varepsilon, \eta) + \Delta = \mathbf{R}^n$. This gives (c) and with $M(\varepsilon) \subset \Lambda'(b\varepsilon)$, it gives (b).

Lemma 5 of Chapter I shows that for small enough η and each $\theta > 1$, $S(\theta, \eta) \cap D = \{0\}$. For each positive C, let Λ_C be the set of all $p_1(d) \in \mathbf{R}$ for which $\mathbf{d} \in D$ and $|p_2(\mathbf{d})| \leqslant C$. Then, for a sufficiently large C, Λ is contained in Λ_C and, if $|m_1| + \cdots + |m_n| \leqslant N$, the set of all sums $m_1\lambda_1 + \cdots + m_n\lambda_n$, $\lambda_j \in \Lambda$, is contained in Λ_{NC}. Since we know that $S(NC, \eta) \cap D = \{0\}$ if η is small enough, we have $|m_1\lambda_1 + \cdots + m_n\lambda_n| \geqslant \eta\,(NC)^{-n+1}$ or $m_1\lambda_1 + \cdots + m_n\lambda_n = 0$ if $|m_1| + \cdots + |m_n| \leqslant N$.

A special case of Theorem V occurs in models of algebraic numbers, which we now study.

8. Harmonious sets of real numbers closed under multiplication

8.1. Let K be a real algebraic number field, $n = d^0\,[K; \mathbf{Q}]$, $\sigma_1 = Id$, $\sigma_2, \ldots, \sigma_n$ the n $\mathbf{Q}$-isomorphisms from K to $\mathbf{C}$, $\mathscr{A}$ the ring of algebraic integers in K and $\omega_1, \ldots, \omega_n$ a basis of $\mathscr{A}$ over $\mathbf{Z}$. With this terminology we have

PROPOSITION 6: *For each positive number A, the set Λ of all λ in $\mathscr{A}$ such* *that* $|\sigma_2(\lambda)| \leqslant A, \ldots, |\sigma_n(\lambda)| \leqslant A$ *is a model and hence is a harmonious set of real numbers.*

Such a Λ *satisfies the equivalent conditions* (a), (b), (c), (d) *and* (e) *of Theorem* V.

We can index the σ_j in such a way that $\sigma_1, \ldots, \sigma_r$ are real and for each j, $r+1 \leqslant j \leqslant r+s$, σ_j and σ_{j+s} is a conjugate pair; we then have $n = r + 2s$. We define $l_1(x_1, \ldots, x_n) = \omega_1 x_1 + \cdots + \omega_n x_n$, $l_j(x_1, \ldots, x_n) = \sigma_j(\omega_1) x_1 + \cdots + \sigma_j(\omega_n) x_n$ for $2 \leqslant j \leqslant r$, $l_j(x_1, \ldots, x_n) = \operatorname{Re} \sigma_j(\omega_1) x_1 + \cdots + \operatorname{Re} \sigma_j(\omega_n) x_n$ for $r+1 \leqslant j \leqslant r+s$ and $l_j(x_1, \ldots, x_n) = \operatorname{Im} \sigma_j(\omega_1) x_1 + \cdots + \operatorname{Im} \sigma_n(\omega_n) x_n$ for $r+s+1 \leqslant j \leqslant n$. Dedekind's lemma shows that $\sigma_1, \ldots, \sigma_n$ are **C**-linearly independent and hence $l_1, \ldots, l_n$ are **R**-linearly independent.

Let Ω be the compact subset of $\mathbf{R}^{n-1}$ defined by $|x_2| \leqslant A, \ldots, |x_r| \leqslant A$, $x_j^2 + x_{j+s}^2 \leqslant A$ for $r+1 \leqslant j \leqslant r+s$. To prove that Λ is a model defined by these l_j and Ω, it remains to show that condition (c) of Section 6 holds: if $\sigma : K \to \mathbf{Q}$ is a **Q**-linear form over K such that $\sigma(\mathscr{A}) \subset \mathbf{Z}$, we have to show that $\sigma, \sigma_2, \ldots, \sigma_n$ are **C**-linearly independent. Assume that for some complex numbers $z_2, \ldots, z_n$ we have $z_2\sigma_2(x) + \cdots + z_n\sigma_n(x) \in \mathbf{Z}$ for each algebraic integer x in K. Choose such an algebraic integer θ of degree n in K and apply these congruences to $x = \theta^k$. We get $z_2\theta_2^k + \cdots + z_n\theta_n^k \in \mathbf{Z}$ for all $k \geqslant 0$ and the proof is now concluded as in Theorem XI of Chapter I.

8.2. PROPOSITION 7: *Let* θ *be a real number greater than* 1 *and* Λ *the set of all finite sums* $\sum_{k \geqslant 0} \varepsilon_k \theta^k$, where each $\varepsilon_k = 0$ *or* 1. *Then the following two statements are equivalent:*

(a) θ *is a Pisot number;*

(b) Λ *is harmonious.*

(a) $\Rightarrow$ (b): Let $\theta_2, \ldots, \theta_n$ be the conjugates of θ different from θ. Then each λ in Λ is an algebraic integer of the field of θ and, for $2 \leqslant j \leqslant n$, $|\sigma_j(\lambda)| \leqslant 1/(1 - |\theta_j|)$. Hence, by Proposition 6, Λ is harmonious.

(b) $\Rightarrow$ (a): There exists a real number $t \neq 0$ such that, for each λ in Λ, $\|t\lambda\| \leqslant \frac{1}{5}$. Put $t\theta^k = p_k + r_k$, where $p_k \in \mathbf{Z}$ and $|r_k| \leqslant \frac{1}{5}$. For each $k \geqslant 0$ and $\varepsilon_j = 0$ or 1, we have $|\varepsilon_0 r_0 + \cdots + \varepsilon_k r_k| \leqslant \frac{1}{5}$; this inequality is true for $k = 0$ and can be proved by induction on k: if $\lambda = \varepsilon_0 + \cdots + \varepsilon_k\theta^k + \varepsilon_{k+1}\theta^{k+1}$, then $t\lambda = q + \varepsilon_0 r_0 + \cdots + \varepsilon_k r_k + \varepsilon_{k+1} r_{k+1} = q + s_{k+1}$, where $q \in \mathbf{Z}$ and $|s_{k+1}| \leqslant \frac{2}{5}$. Hence $|s_{k+1}| = \|t\lambda\|$ and $|s_{k+1}| \leqslant \frac{1}{5}$.

Separating the positive and negative r_k, this implies that $|r_0| + \cdots + |r_k| \leqslant \frac{2}{5}$ and $\sum_0^\infty \|t\theta^k\| < +\infty$. Hence θ is a Pisot number.

8.3. THEOREM VI: *Let K be a real algebraic number field of degree n over* **Q** *and* Λ *the set of all Pisot or Salem numbers of degree n in K. Then* Λ *is contained in* $]1, +\infty[$, *is harmonious and is closed under multiplication. Conversely let M be contained in* $]1, +\infty[$, *harmonious and closed under multiplication. Then there exists a real algebraic number field K of degree n over* **Q** *such that M is contained in the set of all Pisot or Salem numbers of degree n in K.*

Proof: Λ is defined by $\lambda \in \mathscr{A}$, $\lambda > 1$ and $|\sigma_j(\lambda)| \leqslant 1$ for $2 \leqslant j \leqslant n$. Proposition 6 shows that Λ is harmonious and Λ is clearly closed under multiplication.

Conversely let x be in M. The set of all powers of x is harmonious since M is closed under multiplication. Hence x is a Pisot or a Salem number; let $n = d^0\,[x; \mathbf{Q}]$ and $K = \mathbf{Q}[x]$. We shall prove that M is contained in K. We already know that each y in M is algebraic over $\mathbf{Z}$ and hence algebraic over K; let p be the degree of y over K. If $p = 1$ there is nothing to prove. If $p > 1$ there are exactly p K-isomorphisms from $K[y]$ into $\mathbf{C}$. Let τ be one of these isomorphisms such that $z = \tau(y) \neq y$. Then $\tau\,(x^m y) = x^m z$ for each positive integer m; $x^m y$ and $x^m z$ are conjugate over K and hence also conjugate over $\mathbf{Q}$. Since $x > 1$, we can find an m such that $|x^m z| > 1$; in this case $x^m y \in M$ cannot be a Pisot or a Salem number (see [19] for the proofs of the algebraic tools used here).

9. Another definition of models of real numbers

With the notation of Theorem IV, an equivalent definition of models of real numbers can be given with the help of the compact group $\mathfrak{G}$ instead of H. Some fresh notation is needed.

Let

n be an integer, $n \geqslant 1$;

$\mathbf{T}^n$ be the n-dimensional torus;

$\Pi : \mathbf{R}^n \to \mathbf{T}^n$ be the canonical homomorphism whose kernel is $\mathbf{Z}^n$;

$N \subset \mathbf{R}^n$ be an hyperplane in $\mathbf{R}^n$;

$U \subset N$ be a compact subset of N;

$\mathscr{U} = \Pi(U)$ be the corresponding compact subset of $\mathbf{T}^n$;

$\alpha_1, \ldots, \alpha_n$ be n real numbers linearly independent over $\mathbf{Q}$;

$\boldsymbol{\alpha} = (\alpha_1, \ldots, \alpha_n)$ be the corresponding vector in $\mathbf{R}^n$; $\boldsymbol{\alpha}$ does not lie in N;

$h : \mathbf{R} \to \mathbf{R}^n$ be the injective map with dense range defined by $h(t) = (\exp 2\pi i\alpha_1 t, \ldots, \exp 2\pi i\alpha_n t)$.

With this terminology we have

PROPOSITION 8: *The set Λ of all real λ such that $h(\lambda) \in \mathcal{U}$ is a model of real numbers and each model of real numbers can be defined in this way.*

Proof: Using the notation of Theorem IV, $G \times H$ is replaced by $\mathbf{R}^n$, G is replaced by the line whose parametric equations are $x_1 = \alpha_1 t, \ldots, x_n = \alpha_n t$; H is replaced by N and the restriction of Π to G is injective; otherwise, still with the terminology of Theorem IV, $p_2(D)$ would not be dense in H.

The numbers $\alpha_1, \ldots, \alpha_n$ can be defined, in the language of Section 6, by $l_1(\alpha_1, \ldots, \alpha_n) = 1$, $l_j(\alpha_1, \ldots, \alpha_n) = 0$ for $2 \leqslant j \leqslant n$. If Λ is defined by Theorem VI, $\alpha_1, \ldots, \alpha_n$ is a dual basis of a basis $(\omega_1, \ldots, \omega_n)$ of algebraic integers in K.

10. Harmonious sets in p-adic fields

10.1. Let p be a prime, $\mathbf{Q}_p$ the p-adic field and A_p the subset of $\mathbf{Q}_p$ consisting of all ordinary rational numbers of the form mp^n, where m and $n \in \mathbf{Z}$. The p-adic norm $|\ \ |_p$ is defined on A_p by $|mp^n| = p^{-n}$ if $m, n \in \mathbf{Z}$ and p does not divide m. Then $A_p \subset \mathbf{Q} \subset \mathbf{Q}_p$ and $\mathbf{Q}_p$ can be defined as the completion of A_p with respect to $|\ \ |_p$.

The ring A_p is principal. Since $\mathbf{Z}$ is contained in A_p, $J = I \cap \mathbf{Z}$ is an ideal of $\mathbf{Z}$ for each ideal I of A_p. Hence $J = m\mathbf{Z}$ and $I \subset mA_p$. But, for each x in I, there exists a rational integer r such that $xp^r \in \mathbf{Z}$. Hence $x \in mA_p$ and the proof is complete.

Let $\mathbf{Z}_p$ be the closure of $\mathbf{Z}$ in $\mathbf{Q}_p$; $\mathbf{Z}_p$ is an open and closed subgroup of $\mathbf{Q}_p$ defined by $|x|_p \leqslant 1$ and $\mathbf{Z}_p \cap \mathbf{Q} = \mathbf{Z}$. Each element x of $\mathbf{Q}_p$ can be written uniquely as $\xi + r$, where $\xi \in \mathbf{Z}_p$ and $r \in A_p$, $-\frac{1}{2} \leqslant r < \frac{1}{2}$. This decomposition of x is called Artin's decomposition. We put $r = \varrho_p(x)$. For each y in $\mathbf{Q}_p$, the mapping $\chi_y : \mathbf{Q}_p \to \mathbf{T}$ defined by $\chi_y(x) = \exp 2\pi i \varrho_p(xy)$ is a continuous homomorphism and each element of Hom $(\mathbf{Q}_p, \mathbf{T})$ can be written in this way. If x and y belong to A_p, $\exp 2\pi i \varrho_p(xy) = \exp 2\pi i xy$ and by a misuse of language we write $\exp 2\pi i xy$ instead of $\chi_y(x)$. The dual group of $\mathbf{Q}_p$ is $\mathbf{Q}_p$ itself.

Problems of repartition modulo 1 in $\mathbf{R}$ become here problems of repartition modulo $\mathbf{Z}_p$.

The field $\mathbf{Q}_p$ does not contain discrete subgroups. To discretize on $\mathbf{Q}_p$, harmonious subsets of $\mathbf{Q}_p$ will be needed. We shall now learn how to construct these sets.

10.2. On A_p there are two norms: the p-adic one for which A_p is dense in $\mathbf{Q}_p$ and the Archimedean one, denoted by $|\ \ |_\infty$ for which A_p is dense in $\mathbf{R}$. These two points of view of A_p are fundamental for the following

PROPOSITION 9: *For each positive real number c, let Λ_c be the set of all λ in A_p such that $|\lambda|_\infty \leqslant c$. Then Λ_c is harmonious and relatively dense in $\mathbf{Q}_p$.*

Proof: Let D be the subgroup of $\mathbf{Q}_p \times \mathbf{R}$ consisting of all (λ, λ) for $\lambda \in A_p$; D is discrete since $|\lambda|_p \leqslant 1$ and $|\lambda|_\infty \leqslant \frac{1}{2}$ imply $\lambda = 0$. Let C be the compact subset of $\mathbf{R}$ defined by $|t| \leqslant \frac{1}{2}$. Consider $\Lambda + \mathbf{Z}_p \times C$. This set is closed in $\mathbf{Q}_p \times \mathbf{R}$ since Λ is discrete. To prove that $\Lambda + \mathbf{Z}_p \times C = \mathbf{Q}_p \times \mathbf{R}$, we have to show that $\Lambda + \mathbf{Z}_p \times C$ contains $A_p \times A_p$, which is dense in $\mathbf{Q}_p \times \mathbf{R}$. We are led to solve the system $x = \lambda + s$, $y = \lambda + t$ for each x and y in A_p, where $\lambda \in A_p$, $s \in \mathbf{Z}_p$ and $|t| \leqslant \frac{1}{2}$. We replace this system by the equivalent one $x = \lambda + s$, $x - y = s - t$; the second equation can be solved with $s \in \mathbf{Z}$ and $-\frac{1}{2} \leqslant t < \frac{1}{2}$ and the first one then gives λ in A_p.

At this stage, we apply Theorem IV. The harmonious sets Λ_c play exactly the same role in $\mathbf{Q}_p$ to discretize distributions as the subgroups $h\mathbf{Z}$ of $\mathbf{R}$ $(h > 0)$. We refer to Chapter V, Section 8.

10.3. To get all harmonious and relatively dense subsets of $\mathbf{Q}_p$, we have to find all discrete and relatively dense subgroups D of $\mathbf{Q}_p \times \mathbf{R}^n$; here $n \geqslant 1$ since $\mathbf{Q}_p$ does not contain such a D.

We shall regard $\mathbf{Q}_p$ as a module on the principal ring A_p.

PROPOSITION 10: *Let $\omega_1, \ldots, \omega_n$ be n p-adic numbers linearly independent over A_p and let $a_2, \ldots, a_n$ be $n - 1$ p-adic numbers such that $\omega_1 + a_2\omega_2 + \cdots + a_n\omega_n \neq 0$. Let $Y_1, \ldots, Y_n$ be n vectors in $\mathbf{R}^n$ linearly independent over $\mathbf{R}$. For each positive c, a subgroup D of $\mathbf{Q}_p \times \mathbf{R}^n$ is defined by the parametric representation $x = t_1\omega_1 + \cdots + t_n\omega_n$, $Y = t_1Y_1 + \cdots + t_nY_n$, where $t_1, \ldots, t_n \in A_p$ and $|t_2 - a_2t_1|_p \leqslant c, \ldots, |t_n - a_nt_1|_p \leqslant c$.*

Then D is a discrete and relatively dense subgroup of $\mathbf{Q}_p \times \mathbf{R}^n$ and the intersection of D with the subgroup $\{0\} \times \mathbf{R}^n$ of $\mathbf{Q}_p \times \mathbf{R}^n$ is equal to $\{0\}$.

Conversely let D′ be a discrete and relatively dense subgroup of $\mathbf{Q}_p \times \mathbf{R}^n$ such that the intersection of D′ with the subgroup $\{0\} \times \mathbf{R}^n$ of $\mathbf{Q}_p \times \mathbf{R}^n$ is equal to $\{0\}$. Then there exists a subgroup D defined as in the first part of Proposition 10 such that D′ is contained in D.

(a) D is discrete. The inequalities $|x|_p = |t_1\omega_1 + \cdots + t_n\omega_n|_p \leqslant 1$, $|t_2 - a_2t_1|_p \leqslant c, \ldots, |t_n - a_nt_1|_p \leqslant c$ imply $|t_1|_p \leqslant a, \ldots, |t_n|_p \leqslant a$. Let L_a be the set of all $t_1Y_1 + \cdots + t_nY_n$ for $t_1, \ldots, t_n$ in A_p such that

$|t_1|_p \leqslant a, \ldots, |t_n|_p \leqslant a$. Then L_a is a lattice in $\mathbf{R}^n$. The ordinary Euclidean norm in $\mathbf{R}^n$ is written $|\ \ |_\infty$, and we can find a positive ε such that $Y \in L_a$ and $|Y|_\infty \leqslant \varepsilon$ imply $Y = 0$. Hence D is discrete.

(b) D is relatively dense. The subgroup of $\mathbf{Q}_p$ defined by $x = t_1\omega_1 + \cdots + t_n\omega_n$ for those $t_1, \ldots, t_n$ in A_p such that $|t_2 - a_2 t_1|_p \leqslant c, \ldots, |t_n - a_n t_1|_p \leqslant c$ is dense in $\mathbf{Q}_p$: if we want to get $|x - t_1\omega_1 - t_2\omega_2 - \cdots - t_n\omega_n|_p \leqslant 1$, we let t_1 be a number in A_p sufficiently near $(\omega_1 + a_2\omega_2 + \cdots + a_n\omega_n)^{-1} x$, we let t_2 be a number in A_p sufficiently near $a_2 t_1$, and so on.

Starting from this approximation of x, we shall modify $t_1, \ldots, t_n$ to get a worse approximation of x, but a better approximation of a given Y in $\mathbf{R}^n$ by $t_1' Y_1 + \cdots + t_n' Y_n$ ($t_1', \ldots, t_n'$ are the modified $t_1, \ldots, t_n$).

Precisely, we define $\varepsilon > 0$ by $\varepsilon \leqslant c$, $\varepsilon |a_2|_p \leqslant c, \ldots, \varepsilon |a_n|_p \leqslant c$ and replace t_1 by $t_1' = t_1 + r_1, \ldots, t_n$ by $t_n' = t_n + r_n$, where $r_1, \ldots, r_n$ are rational integers such that $|r_1|_p \leqslant \varepsilon, \ldots, |r_n|_p \leqslant \varepsilon$. We get $|t_2' - a_2 t_1'|_p \leqslant c, \ldots, |t_n' - a_n t_1'|_p \leqslant c$ and $|x - t_1'\omega_1 - \cdots - t_n'\omega_n|_p \leqslant d$ (d does not depend on x or on $t_1', \ldots, t_n'$). The set of all corresponding $t_1' Y_1 + \cdots + t_n' Y_n$ has the form $X + L_\varepsilon$; X depends on $t_1, \ldots, t_n$ and L_ε is the lattice defined in the first part of the proof. Since L_ε is relatively dense, we can solve $|Y - t_1' Y_1 - \cdots - t_n' Y_n|_\infty \leqslant a$ for each Y in $\mathbf{R}^n$ by judicious choice of $r_1, \ldots, r_n$ (a depends only on $Y_1, \ldots, Y_n$ and ε). Therefore D is relatively dense.

To get the converse result, the following lemma is needed:

LEMMA 24: *Let H be a subgroup of $\mathbf{Q}_p$ and $f: H \to \mathbf{Q}_p$ an additive function. Assume that $x \in H$ and $|x|_p \leqslant 1$ imply $|f(x)|_p \leqslant 1$. Then there exist a positive b and a p-adic number α such that, for each x in H, $|f(x) - \alpha x|_p \leqslant b$.*

In the proof it will be assumed that H is dense in $\mathbf{Q}_p$. The obvious modifications should be made in the other case.

(a) $|x|_p > 1$ implies $|f(x)|_p \leqslant |x|_p$. In fact, if $p^k = |x|_p$, then $p^k x \in \mathbf{Z}_p$ and since f is additive, $f(p^k x) = p^k f(x)$ ($k \geqslant 1$ since $|x|_p > 1$). We get $|f(p^k x)|_p = |p^k|_p\, |f(x)|_p \leqslant 1$. Therefore $|f(x)|_p \leqslant |x|_p$.

(b) $1 < |y|_p \leqslant |x|_p$ implies

$$\left| \frac{f(x)}{x} - \frac{f(y)}{y} \right|_p \leqslant \frac{1}{|y|_p}.$$

Since $x\mathbf{Z}$ is dense in the ball of radius $|x|_p$, we can find a rational integer m such that $|y - mx|_p \leqslant 1$; this implies $|f(y) - mf(x)|_p \leqslant 1$. The inequality

$|f(x)|_p \leqslant |x|_p$ and an easy computation give

$$\left|\frac{f(x)}{x} - \frac{f(y)}{y}\right|_p \leqslant \frac{1}{|y|_p}.$$

When x tends to infinity $(f(x))/x$ is a Cauchy sequence in $\mathbf{Q}_p$; let α be its limit. We have

$$\left|\alpha - \frac{f(y)}{y}\right|_p \leqslant \frac{1}{|y|_p},$$

which gives Lemma 24.

Returning to Proposition 10, let D_a, for each a in $\mathbf{Q}_p$, be the set of all $d = (x, Y)$ in $\mathbf{Q}_p \times \mathbf{R}^n$, such that $|x|_p \leqslant |a|_p$. Since D is relatively dense, D_a is relatively dense in $a\mathbf{Z}_p \times \mathbf{R}^n$ provided $|a|_p$ is sufficiently large. The projection L of D_a on $\mathbf{R}^n$ is a discrete and relatively dense subgroup of $\mathbf{R}^n$; therefore $L = \mathbf{Z}Y_1 + \cdots + \mathbf{Z}Y_n$. The only element common to D and $\mathbf{Q}_p \times \{0\}$ is 0 since their intersection would be a discrete subgroup of $\mathbf{Q}_p$. On the other hand we have assumed that the only element common to D and $\{0\} \times \mathbf{R}^n$ is 0. Therefore, we can find n p-adic numbers $\omega_1, \ldots, \omega_n$ independent over $\mathbf{Z}$ such that $(\omega_1, Y_1) \in D, \ldots, (\omega_n, Y_n) \in D$.

For each (x, Y) in D, there is a rational integer k such that $(p^k x, p^k Y) \in D_a$. For each (x, Y) in D we can therefore write $x = t_1\omega_1 + \cdots + t_n\omega_n$, $Y = t_1 Y_1 + \cdots + t_n Y_n$ for a suitable sequence $t_1, \ldots, t_n$ of elements of A_p, and the mappings $f_1, \ldots, f_n$ from $H = A_p\omega_1 + \cdots + A_p\omega_n$ to A_p defined by $f_1(x) = t_1, \ldots, f_n(x) = t_n$ are additive. Moreover, if $|x|_p \leqslant |a|_p$, we get $|t_1|_p \leqslant 1, \ldots, |t_n|_p \leqslant 1$. Lemma 24 shows that $|t_1 - \alpha_1 x|_p \leqslant b$, $\ldots, |t_n - \alpha_n x|_p \leqslant b$; if $\alpha_1 = \cdots = \alpha_n = 0$, D could not be relatively dense and if, for example, $\alpha_1 \neq 0$ Proposition 10 is proved.

10.4. Definition 7: *With the notation of Proposition* 10, *let* Λ_c, *for each* $c > 0$, *be the set of all* $x = t_1\omega_1 + \cdots + t_n\omega_n$ *in* $\mathbf{Q}_p$ *such that* $|t_1|_\infty \leqslant c$, $\ldots, |t_n|_\infty \leqslant c$ *and* $|t_2 - a_2 t_1|_p \leqslant c, \ldots, |t_n - a_n t_1|_p \leqslant c$ *with* $t_1 \in A_p$, $\ldots, t_n \in A_p$.

Such a Λ_c *is called a model in* $\mathbf{Q}_p$.

Theorem VII: *Each model in* $\mathbf{Q}_p$ *is harmonious and relatively dense. For each harmonious and relatively dense subset M of* $\mathbf{Q}_p$, *there exists a model* Λ *and a finite subset F of* $\mathbf{Q}_p$ *such that M is contained in* $\Lambda + F$.

Proof: Theorem IV and Proposition 10 give Theorem VII.

11. Counter-examples

PROPOSITION 11: *There exists a metrizable and separable l.c.a. group G with the following properties:*

(a) *G is not compact;*

(b) *G is torsion-free;*

(c) *each harmonious subset Λ of G is finite.*

Proof: Let p be a prime, $p \geqslant 2$, and let $\mathbf{Z}_p$ be the compact group of p-adic integers. The compact subgroup $p\mathbf{Z}_p$ of $\mathbf{Z}_p$ is defined by $|x|_p \leqslant p^{-1}$ and we denote by N the compact group $(p\mathbf{Z}_p)^N$ with the product-topology (convergence means componentwise convergence).

To start with, we construct the group G (without topology): G is the subgroup of $(\mathbf{Z}_p)^N$ defined by the condition that $\omega = (\omega_0, \ldots, \omega_k, \ldots)$ belongs to G if and only if ω_k belongs to $p\mathbf{Z}_p$ for large enough k (i.e. $k \geqslant k(\omega)$, say). Since $\mathbf{Z}_p$ is torsion-free, so is G.

The subgroup N of $(\mathbf{Z}_p)^N$ defined by $\omega_k \in p\mathbf{Z}_p$ for all $k \geqslant 0$ is contained in G and a topology $\mathcal{T}$ on G is given by the following two conditions: the topology $\mathcal{T}$ induces on N its topology as the compact group specified above and N is open in G.

Then G is an l.c.a. group; G is metrizable and separable and G/N is a countable discrete abelian group isomorphic to the direct sum of a countable set of copies of $\mathbf{Z}/p\mathbf{Z}$. Therefore G is not compact.

Let $h \in \operatorname{Hom}(G, N)$ be defined by $h(x) = px$; the kernel of h is $\{0\}$.

If $\Lambda \subset G$ is harmonious, so is $h(\Lambda) = p\Lambda$, and $h(\Lambda)$ is contained in the compact subgroup N of G. Therefore $h(\Lambda)$ is finite and so is Λ.

The dual group Γ of G is the set of all sequences $\gamma = (\gamma_k)_{k \geqslant 0}$, where $\gamma_k \in A_p/\mathbf{Z}$ for all $k \geqslant 0$ and $\gamma_k \in (p^{-1}\mathbf{Z})/\mathbf{Z}$ for sufficiently large k. A fundamental sequence $(\Gamma_n)_{n \geqslant 0}$ of neighbourhoods of 0 for the topology of Γ is defined by $\gamma \in \Gamma_n$ if and only if $\gamma_0 = \cdots = \gamma_{n-1} = 0$ while $\gamma_k \in (p^{-1}\mathbf{Z})/\mathbf{Z}$ for all $k \geqslant n$. Hence each element γ of Γ_n is of order p (see Theorem III in Section 4.1).

12. Pisot-Salem-Chabauty numbers in $\mathbf{Q}_p$

12.1. The field of fractions of the principal ring A_p is $\mathbf{Q}$. Therefore if K is an algebraic extension of $\mathbf{Q}$ of degree n, the ring $\mathcal{A}$ of all algebraic integers over the ring A_p which belong to K is a free A_p-module and $\mathcal{A} = A_p\omega_1 + \cdots + A_p\omega_n$ (see [19], p. 48, th. 1).

Let Ω_p be the algebraic closure ([19], § 2.3, p. 39) of $\mathbf{Q}_p$ and $|\ \ |_p$ be the p-adic norm on Ω_p which extends the given one on $\mathbf{Q}_p$. Then for an element x of Ω_p, the following two properties are equivalent: $|x|_p \leqslant 1$ and, for some $n \geqslant 1$ and $a_1, \ldots, a_n$ in $\mathbf{Z}_p$, $x^n + a_1 x^{n-1} + \cdots + a_n = 0$ (x is an algebraic integer over the ring $\mathbf{Z}_p$). The reader is referred to [3], theorem 2, Section 1, Chapter 4.

Let $\sigma_1, \ldots, \sigma_n$ be the n $\mathbf{Q}$-isomorphisms from K into Ω_p and $\tau_1, \ldots, \tau_n$ the n $\mathbf{Q}$-isomorphisms from K into $\mathbf{C}$.

12.2. In Theorem VI, we studied the set Λ of all Pisot or Salem numbers of maximum degree lying in a real algebraic number field K; Λ was contained in the free $\mathbf{Z}$-module $\mathbf{Z}\omega_1 + \cdots + \mathbf{Z}\omega_n$ of all algebraic integers of K and, for each λ in Λ, we had $|\sigma_2(\lambda)| \leqslant 1, \ldots, |\sigma_n(\lambda)| \leqslant 1$. Hence Λ was a model.

In $\mathbf{Q}_p$, Theorem VII shows that a model is contained in a free A_p-module $A_p\omega_1 + \cdots + A_p\omega_n$. Let K be an algebraic number field of degree n over $\mathbf{Q}$ contained in $\mathbf{Q}_p$. If we try to define the set Λ of all Pisot and Salem numbers in K as a model closed under multiplication, we are led to consider a free A_p-submodule of K closed under multiplication and therefore Λ is contained in the set of all algebraic integers over the ring A_p ([19], Chapter II).

A model in $\mathbf{Q}_p$ is defined by $n-1$ inequalities involving p-adic norms and n conditions on Archimedean norms. If we seek models which stable with respect to multiplication we are led to the following definition:

Definition 8: *Let θ be a p-adic number. We say that θ is a Pisot-Chabauty number if θ is an algebraic integer over the ring A_p, such that $|\sigma_2(\theta)|_p \leqslant 1, \ldots, |\sigma_n(\theta)|_p \leqslant 1$, $|\tau_1(\theta)|_\infty < 1, \ldots, |\tau_n(\theta)|_\infty < 1$, where n is the degree of θ, $K = \mathbf{Q}(\theta) \subset \mathbf{Q}_p$ and $\sigma_1 = Id: K \to \mathbf{Q}_p, \sigma_2, \ldots, \sigma_n$ are the n $\mathbf{Q}$-isomorphisms from K into Ω_p.*

If, with the same notation, θ is an algebraic integer over A_p, and if $|\sigma_2(\theta)|_p \leqslant 1, \ldots, |\sigma_n(\theta)|_p \leqslant 1$, $|\tau_1(\theta)|_\infty \leqslant 1, \ldots, |\tau_n(\theta)|_\infty \leqslant 1$, then θ is said to be a Pisot-Salem-Chabauty number in $\mathbf{Q}_p$.

12.3. Proposition 12. *If θ is a Pisot-Chabauty number in $\mathbf{Q}_p$, $\varrho_p(\theta^k)$ tends to zero as k tends to infinity. If θ is algebraic, if $|\theta|_p > 1$ and if $\varrho_p(\theta^k)$ tends to zero, θ is a Pisot-Chabauty number. If θ is a p-adic number, if $|\theta|_p > 1$ and if, for some p-adic number $\lambda \neq 0$, $\sum_0^\infty (\varrho_p(\lambda\theta^k))^2 < +\infty$, θ is a Pisot-Chabauty number.*

Assume θ is a Pisot-Chabauty number. Then $\sigma_2(\theta), \ldots, \sigma_n(\theta)$ are integers over the ring $\mathbf{Z}_p$. Since θ is integral over A_p, $\theta^k + (\sigma_2(\theta))^k + \cdots + (\sigma_n(\theta))^k \in A_p$ for each positive k. Hence $(\sigma_2(\theta))^k + \cdots + (\sigma_n(\theta))^k$

belongs to $\mathbf{Q}_p$ and is integral over $\mathbf{Z}_p$. Since $\mathbf{Q}_p$ is integrally closed, $(\sigma_2(\theta))^k + \cdots + (\sigma_n(\theta))^k \in \mathbf{Z}_p$ ([19], § 2.2, p. 36). We have $\theta^k + (\sigma_2(\theta))^k + \cdots + (\sigma_n(\theta))^k = (\tau_1(\theta))^k + \cdots + (\tau_n(\theta))^k$; A_p is here a subring of $\mathbf{Q}_p$ and of $\mathbf{C}$. Hence the archimidean norm of this sum tends to zero as k tends to infinity and we can write $\theta^k = \xi_k + r_k$ $(k \geqslant 0)$ where $\xi_k \in \mathbf{Z}_p$ and $r_k \to 0$ $(k \to +\infty)$. Moreover $\sum_0^\infty |r_k| < +\infty$.

Next, assume θ is algebraic and $\varrho_p(\lambda\theta^k) \to 0$. There exist an integer $n \geqslant 1$ and $n+1$ rational integers $a_0, \ldots, a_n$ such that $a_0 + \cdots + a_n\theta^n = 0$; hence for all $k \geqslant 0$, $a_0\theta^k + \cdots + a_n\theta^{k+n} = 0$. Writing $\lambda\theta^k = \xi_k + r_k$ with $\xi_k \in \mathbf{Z}_p$, $-\frac{1}{2} \leqslant r_k < \frac{1}{2}$, $r_k \in A_p$ and $s_k = a_0\xi_k + \cdots + a_n\xi_{k+n}$, we observe that $s_k \in \mathbf{Z}_p$ for all $k \geqslant 0$ and at the same time $s_k = -a_0 r_k - \cdots - a_n r_{k+n}$. Hence $s_k \in \mathbf{Z}_p \cap A_p = \mathbf{Z}$ and $s_k = 0$ for $k \geqslant k_0$.

On each principal ring, Fatou's lemma is valid (Chapter I, Section 6.2).

Since A_p is principal, we can find two relatively prime elements $P(X)$ and $Q(X)$ of $A_p[X]$ such that $Q(0) = 1$ and $\sum_{k \geqslant 0} r_k X^k = P(X)/Q(X)$. Since $|r_k|_\infty \to 0$ as $k \to +\infty$, the rational fraction $P(X)/Q(X)$ does not have pole in the closed disc $|z|_\infty \leqslant 1$. Writing $r_k = \lambda\theta^k - \xi_k$, we get $P(X)/Q(X) = \lambda/(1-\theta X) - \sum_{k \geqslant 0} \xi_k X^k$. Since $|\xi_k|_p \leqslant 1$, the series $\sum_{k \geqslant 0} \xi_k X^k$ is convergent if $X \in \Omega_p$, $|X|_p < 1$; therefore $1/\theta$ is the unique root of Q in Ω_p whose absolute value is less than one. Hence θ is a Pisot number.

$-\sum_0^\infty (\varrho_k(\lambda\theta^k))^2 < +\infty$. We put $\lambda\theta^k = \xi_k + r_k$ and $\eta_k = r_k - \theta r_{k-1} = \theta\xi_{k-1} - \xi_k$ $(k \geqslant 1)$. Then $|\eta_k|_p \leqslant |\theta|_p$ and $|r_k|_p \leqslant |\lambda|_p\, |\theta|_p^k$. Let D_n be the determinant

$$\begin{vmatrix} r_0 & \cdots & r_n \\ \cdots & \cdots & \cdots \\ r_n & \cdots & r_{2n} \end{vmatrix}.$$

If we show that for $n \geqslant n_0$ $D_n = 0$, we can (as Chapter I, Section 6.5) find an integer n (in fact the smallest n for which $D_n = 0$) and $n+1$ rational integers $a_0, \ldots, a_n$ such that $a_0 r_k + \cdots + a_n r_{k+n} = 0$ for all $k \geqslant 0$. This implies $\lambda\theta^k(a_0 + \cdots + a_n\theta^n) \in \mathbf{Z}_p$ for all $k \geqslant 0$ and since $|\theta|_p > 1$, we get $a_0 + \cdots + a_n\theta^n = 0$. Therefore θ is algebraic and is a Pisot number. To prove that $D_n = 0$ for large enough n, the following lemma will be needed:

LEMMA 25: *If $x \in A_p$ and $|x|_p\, |x|_\infty < 1$, then $x = 0$.*

Proof: Write $x = qp^m$, where p does not divide q; $|x|_p = p^{-m}$, $|x|_\infty = |q|\, |p^m|$ and $|x|_p\, |x|_\infty < 1$ imply $|q| < 1$, hence $q = 0$ and $x = 0$.

Since $\sum_0^\infty r_k^2 < +\infty$, Hadamard's inequality (Chapter I, Section 6.4) shows that, for all positive ε, $D_n = 0(\varepsilon^n)$ as $n \to +\infty$. To get an estimate of $|D_n|_p$ we substract from each column of D_n the preceding one multiplied by θ; we find

$$D_n = \begin{vmatrix} r_0 & \eta_1 & \cdots & \eta_n \\ \cdot\cdot & \cdots & \cdots & \cdots \\ r_n & \eta_{n+1} & \cdots & \eta_{2n} \end{vmatrix}.$$

We expand this new D_n as a sum (or difference) of $(n + 1)!$ elementary products of $n + 1$ terms chosen in D_n; the p-adic norm of each of these products is bounded by $|\lambda|_p\,|\theta|_p^{2n}$ and $|D_n|_p \leqslant |\lambda|_p\,|\theta|_p^{2n}$. Hence $|D_n|_\infty\,|D_n|_p < 1$ for n sufficiently large and since D_n and each r_k belong to A_p, Lemma 25 can be applied [35].

12.4. Let K be an algebraic number field contained in $\mathbf{Q}_p$ and let Λ be the set of all Pisot-Salem-Chabauty numbers θ in K whose degree is equal to the degree of K over $\mathbf{Q}$. Just as in the real case, we get

THEOREM VIII: *The set Λ is harmonious and closed under multiplication. Conversely if M is an harmonious subset of $\mathbf{Q}_p$ closed under multiplication, M is contained in $F\Lambda$ where F is the finite set of roots of unity lying in K and Λ is defined as in the first part of the theorem.*

$F\Lambda$ is the set of all products $f\lambda$ where $f \in F$, $\lambda \in \Lambda$.

The first part of Theorem VIII is an obvious application of Proposition 10. To get the second part, we proceed exactly as in the real case; for more details the reader is referred to [89]. The set of roots of unity in $\mathbf{Q}_p$ is found by the method explained in [20], Chapter II, Section 3.

12.5. As in the real case, we have the following characterization of Pisot-Chabauty numbers: *let θ be a p-adic number and Λ the set of all finite sums $\sum_{k\geqslant 0}\varepsilon_k\theta^k$ where $\varepsilon_k = 0$ or 1. Then Λ is harmonious if and only if θ is a Pisot-Chabauty number.*

Firstly, assume θ is a Pisot-Chabauty number. Let K be the field of θ and $\mathscr{A} = A_p\omega_1 + \cdots + A_p\omega_n$ be the ring of all elements of K which are integral over A_p; $\sigma_1 : K \to \mathbf{Q}_p$ is the identity mapping, $\sigma_2, \ldots, \sigma_n$ the $n - 1$ other $\mathbf{Q}$-isomorphisms from K into Ω_p and $\tau_1, \ldots, \tau_n$ are the n $\mathbf{Q}$-isomorphisms from K into $\mathbf{C}$.

Then each λ in Λ belongs to $\mathscr{A}$; $|\sigma_2(\lambda)|_p \leqslant \sup\limits_{k\geqslant 0} |\sigma_2(\theta^k)|_p \leqslant 1$, and so on; $|\tau_1(\lambda)|_\infty \leqslant 1|(1 - |\tau_1(\theta)|_\infty), \ldots, |\tau_n(\lambda)|_\infty \leqslant 1/(1 - |\tau_n(\theta)|_\infty)$. We apply Theorem VII.

On the other hand, if Λ is harmonious there is a $\mu \neq 0$ in $\mathbf{Q}_p$ such that $|\varrho_p(\mu\lambda)| \leqslant \frac{1}{5}$ for each λ in Λ. We write $r_k = \varrho_p(\mu\theta^k)$. Then, for all finite sequences ε_k, $k \geqslant 0$, where each $\varepsilon_k = 0$ or 1, $\|\sum_{k\geqslant 0}\varepsilon_k r_k\| \leqslant \frac{1}{5}$ ($\|x\|$ is the distance from the real number x to the nearest integer). As in Proposition 7, we get $\sum_{k\geqslant 0} |r_k| \leqslant \frac{2}{5}$ and Proposition 12 gives the result.

13. Adeles and harmonious sets closed under multiplication

13.1. The introduction of adeles provides a remarkable interpretation of harmonious subsets of $\mathbf{R}$ and $\mathbf{Q}_p$ closed under multiplication.

The reader is referred to [3] and [22] throughout this section.

A local field will be a commutative non discrete locally compact field. We will only be concerned with local fields of characteristic 0 in this section; they are of two types, namely $\mathbf{R}$ or $\mathbf{C}$, or finite algebraic extensions K of a p-adic field $\mathbf{Q}_p$ (from now on, K is no longer an algebraic number field as was the case in the preceding sections).

The maximal compact subring of $\mathbf{Q}_p$ is $\mathbf{Z}_p$, while the maximal compact subring of a finite algebraic extension K of a p-adic field $\mathbf{Q}_p$ is the ring of all algebraic integers over $\mathbf{Z}_p$.

If the local field K is $\mathbf{R}$ or $\mathbf{C}$, $|\ \ |_\infty$ will denote the usual absolute value. On each finite algebraic extension K of $\mathbf{Q}_p$, the p-adic valuation (of $\mathbf{Q}_p$) can be extended in a unique way and the maximal compact subring of K becomes the unit ball of K.

Let k be an algebraic number field; k is an algebraic extension of $\mathbf{Q}$ of finite degree n.

A completion (π, K) of k is an ordered pair consisting of a local field K and an isomorphic embedding π of k into K whose image $\pi(k)$ is dense in K.

Two completions are called equivalent if they only differ in that K is replaced by an isomorphic copy K' of K.

A place v of k is an equivalence class of completions.

It may easily be checked that all places v of an algebraic number field k constitute a set denoted by P.

From now on, if v is a place of k, the corresponding local field K will be written k_v and the valuation (or the absolute value) on K will be denoted by $|\ \ |_v$.

If $K = \mathbf{R}$ or $K = \mathbf{C}$, the corresponding places are called infinite places and the corresponding absolute value will be denoted by $|\ \ |_\infty$. All other places are called finite places.

If v is a finite place of k, $\mathfrak{O}_v$ will be the maximal compact subring of the local field k_v; $\mathfrak{O}_v$ is defined by $|x|_v \leqslant 1$.

The adeles ring k_A of k is constructed inductively. For each finite subset F of P containing all infinite places of k, $k_A(F) = \prod_{v \in F} k_v \prod_{v \notin F} \mathfrak{O}_v$ is a locally compact ring.

As a set, k_A is the inductive limit of all $k_{\mathbf{A}}(F)$ while $k_{\mathbf{A}}$ is topologized by the condition that, for any finite subsets F and F' of P, $k_{\mathbf{A}}(F)$ is an open subgroup of $k_{\mathbf{A}}(F')$ whenever F is contained in F'.

An element $X = (X_v)_{v \in P}$ of $k_{\mathbf{A}}$ is an arbitrary sequence of elements of k_v such that $X_v \in \mathfrak{O}_v$ for all places v of P with the possible exception of a finite number of places.

For each x in k, there is only a finite number of finite places $v = (\pi, K)$ such that $\pi(x)$ does not belong to the maximal compact subring of K. Hence a mapping $\varphi: k \to k_{\mathbf{A}}$ can be defined by the rule: for each x in k, $\varphi(x)$ is the family of all $(\pi_v(x))_{v \in P}$.

It may easily be checked that φ is an isomorphic embedding of k in $k_{\mathbf{A}}$, and so, with a slight abuse of language, we henceforth regard k as a subgroup of $k_{\mathbf{A}}$. Then k is discrete in $k_{\mathbf{A}}$ and k is relatively dense in $k_{\mathbf{A}}$.

Each k_v can be embedded in $k_{\mathbf{A}}$: to each x in k_v there corresponds an element $(X_w)_{w \in P}$ of $k_{\mathbf{A}}$ given by $X_w = x$ if $w = v$ and $X_w = 0$ in all other cases.

From now on each k_v will be regarded as a subring of $k_{\mathbf{A}}$: such closed subrings are called quasifactors of $k_{\mathbf{A}}$ and, for each such quasi factor k_v and each X in $k_{\mathbf{A}}$, the component X_v of X whose subscript is v will be called the projection of X onto k_v.

DEFINITION 9: *Let k be an algebraic number field and (π_v, k_v) a completion of k. We shall write $K = k_v$.*

A standard subset Λ of K is the set of all projections on k_v of all x in k such that, for all places $w \neq v$ in P, $|x|_w \leqslant 1$.

13.2. It is remarkable that standard sets can be defined by Diophantine approximation properties.

THEOREM IX: *Let K be a local field of characteristic 0 and let Λ be a subset of K. If such subsets Λ are ordered by inclusion, then the following two properties of Λ are equivalent:*

(a) *Λ is a standard subset of K;*

(b) *Λ is maximal among the class of those harmonious subsets of K which are closed under multiplication.*

13.3. Proof of Theorem IX when $K = \mathbf{R}$

First of all, standard subsets of $\mathbf{R}$ must be specified. Since $\mathbf{R}$ is viewed as a completion of an algebraic number field k, it may be assumed that k is contained in $\mathbf{R}$. Elements λ of Λ are defined by three conditions:

(a) $\lambda \in k$;

(b) for all primes p and all isomorphic embeddings $\pi : k \to K$ of k into a dense subset of a finite algebraic extension K of $\mathbf{Q}_p$, $|\pi(\lambda)|_p \leqslant 1$.

(c) if $\sigma_1, \sigma_2, \ldots, \sigma_n$ are the n $\mathbf{Q}$-isomorphisms from k into $\mathbf{C}$ ordered in such a way that σ_1 is the identity mapping from k into $\mathbf{R}$, then $|\sigma_2(\lambda)| \leqslant 1, \ldots, |\sigma_n(\lambda)| \leqslant 1$.

Let $P(X)$ be the minimal polynomial of the algebraic number λ, a_0 the leading coefficient of P, and m the degree of P. Let Ω_p be an algebraic closure of $\mathbf{Q}_p$ and $z_1, \ldots, z_m$ the roots of $P(X) = 0$ in Ω_p. Then, for each j, $1 \leqslant j \leqslant m$, there exists an isomorphic embedding $\tau_j : k \to \mathbf{Q}_p(z_j)$ such that $\tau_j(\lambda) = z_j$.

Hence (b) implies that all roots z_j of $P(X)$ in Ω_p must be algebraic integers over $\mathbf{Z}_p$.

Classical relations between symmetric functions of roots and coefficients $a_0, a_1, \ldots, a_m$ of P imply that, for $1 \leqslant j \leqslant m$, $a_j/a_0 \in \mathbf{Z}_p$. Since $a_0, a_1, \ldots, a_m$ are assumed to be relatively prime, p does not divide a_0.

But (b) must be valid for all primes $p \geqslant 2$; therefore $a_0 = 1$ and λ is necessarily an algebraic integer.

A standard subset Λ of $\mathbf{R}$ may be defined in the following way: λ belongs to Λ if and only if $\pm\lambda = 1$ or if there exists a real algebraic field k and $\pm\lambda$ is a Pisot or a Salem number of k whose degree equals the degree of k over $\mathbf{Q}$.

Theorem IX is simply a restatement of Theorem VI.

13.4. Standard subsets of $\mathbf{C}$

The same argument may be used to identify standard subsets of $\mathbf{C}$. There exists a complex algebraic number field k and λ belongs to Λ if and only if λ is a root of 1 lying in k or if λ is a Pisot or a Salem number of k whose degree equals the degree of k over $\mathbf{Q}$ (such Pisot or Salem numbers are algebraic integers over $\mathbf{Z}$ all of whose conjugates μ, with the exception of λ and $\bar{\lambda}$, satisfy $|\mu|_\infty \leqslant 1$).

A proof of Theorem IX for this case can be found in [42].

13.5. Standard subsets of $\mathbf{Q}_p$

Let Ω_p be an algebraic closure of $\mathbf{Q}_p$. A standard subset Λ of $\mathbf{Q}_p$ is then defined in the following way: there exists an algebraic number field k of degree n over $\mathbf{Q}$, and an isomorphic embedding of k in $\mathbf{Q}_p$ and then Λ is the set of all λ in k such that

(a) the leading coefficient a_0 of the minimal polynomial $P(X) = a_0X^m + a_1X^{m-1} + \cdots + a_m$ of λ is a power of p ($a_0 = p^h$, $h \geqslant 0$);

(b) all other roots $\lambda_2, \ldots, \lambda_m$ of P in Ω_p satisfy $|\lambda_j|_p \leqslant 1$ $(2 \leqslant j \leqslant m)$;

(c) all roots $\zeta_1, \ldots, \zeta_m$ of P in $\mathbf{C}$ satisfy $|\zeta_j| \leqslant 1$ $(1 \leqslant j \leqslant m)$. It may easily be proved that such a λ is either a root of 1 lying in k or a Pisot-Salem-Chabauty number of k whose degree (over $\mathbf{Q}$) is n.

In this case, a proof of Theorem IX may be found in [89].

13.6. Why is a standard subset of a local field of characteristic 0 harmonious? The introduction of the adele ring $k_{\mathbf{A}}$ of k gives a very easy proof. With the notation of Section 5, G is our local field $K = k_v$, H is the incomplete adele ring $k'_{\mathbf{A}}$ constructed from all places $w \neq v$, and D is the discrete and relatively dense subgroup k of $k_{\mathbf{A}} = k_v \times k'_{\mathbf{A}}$. Our standard subset Λ of K was defined to be the set of all projections onto $K = k_v$ of all $d \in k$ whose projections onto $k'_{\mathbf{A}}$ belonged to the compact set $\prod_{w \neq v} \mathfrak{O}_w$. Hence Λ is a model.

The proof of Theorem IX in the general case may be found in [89].

14. Characterization of harmonious sets by additive properties

14.1. Let G be an l.c.a. group and Λ a subset of G. We say that Λ is discrete if for each λ in Λ there exists a neighbourhood V of λ in G whose intersection with Λ contains only λ. We say that Λ is relatively dense if there exists a compact subset K of G such that any translate of K intersects Λ.

Finally for each subset Λ of G, $\Lambda - \Lambda$ will denote the set of all differences $\lambda_1 - \lambda_2$, $\lambda_1 \in \Lambda$, $\lambda_2 \in \Lambda$.

Using this notation, we can state

THEOREM X: *Let G be an l.c.a. group and Λ a subset of G. The following two properties of Λ are equivalent:*

(a) *Λ is discrete in G, Λ is relatively dense in G and there is a finite subset F of G such that $\Lambda - \Lambda$ is contained in $\Lambda + F$;*

(b) *Λ is relatively dense in G and Λ is harmonious.*

Remarks: If $F = \{0\}$ and if $\Lambda - \Lambda$ is contained in $\Lambda + F, \Lambda$ is a subgroup of G. Theorem X shows, incidentally, that relatively dense harmonious sets are a natural generalization of discrete subgroups.

14.2. The proof of (b) $\Rightarrow$ (a) is obvious. Replacing Λ by $x_0 + \Lambda$ it may be assumed that $0 \in \Lambda$; such a translation does not alter property (a). Let $\Lambda_1 = \Lambda - \Lambda$. Since Λ is relatively dense, there is a compact subset K of G such that each λ_1 in Λ_1 can be written $\lambda_1 = \lambda + r$, where $\lambda \in \Lambda$ and $r \in K$; hence $r \in (\Lambda - \Lambda - \Lambda) \cap K = F$. But Λ and hence also $\Lambda - \Lambda - \Lambda$ are harmonious subsets of G; therefore F is a finite set and $\Lambda - \Lambda \subset \Lambda + F$.

14.3. To prove that (a) implies (b) the following lemma will be used:

LEMMA 26: *Let G and G' be two l.c.a. groups, Λ a relatively dense subset of G, G_1 the group generated by Λ in G (G_1 may not be a closed subgroup of G) and $h: G_1 \to G'$ a homomorphism.*

Assume that a neighbourhood V of 0 in G, a neighbourhood W of 0 in G' and a compact subset B of G' can be found such that

$$h(\Lambda) \quad \text{is contained in } B \tag{2.1}$$

$$x \in G_1 \cap V \quad \text{and} \quad h(x) \in W \quad \text{imply} \quad x = 0. \tag{2.2}$$

Then Λ is harmonious.

By Proposition 3, it suffices to prove that Λ is contained in a model. Definition 4 will be applied. Replacing G' by the closure H of $h(G_1)$ in G', it may be assumed that $h(G_1)$ is dense in G'; obviously this substitution does not alter (2.1) and (2.2).

Let D be the subgroup of $G \times H$ consisting of all pairs $(t, h(t))$, where t runs over G_1. To show that G produces H, it suffices now to prove that conditions (a), (b), (c) and (d) of Definition 4 are satisfied.

(a) D is discrete: obvious by (2.2);

(b) D is relatively dense in $G \times H$: let K be a compact subset of G such that each translate of K intersects Λ and let L be any compact neighbourhood of 0 in H. Since $h(G_1)$ is dense in H, there exists, for any y in H, a t in G_1 such that $y \in h(t) + L$. On the other hand, by the definition of K, a λ in Λ and an r in K can be found for each x in G, such that $x - t = \lambda + r$. Then $t' = t + \lambda$ has the following two properties: $x \in t' + K$ and $y \in h(t') + L - B$.

(a) $(0, y) \in D$ implies $y = 0$. Obvious: $y = h(0) = 0$;

(b) the second projection of D is dense in H: this was the definition of H.

14.4. The key of the proof is given by the following result about approximate additivity:

For each $m \geqslant 0$, let $\mathbf{R}^m$ be the m-dimensional Euclidean space ($\mathbf{R}^0 = \{0\}$). If $\mathbf{x} = (x_1, \ldots, x_m)$, $|\mathbf{x}| = \sup_{1 \leqslant j \leqslant m} |x_j|$.

PROPOSITION 13: *Let $\varphi : \mathbf{Z} \to \mathbf{R}^m$ be a mapping and C a positive constant such that for each j and k in $\mathbf{Z}$.*

$$|\varphi(j + k) - \varphi(j) - \varphi(k)| \leqslant C. \tag{2.3}$$

Then an $\mathbf{x}$ can be found in $\mathbf{R}^m$ such that, for each j in $\mathbf{Z}$ and the same con stant C

$$|\varphi(j) - j\mathbf{x}| \leqslant C.$$

Writing $\varphi = (\varphi_1, \ldots, \varphi_m)$, (2.3) holds for each φ_j by definition of the distance in $\mathbf{R}^m$. We may thus restrict our attention to the case $m = 1$.

As a first step, it will be proved that $j^{-1}\varphi(j)$ tends to a limit x as $|j| \to +\infty$.

Applying (2.3), inductively, we get $|\varphi(nA) - n\varphi(A)| \leqslant nC$ for each $n \geqslant 1$ and $A > 0$. Hence

$$\left|\frac{\varphi(nA)}{nA} - \frac{\varphi(A)}{A}\right| \leqslant \frac{C}{A}. \tag{2.4}$$

If $0 \leqslant k \leqslant A - 1$, we get $|\varphi(nA + k) - \varphi(nA) - \varphi(k)| \leqslant C$. Hence

$$\left|\frac{\varphi(nA + k)}{nA + k} - \frac{\varphi(nA)}{nA}\right| \leqslant \frac{2C + |\varphi(k)| + |\varphi(A)|}{nA}. \tag{2.5}$$

For each positive ε, there is an A such that $C/A \leqslant \varepsilon/3$; now regarding A as fixed, a positive N can be found such that

$$\frac{2C + |\varphi(k)| + |\varphi(A)|}{NA} \leqslant \frac{\varepsilon}{3}$$

for each $0 \leqslant k \leqslant A - 1$. If $m \geqslant NA$ and $m' \geqslant NA$ (2.4) and (2.5) imply

$$\left|\frac{\varphi(m)}{m} - \frac{\varphi(m')}{m'}\right| \leqslant \varepsilon.$$

The Cauchy sequence $(\varphi(j))/j$ tends to a limit x as $j \to +\infty$. Since $|\varphi(-j) + \varphi(j) - \varphi(0)| \leqslant C$, the same holds as $j \to -\infty$.

Let $\psi(j) = \varphi(j) - jx$. Then $\psi(j)$ satisfies (2.3) and $\psi(j) = o(j)$ as $|j| \to +\infty$. It will be proved that $|\psi| \leqslant C$.

Assume that a positive ε and an integer j exist such that $|\psi(j)| \geqslant C(1+\varepsilon)$. Replacing ψ by $-\psi$ if necessary, it may be assumed that $\psi(j) \geqslant C(1+\varepsilon)$. Using (2.3) inductively, we get $\psi(nj) \geqslant C + nC\varepsilon$ and $\psi(nj)$ is not $o(nj)$ as n tends to infinity (if $j = 0$, the last inequality obviously cannot hold).

The following result can be proved by the same sort of argument.

PROPOSITION 14: *Let $\varphi : \mathbf{R}^n \to \mathbf{R}^m$ be a continuous mapping such that, for a constant C and every $\mathbf{x}$ and $\mathbf{y}$ in $\mathbf{R}^n$ $|\varphi(\mathbf{x}+\mathbf{y}) - \varphi(\mathbf{x}) - \varphi(\mathbf{y})| \leqslant C$. Then $\varphi(\mathbf{x}) = l(\mathbf{x}) + \psi(\mathbf{x})$ where $l : \mathbf{R}^n \to \mathbf{R}^m$ is linear and $|\psi(\mathbf{x})| \leqslant C$.*

14.5. The proof of Theorem X for the case $G = \mathbf{R}^n$

If Λ is discrete and is relatively dense in G and if $\Lambda - \Lambda \subset \Lambda + F$, then any sum or difference $\pm\Lambda \pm \Lambda \pm \cdots \pm \Lambda$ has the same properties. Moreover $\pm\Lambda \pm \Lambda \pm \cdots \pm \Lambda$ is uniformly discrete in the following sense: for ε sufficiently small, a ball of radius ε intersects this set in, at most, a single point.

Replacing Λ by $\Lambda - \Lambda$, it may be assumed that $0 \in \Lambda$, $\Lambda = -\Lambda$; if $\Lambda - \Lambda$ is proved to be harmonious, then $\Lambda - \lambda$ will be harmonious for each λ in Λ, and hence so will Λ (Theorem II).

Let G_1 be the subgroup of G ($\mathbf{R}^n$ in this case) generated by Λ. To show that Λ is harmonious, we construct, using Proposition 14, a homomorphism $h : G_1 \to \mathbf{R}^m$ satisfying (2.1) and (2.2).

To start with, the structure of G_1 will be described when $G = \mathbf{R}^n$. Let $B(a)$ be the ball of $\mathbf{R}^n$ defined by $|\mathbf{x}| \leqslant a$. If a is large enough, each translate of $B(a)$ intersects Λ.

Let $F_1 = B(a) \cap (\Lambda - \Lambda)$. Starting from 0 it may be checked, by an obvious induction that Λ is contained in the $\mathbf{Z}$-module G_1 generated by F_1.

Since F_1 is finite, the structure theorem of modules on principal rings shows that G_1 is a direct sum $\mathbf{Z}\boldsymbol{\omega}_1 + \cdots + \mathbf{Z}\boldsymbol{\omega}_m$.

We now describe a mapping $\varphi : \mathbf{R}^n \to \mathbf{R}^m$ to which Proposition 14 will be applied: the linear part l of φ will be used to construct the homomorphism $h : G_1 \to \mathbf{R}^m$.

On $\Lambda + \Lambda$, φ is defined by: $\varphi(\mathbf{x}) = (p_1, \ldots, p_m)$ if $\mathbf{x} = p_1\omega_1 + \cdots + p_m\omega_m$ ($p_j \in \mathbf{Z}$, $1 \leqslant j \leqslant m$).

Let $\mathbf{x}$ and $\mathbf{y}$ be two points of $\Lambda + \Lambda$ and let $B = B(a)$ be the ball $|\mathbf{x}| \leqslant a$ of $\mathbf{R}^n$. If $\mathbf{y} - \mathbf{x} \in 2B$, $\mathbf{y} - \mathbf{x}$ belongs to the finite set ($\Lambda + \Lambda - \Lambda$

$-\Lambda) \cap 2B$; hence $\varphi(\mathbf{y}) - \varphi(\mathbf{x})$ lies in a finite subset of $\mathbf{Z}^m$. It follows that, for all $\mathbf{x}$ and $\mathbf{y}$ in $\Lambda + \Lambda$,

$$|\varphi(\mathbf{y}) - \varphi(\mathbf{x})| \leqslant c\,|\mathbf{y} - \mathbf{x}|. \tag{2.6}$$

A paving of $\mathbf{R}^n$ with $n + 1$-simplexes can be constructed in such a way that the vertices belong to $\Lambda + \Lambda$ and the distance between two vertices of the same simplex is less than $2a$.

Using this paving, we extend φ in the following way: the extension $\bar{\varphi}$ equals φ on $\Lambda + \Lambda$ and is affine on each simplex of the paving.

The following two properties of $\bar{\varphi}$ are immediate consequences of those of φ:

$$\text{for each } \lambda \text{ and } \lambda' \text{ in } \Lambda, \quad \bar{\varphi}(\lambda + \lambda') = \bar{\varphi}(\lambda) + \bar{\varphi}(\lambda') \tag{2.7}$$

$$\text{for each } \mathbf{x} \text{ and } \mathbf{x}' \text{ in } \mathbf{R}^n, \quad |\bar{\varphi}(\mathbf{x}') - \bar{\varphi}(\mathbf{x})| \leqslant c\,|\mathbf{x}' - \mathbf{x}|. \tag{2.8}$$

Each $\mathbf{x}$ in $\mathbf{R}^n$ can be written $\lambda(\mathbf{x}) + r(\mathbf{x})$, where $\lambda(\mathbf{x}) \in \Lambda$ and $|r(\mathbf{x})| \leqslant a$; (2.7) and (2.8) give

$$|\bar{\varphi}(\mathbf{x} + \mathbf{x}') - \bar{\varphi}(\mathbf{x}) - \bar{\varphi}(\mathbf{x}')| \leqslant C \tag{2.9}$$

for each $\mathbf{x}$ and $\mathbf{x}'$ in $\mathbf{R}^n$.

By Proposition 14, $\varphi(\mathbf{x}) = l(\mathbf{x}) + \psi(\mathbf{x})$, where $l\colon \mathbf{R}^n \to \mathbf{R}^m$ is linear and $\psi\colon \mathbf{R}^n \to \mathbf{R}^m$ is bounded.

We write $l = (l_1, \ldots, l_m)$, where each l_j is a linear form on $\mathbf{R}^n$; there is a constant C such that, for each $\lambda = p_1\boldsymbol{\omega}_1 + \cdots + p_m\boldsymbol{\omega}_m$ in Λ, $|\varphi(\lambda) - l(\lambda)| \leqslant C$, which can also be written

$$|p_j - l_j(\lambda)| \leqslant C, \quad 1 \leqslant j \leqslant m. \tag{2.10}$$

Let $h_1 : G_1 \to \mathbf{R}^m$ be the homomorphism defined by the formula (2.10) $h_1(\mathbf{t}) = (p_1 - l_1(\mathbf{t}), \ldots, p_m - l_m(\mathbf{t}))$ for each $\mathbf{t} = p_1\boldsymbol{\omega}_1 + \cdots + p_m\boldsymbol{\omega}_m$.

Then, if ε is small enough, $|\mathbf{t}| \leqslant \varepsilon$ and $|h_1(\mathbf{t})| \leqslant \varepsilon$ imply $\mathbf{t} = 0$: if $|l_1(\mathbf{t})| < \frac{1}{2}, \ldots, |l_m(\mathbf{t})| < \frac{1}{2}$ and if $|p_1 - l_1(\mathbf{t})| < \frac{1}{2}, \ldots, |p_m - l_m(\mathbf{t})| < \frac{1}{2}$, we get $|p_1| < 1, \ldots, |p_m| < 1$; the rational integers $p_1, \ldots, p_m$ vanish and so does $\mathbf{t}$.

Lemma 26 and (2.10) show that Λ is harmonious.

14.6. The proof of Theorem X for the case $G = \mathbf{R}^n \times K$, *where K is a compact abelian group*

In outline the proof is exactly the same. We seek an integer $m \geqslant 0$ and a homomorphism h_1 from the group G_1 generated by Λ into $\mathbf{R}^m$ with the two properties: $h_1(\Lambda)$ is bounded in $\mathbf{R}^m$ and for a suitable neighbourhood

V of 0 in G and a suitable positive ε, $t \in G_1 \cap V$ and $|h_1(t)| \leqslant \varepsilon$ imply $t = 0$. Such an h_1 will be called fit.

G_1 must first be described. Since Λ is relatively dense in $\mathbf{R}^n \times K$ and since $\Lambda - \Lambda$ is uniformly discrete, the argument used in Section 14.5 applies here: G_1 is generated by a finite subset F_1 of G.

A standard structure theorem gives: $G_1 \cong \mathbf{Z}^m \times \Phi$ where Φ is a finite subgroup of K ([19], p. 27, Corollaire 1 of Th. 1).

Let $\pi : G \to \bar{G} = G/\Phi$ be the canonical homorphism whose kernel is the group Φ, $\bar{\Lambda} = \pi(\Lambda)$ and $\bar{F} = \pi(F)$. Obviously $\bar{\Lambda} - \bar{\Lambda} \subset \bar{\Lambda} + \bar{F}$ and $\bar{\Lambda}$ is relatively dense in G; if we construct a fit $\bar{h}_1 : \bar{G} \to \mathbf{R}^m$, then $h_1 = \bar{h}_1 \circ \pi$ will be also a fit homomorphism (since Φ is finite).

It suffices to work with $\bar{G} = \mathbf{R}^n \times (K/\Phi)$ instead of G, in this case the group generated by $\bar{\Lambda}$ is isomorphic to $\mathbf{Z}^m$. From now on, we write G instead of $\bar{G}$ and Λ instead of $\bar{\Lambda}$ and it may be assumed that the group G_1 generated by Λ is a direct sum $\mathbf{Z}\omega_1 + \cdots + \mathbf{Z}\omega_m$. We also write K instead of $\bar{K} = K/\Phi$.

Each x in G can be written $x = (\mathbf{x}', y)$, where $\mathbf{x}' \in \mathbf{R}^n$, $y \in K$; in all of what follows, 'prime' will denote projection on $\mathbf{R}^n$.

Let F_1 be the finite set $(\Lambda + \Lambda - \Lambda - \Lambda) \cap (\{0\} \times K)$; then $(\mathbf{x}', y) \in \Lambda + \Lambda$ and $(\mathbf{x}', z) \in \Lambda + \Lambda$ imply $z - y \in F_1$. Note that F_1 is contained in the group G_1 generated by Λ.

Let Λ' be the projection of Λ on $\mathbf{R}^n$ and let $\varphi : \Lambda' + \Lambda' \to \mathbf{Z}^m$ be any mapping with the following property: for each $\mathbf{x}'$ in $\Lambda' + \Lambda'$, the equality $\varphi(\mathbf{x}') = \mathbf{p} = (p_1, \ldots, p_m)$ implies that $x = p_1\omega_1 + \cdots + p_m\omega_m$ belongs to $\Lambda + \Lambda$ and that $\mathbf{x}'$ is the projection of x on $\mathbf{R}^n$.

Let φ_1 and φ_2 be two such mappings φ. Then for each $\mathbf{x}'$ in Λ', $\varphi_1(\mathbf{x}') - \varphi_2(\mathbf{x}') \in F_1$.

Let $\bar{\varphi}$ be a piecewise affine extension of φ; $\bar{\bar{\varphi}} : \mathbf{R}^n \to \mathbf{R}^m$ is given by the procedure of Section 14.5. Let A be the set of all $\mathbf{p} \in \mathbf{Z}^m$ for which $p_1\omega_1 + \cdots + p_m\omega_m \in F_1$. Then all such mappings $\bar{\varphi}$ have the following properties:

$$\text{for each } \mathbf{u} \text{ and } \mathbf{v} \text{ in } \Lambda', \quad \bar{\varphi}(\mathbf{u} + \mathbf{v}) - \bar{\varphi}(\mathbf{u}) - \bar{\varphi}(\mathbf{v}) \in A \tag{2.11}$$

$$\text{for each } \mathbf{u} \text{ and } \mathbf{v} \text{ in } \mathbf{R}^n, \quad |\bar{\varphi}(\mathbf{v}) - \bar{\varphi}(\mathbf{u})| \leqslant c\,|\mathbf{v} - \mathbf{u}| \tag{2.12}$$

and two such $\bar{\varphi}$ differ by a bounded function.

By proposition 14, $\bar{\varphi} = l + \psi$ where $l : \mathbf{R}^n \to \mathbf{R}^m$ is linear and ψ is bounded. We write $l = (l_1, \ldots, l_m)$ and, for some constant $C > 0$ and any $\lambda = (\lambda', y)$, λ' and $l(\lambda')$ are connected by the relations

$$\lambda = p_1\omega_1 + \cdots + p_m\omega_m \text{ and } |l_j(\lambda') - p_j| \leqslant C,\ 1 \leqslant j \leqslant m. \quad (2.13)$$

Let $h_1 : G_1 \to \mathbf{R}^m$ be the homomorphism defined by $h_1(t) = (l_1(\mathbf{t}') - p_1, \ldots, l_m(\mathbf{t}') - p_m)$, where $t = p_1\omega_1 + \cdots + p_m\omega_m$ and $\mathbf{t}'$ is the projection of t on $\mathbf{R}^n$.

As in Section 14.5 it may be checked that h_1 is fit; hence Λ is harmonious.

14.7. The proof of Theorem X for the general case

Let G be an l.c.a. group, Λ a discrete subset of G, L a compact subset of G and F a finite subset of G. We assume that $\Lambda + L = G$ and $\Lambda - \Lambda \subset \Lambda + F$.

Let V be a compact neighbourhood of 0 containing L and F. The open subgroup $\mathfrak{G}$ of G generated by V is a product $\mathbf{R}^n \times \mathbf{Z}^p \times K$, where $n \geqslant 0$, $p \geqslant 0$ and K is a compact abelian group; see [7], Ch. II th. 9.8.

Let $\Lambda_1 = \Lambda \cap \mathfrak{G}$ and G_1 be the subgroup, without topology, of G generated by $\Lambda_1 \cup F$; since $\mathfrak{G}$ is a relatively dense subgroup of $\mathbf{R}^{n+p} \times K$, an integer $m \geqslant 0$ and a fit $h_1 \in \text{Hom}(G_1, \mathbf{R}^m)$ can be found such that $h_1(\Lambda_1)$ is bounded in $\mathbf{R}^m$ (see Section 14.6).

A constant $a > 0$, fixed from now on, is chosen with the two properties (i) if $B \subset \mathbf{R}^m$ is the ball $|\mathbf{x}| \leqslant a$, $h_1(\Lambda_1)$ and $h_1(F)$ are contained in B, and (ii) for each x in the finite set $(\Lambda + F + F) \cap (K + K - K)$, $h_1(x) \in B$.

Let H be the subgroup without topology, of G generated by Λ. Using $\Lambda = -\Lambda$ and $\Lambda - \Lambda \subset \Lambda + F$, we have $\Lambda + \cdots + \Lambda$ (n times) $\subset \Lambda + F + \cdots + F (n - 1$ times). Hence $H \subset \Lambda + G_1$ and $H \cap \mathfrak{G} = G_1$.

LEMMA 27: *A homomorphism $h : H \to \mathbf{R}^m$ can be found with the following two properties*

$$h(\Lambda) \textit{ is contained in } B \quad (2.14)$$

$$h \textit{ restricted to } G_1 \textit{ is } h_1. \quad (2.15)$$

Taking Lemma 27 for granted, h is fit since each sufficiently small x in G belongs to the open subgroup $\mathfrak{G}$ and then $h(x) = h_1(x)$. Thus Lemma 26 implies that Λ is harmonious.

The proof of Lemma 27 uses a standard transfinite induction of which the modern treatment is by Zorn's lemma.

Consider the set of pairs $(H_\alpha, h_\alpha)_{\alpha \in A}$, where for each $\alpha \in A$, H_α is a subgroup of H containing G_1 and $h_\alpha : H_\alpha \to \mathbf{R}^m$ is a homomorphism with the two properties

$$h_\alpha(\Lambda \cap H_\alpha) \quad \text{is contained in } B \tag{2.16}$$

$$h_\alpha \text{ restricted to } G_1 \text{ is } h_1. \tag{2.17}$$

We order A by $\alpha < \beta$ if $H_\alpha \subset H_\beta$ and h_β restricted to H_α is h_α. For each totally ordered subset B of A, the least upper bound γ of all β in B is well determined: H_γ is the union of all the H_β and $h_\gamma: H_\gamma \to \mathbf{R}^m$ is uniquely defined by the consistency rule (2.17).

Let α be a maximal element of A. It will be proved that $H_\alpha = H$ and the proof of Lemma 27 will be complete.

If $H_\alpha \neq H$, there is an $x \in \Lambda$, $x \notin H_\alpha$ since H is the group generated by Λ. Two cases may occur

Case 1: There is a least integer $q \geqslant 1$ such that $qx \in H_\alpha$. We define $H_\beta = \mathbf{Z}x + H_\alpha$ and $h_\beta: H_\beta \to \mathbf{R}^m$ by $h_\beta(qx) = qh_\beta(x)$; it is a simple exercise to check to consistency of this definition of the homomorphism h_β.

We have to prove (2.16). Let $y \in \Lambda \cap H_\beta$; then $qy \in H_\alpha$ and, using $\Lambda = -\Lambda$ (Section 14.5) and $\Lambda - \Lambda \subset \Lambda + F$, we can write $qy \in q\Lambda \subset \Lambda + \cdots + \Lambda$ (q times), which is contained in $\Lambda + F + \cdots + F$ (q times); hence $qy = \lambda + z$ where $\lambda \in \Lambda$ and $z \in F + \cdots + F$ ($q - 1$ times).

Since F is contained in G_1, z belongs to G_1 and $h_\beta(z)$ equals $h_1(z) \in (q-1)B$. On the other hand, $\lambda = qy - z$ belongs to $\Lambda \cap H_\alpha$ and the induction hypothesis gives $h_\beta(\lambda) = h_\alpha(\lambda) \in B$. Hence $h_\beta(qy) \in qB$ and $h_\beta(y) \in B$.

Case 2: No such $q \geqslant 1$ exists.

We define $H_\beta = \mathbf{Z}x + H_\alpha$ and $h_\beta: H_\beta \to \mathbf{R}^m$ by $h_\beta(x) = \mathbf{u}$. Although $\mathbf{u}$ is not yet fixed, h_β is, in any case, a homomorphism. It will be seen that a suitable choice of $\mathbf{u}$ gives (2.16). Since Λ is relatively dense in G, we have for each $j \in \mathbf{Z}$,

$$jx = \lambda_j + r_j, \quad \text{where} \quad \lambda_j \in \Lambda \quad \text{and} \quad r_j \in L. \tag{2.18}$$

Since $x \in \Lambda$ and $-x \in \Lambda$ ($\Lambda = -\Lambda$), we may choose $r_1 = r_{-1} = 0$. It will be proved by induction on j that each r_j belongs to G_1 For each j and k in $\mathbf{Z}$, we have $\lambda_{j+k} - \lambda_j - \lambda_k = r_j + r_k - r_{j+k}$ The left-hand side belongs to $\Lambda - \Lambda - \Lambda \subset \Lambda + F + F$ and the right-hand side belongs to $L + L - L$.

We recall that the group $\mathfrak{G}$ contains F and K and that G_1 is generated by F and $\Lambda_1 = \Lambda \cap \mathfrak{G}$. Hence $S = (\Lambda + F + F) \cap (L + L - L)$ is a finite subset of G_1 and for each pair of rational integers j and k, $r_j + r_k$

$- r_{j+k} \in S$ Starting with $r_1 = 0$, we get $r_j \in G_1$ for all $j \geqslant 1$ and starting with $r_{-1} = 0$ we get $r_j \in G_1$ for all other rational integers.

By our choice of a, $h_1(S)$ is contained in the ball $|x| \leqslant a$ of $\mathbf{R}^m$. We put $\varrho_j = h_1(r_j)$ to get $|\varrho_{j+k} - \varrho_j - \varrho_k| \leqslant a$ and Proposition 13 gives $\varrho_j = j\mathbf{u} + \psi(j)$, where $|\psi(j)| \leqslant a$; $\mathbf{u}$ is now defined and we put $h_\beta(x) = \mathbf{u}$. Here it is essential to retain the same constant a. We have $h_\beta(\lambda_j) = h_\beta(jx) + h_\beta(\lambda_j - jx) = ju - \varrho_j = -\psi(j)$; hence $|h_\beta(\lambda_j)| \leqslant a$.

If $y \in H_\beta \cap \Lambda$, it will be proved initially that $|h_\beta(y)| \leqslant 3a$; a more careful examination will show that $|h_\beta(y)| \leqslant a$.

Writing $y = jx + r$, $r \in H_\alpha$ and $\lambda_j = jx - r_j$, we get $y = \lambda_j + r + r_j$; $r + r_j \in H_\alpha$ because r and r_j each $\in H_\alpha$, and $r + r_j$ belongs to $\Lambda - \Lambda$, as $y - \lambda_j$ does and $r + r_j = y - \lambda_j$; hence $r + r_j \in H_\alpha \cap (\Lambda + F) = H_\alpha \cap \Lambda + F$, as $F \subset H_\alpha$. We get $|h_\beta(r + r_j)| = |h_\alpha(r + r_j)| \leqslant 2a$, since $h_\alpha(H_\alpha \cap \Lambda) \subset B$ and $h_\alpha(F) \subset B$. But $|h_\beta(\lambda_j)| \leqslant a$. Hence $|h_\beta(y)| \leqslant 3a$.

The treatment of y follows that of x, above. For each $k \geqslant 1$, we write $ky = \lambda_k + s_k$, where $\lambda_k \in \Lambda$, $s_k \in L \cap G_1$. Proposition 13 gives $h_1(s_k) = k\mathbf{v} + w_k$, where $\mathbf{v} \in \mathbf{R}^m$ and $|w_k| \leqslant a$.

Then $h_\beta(ky) = h_\beta(\lambda_k) + h_\beta(s_k) = h_\beta(\lambda_k) + h_1(s_k) = h_\beta(\lambda_k) + k\mathbf{v} + w_k$ or

$$k\,(h_\beta(y) - \mathbf{v}) = h_\beta(\lambda_k) + w_k. \tag{2.19}$$

Our first estimate gives $|h_\beta(\lambda_k)| \leqslant 3a$ ($\lambda_k \in \Lambda \cap H_\beta$) and so the right-hand side of (2.19) is bounded in $\mathbf{R}^m$. Hence so is the left-hand side, and we get

$$h_\beta(y) = \mathbf{v} \quad \text{and} \quad h_\beta(\lambda_k) = -w_k. \tag{2.20}$$

For $k = 1$, we have chosen $y = \lambda_1$ and we get $|h_\beta(y)| = |h_\beta(\lambda_1)| \leqslant a$. This concludes the proof of Lemma 27.

Applications

1. Let θ be a real number greater than 1. Assume that there exists a function $f: \mathbf{N} \to (0, +\infty)$ such that, for each integer $m \geqslant 0$ and each sequence $p_0, \ldots, p_m$ of rational integers, $|\sum_0^m p_j\theta^j| \geqslant f(\sup_{0 \leqslant j \leqslant m} |p_j|) > 0$. Then θ is a Pisot number and each Pisot number has this property.

Proof: For each integer $N \geqslant 1$ let Λ_N be the set of all finite sums $\sum_0^m p_j\theta^j$, where $|p_j| \leqslant N$. Then, for $N \geqslant \theta$, Λ_N is relatively dense and for each N and each $\lambda \in \Lambda_N$, $|\lambda| \geqslant f(N)$. Moreover $\Lambda_N \pm \Lambda_N$ is contained in Λ_{2N}. These three properties imply that, for $N \geqslant \theta$, $\Lambda_N - \Lambda_N \subset \Lambda_N + F$, where F is finite. Hence Λ_N is harmonious and θ is a Pisot number.

If θ is a Pisot number, the sets Λ_N are harmonious for all $N \geqslant 1$. Proposition 2 gives the result.

2. Let θ be a p-adic number such that $|\theta|_p > 1$. Assume that there exists a function $f: \mathbf{N} \to (0, +\infty)$ such that, for each integer $m \geqslant 0$ and each sequence $(p_0, \ldots, p_m)$ of rational integers $|\sum_0^m p_j \theta^j|_p \geqslant f(\sup_{0 \leqslant j \leqslant m} |p_j|)$. Then θ is a Pisot-Chabauty number and each Pisot- Chabauty number has this property. The proof is similar to the one just given.

15. Notes

The algebraic tools used in Section 13 may be found in [3] and in [22].

The completions of an algebraic number field k are $\mathbf{R}$, $\mathbf{C}$ or finite extensions $K_{\mathfrak{p}}$ of the $\mathbf{Q}_p$ associated with prime ideals $\mathfrak{p}$ of the ring of algebraic integers of k. The reader is referred to [3] for an excellent discussion.

A description of the multiplicative group of $\mathbf{Q}_p$ may be found in [20].

CHAPTER III

UNIQUENESS OF REPRESENTATION BY TRIGONOMETRIC SERIES

With any sequence $(c_k)_{k \in \mathbf{Z}}$ of complex numbers, we associate a formal series $S = \sum_{-\infty}^{+\infty} c_k e^{ikx}$; we put $S_n(x) = \sum_{-n}^{n} c_k e^{ikx}$. We say that the trigonometric series S converges to a finite function $f: \Omega \to \mathbf{C}$, where Ω is an open subset of $[0, 2\pi]$, if $S_n(x) \to f(x)$ for each x in Ω.

It is indeed remarkable that a trigonometric series may converge to an integrable function f almost everywhere on $[0, 2\pi]$ without being the Fourier series of this function. The explanation of this fact comes from the study of the exceptional set E of $x \in [0, 2\pi]$ for which $S_n(x)$ does not tend to $f(x)$; E is called a set of multiplicity.

If E is a compact set of real numbers, E is a set of multiplicity if and only if E carries a distribution T whose Fourier transform vanishes at infinity (Section 1).

For symmetric compact sets E of Cantor type, almost complete results are given in Sections 3, 4 and 5.

The p-adic case is studied in Section 6.

The models defined in Chapter II play a fundamental role in the proofs.

1. Riemann's theory

1.1. Three equivalent ways of defining a set of multiplicity are given in the following result:

THEOREM I: *Let E be a compact subset of $[0, 2\pi]$ and $[0, 2\pi] \backslash E$ be complement of E in $[0, 2\pi]$. Then the following three properties of E are equivalent:*

(a) *There exists a finite function $f: [0, 2\pi] \to \mathbf{C}$ in $L^1\,[0, 2\pi]$ and a trigonometric series S such that $S_n(x) \to f(x)$ for all x in $[0, 2\pi] \backslash E$ while S is not the Fourier series of f;*

(b) *There exists a trigonometric series S such that $S_n(x) \to 0$ $(n \to +\infty)$ for each x in $[0, 2\pi] \backslash E$ while not all coefficients of S vanish;*

(c) *There exists a distribution T carried by E which is not the zero distribution and whose Fourier transform vanishes at infinity.*

1.2. Before giving the proof of Theorem I, let us make some comments and give some explanations. If E possesses one of the equivalent properties (a), (b) or (c), E is called a set of multiplicity. If E is not a set of multiplicity, E is called a set of uniqueness. If for each $x \in [0, 2\pi]\backslash E$, a trigonometric series S converges to an integrable function f, and if E is a set of uniqueness, then S is the Fourier series of f.

A trigonometric series is a formal series $\sum c_k e^{ikx}$.

The Fourier transform $\hat{T}$ of a distribution T with compact support is a complex valued continuous function defined on $\mathbf{R}$ by $\hat{T}(x) = \langle T, e^{-ixt}\rangle$ which we loosely write $\int_{\mathbf{R}} e^{-ixt} T(t)\, dt$.

1.3. Now to the proof. Obviously (b) ⇒ (a). As we shall see, the proof of the implication (c) ⇒ (b) is very easy: the required S is the Fourier series of the distribution T. The proof of the implicationt (a) ⇒ (c) is deeper.

If $E = [0, 2\pi]$, there is nothing to prove. Obviously the problem is translation-invariant and *from now on it will be assumed that E is a compact subset of the open interval* $[0, 2\pi]$.

The Fourier series of a distribution T carried by E is the formal series

$$S(T; x) = \sum_{-\infty}^{+\infty} c_k e^{ikx}, \quad \text{where} \quad c_k = \frac{1}{2\pi}\int_0^{2\pi} e^{-ikt} T(t)\, dt.$$

1.4. The pseudofunctions

DEFINITION 1: *A distribution T carried by a compact subset of the open interval* $]0, 2\pi[$ *is called a pseudo-function if* $\lim_{|k|\to\infty} \hat{T}(k) = 0$.

DEFINITION 2: *A function* $\alpha : [0, 2\pi] \to \mathbf{C}$ *belongs to A if the Fourier series of* α *is absolutely convergent.*

LEMMA 1: *If T is a pseudofunction and if* α *belongs to A, the product* αT *is a pseudofunction.*

Under the Fourier transform, T becomes an element of c_0 and α an element of l^1. The product αT becomes the convolution product of an element of c_0 and an element of l^1; such a convolution belongs to c_0.

LEMMA 2: *If T is a pseudofunction and if* $S_n(T, x) = \sum_{-n}^{n} \hat{T}(k)\, e^{ikx}$, *then* $S_n(T, x) \to 0$ as $n \to +\infty$, *for each x not belonging to the support of T.*

Proof: We put

$$D_n(t) = \sum_{-n}^{n} e^{ikt} = \frac{\sin\left(n+\frac{1}{2}\right)t}{\sin\left(\frac{t}{2}\right)} = \sin nt \, \frac{\cos\frac{t}{2}}{\sin\frac{t}{2}} + \cos nt .$$

Let T be a pseudo-function carried by a compact subset E of the open interval $]0, 2\pi[$. Then for each x in $[0, 2\pi]$,

$$S_n(T, x) = \int_0^{2\pi} D_n(x-t)\, T(t)\, dt .$$

If x does not belong to E, we can find an indefinitely differentiable 2π-periodic function α such that, on a neighborhood of E,

$$\alpha(t-x) = \frac{\cos\frac{t-x}{2}}{\sin\frac{t-x}{2}} .$$

Then
$S_n(T, x) = \int_0^{2\pi} \sin n(x-t)\, \alpha(x-t)\, T(t)\, dt + \int_0^{2\pi} \cos n(x-t)\, T(t)\, dt.$
Since T is a pseudofunction, the second integral tends to 0, and by Lemma 1, so does the first one.

1.5. *Proof of* (c) ⇒ (b)

Let T be the pseudo-function carried by E and $\sum_{-\infty}^{+\infty} c_k e^{ikx} = S$ the Fourier series of T. Then if $T \neq 0$, the sequence $(c_k)_{k \in \mathbf{Z}}$ is not the zero sequence. By Lemma 2, $S_n(x)$ tends to 0 for each x in $[0, 2\pi] \setminus E$ and (b) is proved.

1.6. *Sketch of the proof of the implication* (a) ⇒ (c)

Let $S = \sum_{-\infty}^{+\infty} c_k e^{ikx}$ be the given trigonometric series and F the 2π-periodic continuous function whose Fourier series is obtained by integrating the trigonometric series $\sum_{-\infty}^{+\infty} c_k e^{ikx}$ twice: we may assume that $c_0 = 0$ and the Fourier series of F is $\sum_{-\infty}^{+\infty} -c_k/k^2\, e^{ikx}$. We define $\Delta^2(F; x, h)$ by $(F(x+h) + F(x-h) - 2F(x))/h^2$ and $D^2(F; x) = \lim_{h\to 0} \Delta^2(F; x, h)$, $\underline{D}^2 = \liminf_{h\to 0} \Delta^2(F; x, h)$, $\overline{D}^2 = \limsup_{h\to 0} \Delta^2(F; x, h)$.

Then $D^2(F) = f$ on $[0, 2\pi]\backslash E$ (Lemma 3). This implies that

$$F(x) = \int_a^x dy \int_a^y f(t)\, dt + Ax + B$$

on each interval $]a, b[$ contained in $[0, 2\pi]\backslash E$.

The distribution T whose Fourier series is $\sum_{-\infty}^{+\infty} c_k e^{ikx}$ is then the sum $f + T_1$, where T_1 is a pseudofunction carried by E.

1.7. Definition of the distribution

Let $S = \sum_{-\infty}^{+\infty} c_k e^{ikx}$ be the given trigonometric series. If $E = [0, 2\pi]$ there is nothing to prove and from now on we shall assume that E is contained in the open interval $]0, 2\pi[$. If $[0, 2\pi]\backslash E$ contains an interval $[a, b]$ the convergence of S_n on $[a, b]$ implies that $c_n e^{inx} + c_{-n} e^{-inx}$ tends to 0 on $[a, b]$; multiplying by e^{inx}, we see that $c_{-n} + c_n e^{2inx}$ tends to 0 on $[a, b]$. Hence $c_n (e^{2inx} - e^{2ina})$ tends to 0. If c_n does not tend to 0, $|c_n| \geqslant \varepsilon$ for an infinite set J of integers; then $\varepsilon\, |e^{2inx} - e^{2ina}| \leqslant |c_n|\, |e^{2inx} - e^{2ina}|$ tends to 0 ($n \in J$) and so does $f_n(x) = e^{2inx} - e^{2ina}$ on the whole interval $[a, b]$. Since $|f_n(x)| \leqslant 2$, Lebesgue's dominated convergence theorem gives $\int_a^b f_n(x)\, dx \to 0$ ($n \in J$) and a simple computation gives $\int_a^b e^{2inx} dx \to 0$. On the other hand $e^{2ina} \int_a^b dx = (b - a)\, e^{2ina}$ does not tend to 0. It follows immediately that $c_n \to 0$ when $|n| \to +\infty$. Since $c_n \to 0$, we can define a distribution T by means of the formal series S and T is a pseudofunction.

We may assume that $c_0 = 0$ and, integrating T twice, we obtain a continuous function F whose Fourier series is $\sum_{-\infty}^{+\infty} -c_k/k^2\, e^{ikx}$.

1.8. $D^2(F) = f$ on $[0, 2\pi]\backslash E$

We have $\Delta^2(F; x, h) = \sum_{-\infty}^{+\infty} c_k e^{ikx} \lambda_k(h)$, where $\lambda_k(h) = 2(1 - \cos kh)/k^2h^2$. Let $d_k = c_k e^{ikx}$ and $\mu_k(h) = \lambda_{k+1}(h) - \lambda_k(h)$.

We apply the following lemma:

LEMMA 3: *If* $\lim_{h\to 0} \lambda_k(h) = 1$ *for each fixed rational integer k and if there is a constant A, independent of h, such that* $\sum_{-\infty}^{+\infty} |\mu_k(h)| \leqslant A$, *then for each convergent series* $\sum_{-\infty}^{+\infty} d_k$, $\lim_{h\to 0} \sum_{-\infty}^{+\infty} \lambda_k(h)\, d_k$ *exists and equals* $\sum_{-\infty}^{+\infty} d_k$.

Proof: Writing $d_{|k|} = d_k + d_{-k}$, it may be assumed we sum only over $k \geqslant 1$. We put $D_n = \sum_n^{\infty} d_k$ and we get $\sum_{k\geqslant 1} d_k \lambda_k(h) = D_1 \lambda_{1,h} + \sum_2^{\infty} D_k \mu_k(h)$. The first term tends to $\sum_{-\infty}^{+\infty} d_k$, while the second tends to 0: for each positive ε, there exists an n such that $|D_k| \leqslant \varepsilon/2A$ for $k \geqslant n$ and this implies $|\sum_{n+1}^{\infty} D_k \mu_k(h)| \leqslant \varepsilon/2$ while $\sum_2^n D_k \mu_k(h)$ clearly tends to 0 with h.

In our case the inequality $\sum_{k\geqslant 1} |\lambda_{k+1}(h) - \lambda_k(h)| \leqslant A$ can be proved in the following way. Let ω be the function $2(1 - \cos x)/x^2$; its derivative ω' is continuous and belongs to L^1, as a trivial computation shows. Then $\lambda_{k+1}(h) - \lambda_k(h) = \omega((k+1)h) - \omega(kh) = \int_{kh}^{(k+1)h} \omega'(t)\,dt$ and $\sum_{-\infty}^{+\infty} |\mu_k(h)| \leqslant \sum_{-\infty}^{+\infty} \int_{kh}^{(k+1)h} |\omega'(t)|\,dt = \int_{-\infty}^{+\infty} |\omega'(t)|\,dt < +\infty$.

1.9. $F(x) = \int_a^x dy \int_a^y f(t)\,dt + Ax + B$ *on each interval* $]a, b[$ *contained in* $[0, 2\pi]\backslash E$.

The proof depends on the following result:

PROPOSITION 1: *Let* $]a, b[$ *be an open interval of real numbers,* f *a finite integrable function on* $]a, b[$, *and* F *a continuous function on the closed interval* $[a, b]$. *Assume that* $D^2F(x) = f(x)$ *for each* x *in* $[a, b]$. *Then* $F(x) = \int_a^x dy \int_a^y f(t)\,dt + Cx + D$.

The proof of Proposition 1 may be reduced to the case where F and f are both real. The complex case is then dealt with by considering the real and imaginary parts of F separately. We need a number of lemmas.

LEMMA 4: *Let* $F\colon [a, b] \to \mathbf{R}$ *be continuous and such that* $\bar{D}^2F \geqslant 0$ *(the value* $+\infty$ *is allowed) on* $[a, b]$. *Then* F *is convex on* $[a, b]$.

Proof: For each positive ε, let $F_\varepsilon(t) = F(t) + \varepsilon t^2$. If each F_ε is proved to be convex, this will also be true of their uniform limit F. We have $\bar{D}^2F_\varepsilon \geqslant 2\varepsilon$. Let $a < x_1 < x_2 < b$, $M_1 = (x_1, F_\varepsilon(x_1))$, $M_2 = (x_2, F_\varepsilon(x_2))$ and let $l(x) = \alpha x + \beta$ be the equation of M_1M_2. It must be checked that, for each x in (x_1, x_2), $G(x) = F_\varepsilon(x) - l(x) \leqslant 0$. The continuous function G, restricted to $[x_1, x_2]$ attains its maximum at some point x_3 of $[x_1, x_2]$. If $x_3 = x_1$ or $x_3 = x_2$, then $G(x) \leqslant 0$ on the whole of $[x_1, x_2]$. If $x_1 < x_3 < x_2$, then $\bar{D}^2F_\varepsilon(x_3) = \bar{D}^2G(x_3) \leqslant 0$, which contradicts $\bar{D}^2F_\varepsilon \geqslant 2\varepsilon$. Hence F_ε is convex.

LEMMA 5: *For each function* $f\colon [a, b] \to \mathbf{R}$ *which is real and integrable, there exists an increasing sequence* φ_n, $n \geqslant 1$, *of integrable upper semi-continuous functions and a decreasing sequence* ψ_n, $n \geqslant 1$, *of integrable lower semi-continuous functions such that* $\varphi_n \leqslant f \leqslant \psi_n$ *and* $\int_a^b (\psi_n(x) - \varphi_n(x))\,dx \to 0$ $(n \to +\infty)$.

This is a classical result from integration theory. Note that the value $+\infty$ is admitted for ψ_n and $-\infty$ for φ_n.

LEMMA 6: *If* φ *is upper semi-continuous and integrable on* $[a, b]$, *let* $\Phi(x) = \int_a^x dy \int_a^y \varphi(t)\,dt$. *Then, for each* x_0 *in* (a, b),

$$\bar{D}^2\Phi(x_0) \leqslant \varphi(x_0)$$

and, with the same notation, if ψ is lower semi-continuous and integrable on $[a, b]$

$$\underline{D}^2\Psi(x_0) \geqslant \psi(x_0).$$

Proof: For each positive ε, a positive η can be found such that $|x - x_0| \leqslant \eta$ implies $\varphi(x) \leqslant \varphi(x_0) + \varepsilon$. The two functions $\Phi(x)$ and

$$\int_{x_0}^{x} dy \int_{x_0}^{y} \varphi(t)\,dt$$

differ only by a linear term and have the same generalized second derivative. Hence we may assume that $\Phi(x) = \int_{x_0}^{x} dy \int_{x_0}^{y} \varphi(t)\,dt$; if $|x - x_0| \leqslant \eta, \Phi(x) \leqslant (x - x_0)^2 (\varphi(x_0) + \varepsilon)/2$, which implies $\bar{D}^2\Phi(x_0) \leqslant \varphi(x_0) + \varepsilon$ for each positive ε. If $\varphi(x_0) = -\infty$, the obvious modifications to be made in the proof are left to the reader.

Proof of Proposition 1: Let Φ_n and Ψ_n be defined in terms of φ_n and ψ_n in the same way that Φ was defined in Lemma 6. Let $G_n = F - \Phi_n$. Then, by Lemma 6, $\bar{D}^2 G_n = D^2 F - \bar{D}^2\Phi_n \geqslant f - \varphi_n \geqslant 0$. Hence G_n is convex and so is the uniform limit $F(x) - \int_a^x dy \int_a^y f(t)\,dt$ of the G_n, $n \geqslant 1$. Using the same argument, we prove that $\Psi_n - F$ is convex. Hence $F(x) - \int_a^x dy \int_a^y f(t)\,dt$ is both convex and concave. Proposition 1 is therefore proved.

1.10. Conclusion of the proof of (a) $\Rightarrow$ (c)

We define T_1 by $T - f$; T_1 is the second distributional derivative of $F(x) - \int_a^x dy \int_a^y f(t)\,dt$, where a is a fixed point of $[0, 2\pi]$. On each interval I contained in $[0, 2\pi]\backslash E$, this difference is linear and $T_1 = 0$. Hence T_1 is supported by E. Since the Fourier coefficients of T and of f tend to 0, so do those of T_1 and (c) is proved.

1.11. Equivalence of a periodic and a non-periodic problem

THEOREM II: *Let E be a compact subset of the open interval* $]0, 2\pi[$ *and T a distribution carried by E. Then the following two properties of T are equivalent:*

(a) $\lim_{|k| \to +\infty} \int_0^{2\pi} e^{-ikt} T(t)\,dt = 0$;

(b) $\lim_{|x| \to +\infty} \int_0^{2\pi} e^{-ixt} T(t)\,dt = 0$.

Obviously (b) $\Rightarrow$ (a). It remains to prove that (a) $\Rightarrow$ (b). Let $\alpha: [0, 2\pi] \to [0, 1]$ be a C^∞ function such that $\alpha = 1$ on a neighbourhood of E and

$\alpha = 0$ on a neighbourhood of 0 and of 2π. If $0 \leqslant u \leqslant 1$, we define χ_u on $[0, 2\pi]$ by $\chi_u(t) = (e^{-iut} - 1)\,\alpha(t)$; $\chi_u(t)$ is uniformly bounded in u and all derivatives of χ_u are uniformly bounded. Hence the Fourier series of χ_u is $\sum_{-\infty}^{+\infty} a_k(u)\, e^{-ikt}$, where $|a_k(u)| \leqslant a_k$ and $\sum_{-\infty}^{+\infty} a_k < +\infty$.

If $k \leqslant x \leqslant k + 1$, we write $x = k + u$ and $e^{-ixt} - e^{-ikt} = e^{-ikt}\chi_u(t)$ on a neighbourhood of E. Then

$$\left|\frac{1}{2\pi}\int_0^{2\pi} e^{-ikt}\chi_u(t)\,T(t)\,dt\right| = \left|\sum_{-\infty}^{+\infty} \hat{T}(j + k)\,a_j(u)\right| \leqslant \sum_{-\infty}^{+\infty} |\hat{T}(j + k)||a_j|,$$

which tends to 0 as k or x tends to infinity for the reason given in Lemma 1 (convolution of c_0 and l^1).

1.12. The spectrum of a complex valued bounded continuous function

Let $\mathscr{S}$ be the space of complex valued indefinitely differentiable functions all of whose derivatives, together with the function itself decay rapidly at infinity. If α belongs to $\mathscr{S}$, so does the Fourier transform $\hat{\alpha}(x) = \int_{-\infty}^{+\infty} e^{-ixt}\alpha(t)\,dt$ of α [21].

The topology of $\mathscr{S}$ is defined by a family of norms

$$\|\alpha\|_{m,n} = \sup_{\mathbf{R}}\,(1 + |t|)^m |D^n\alpha\,(t)|$$

under which $\mathscr{S}$ is a Frechet space and $\alpha \rightarrow \hat{\alpha}$ is an isomorphism of $\mathscr{S}$.

The Fourier transform of a complex valued bounded continuous function φ is the distribution T defined by $\int_{\mathbf{R}} \alpha(x)\,T(x)\,dx = \int_{\mathbf{R}} \hat{\alpha}(t)\,\varphi(t)\,dt$ for each α in $\mathscr{S}$.

The spectrum of φ is, by definition, the support of the Fourier transform of φ. We can thus give another definition of sets of multiplicity.

DEFINITION 3: *A compact set E of real numbers is a set of multiplicity if there exists a complex valued continuous function $\varphi\,.\,\mathbf{R} \rightarrow \mathbf{C}$ vanishing at infinity whose spectrum lies in E (we exclude the zero function).*

2. Symmetric sets

2.1. We shall now define a family of compact sets of real numbers, the so-called symmetric sets about which striking results have been obtained on the question of uniqueness.

(*a*) Dissection of type ξ on $[a, b]$. If $[a, b]$ is an interval of real numbers and if $0 < \xi < \frac{1}{2}$, we say that we perform the dissection ξ on $[a, b]$ if we replace $[a, b]$ by the union of the two closed intervals $[a, a_1]$ and $[b_1, b]$ of length $(b - a)\,\xi$. These two intervals are disjoint if $\xi < \frac{1}{2}$.

(b) *Sets* $E(\xi)$. For each ξ in $]0, \frac{1}{2}[$ and each interval $[a, b]$, we define a set $E(\xi)$ of Cantor type in the following way: we perform a dissection of type ξ on $[a, b]$ and obtain the closed union E_1 of two intervals of length $(b - a)\,\xi$. On each interval of E_1, we perform another dissection of type ξ and obtain the union E_2 of four intervals of length $(b - a)\,\xi^2$. At the kth step, we obtain a closed union E_k of 2^k intervals of length $(b - a)\,\xi^k$.

We define $E(\xi)$ to be the intersection of all the E_k, $k \geqslant 1$.

Obviously the Lebesgue measure of E is 0. For the same ξ, all the $E(\xi)$ are homothetical to any one of them.

(c) *Sets* $E(\xi_1, \xi_2, \ldots)$. Let $(\xi_k)_{k \geqslant 1}$ be a sequence of real numbers contained in $]0, \frac{1}{2}[$.

The set $E(\xi_1, \xi_2, \ldots)$ is defined in the same way as $E(\xi)$ but to pass from E_k to E_{k+1} we perform a dissection of type ξ_{k+1} on each interval of E_k. The measure of the intersection E of all the E_k is the infinite product $(b - a) \prod_1^\infty (2\xi_k)$.

A formula for $E(\xi)$ *and* $E(\xi_1, \xi_2, \ldots)$: Let $l = b - a$ and $t_k = \xi_1 \cdots \xi_{k-1} (1 - \xi_k)$. Then the elements x of $E(\xi_1, \xi_2, \ldots)$ are given by all convergent series $a + l \sum_1^\infty \varepsilon_k t_k$ for all sequences $(\varepsilon_k)_{k \geqslant 1}$, where each $\varepsilon_k = 0$ or 1.

Definition 3 shows that if E is a set of uniqueness (or of multiplicity) so is each translate of E and each set homothetical to E.

Hence, from now on, $E(\xi_1, \xi_2, \ldots)$ will be defined by taking $a = 0$, $l = 1$. Then $x \in E(\xi_1, \xi_2, \ldots)$ means $x = \sum_1^\infty \varepsilon_k t_k$, $\varepsilon_k = 0$ or 1, $t_k = \xi_1 \cdots \xi_{k-1} (1 - \xi_k)$.

In the case where all $\xi_k = \xi$, $k \geqslant 1$, we shall assume that $a = 0$ and $l = \xi/(1 - \xi)$. From now on, we put $\xi = \theta^{-1}$, $\theta > 2$ and we write $E(\theta)$ instead of this particular $E(\xi)$. No confusion is possible, since $\theta > 2$ cannot be a dissection ratio.

Then $x \in E(\theta)$ means

$$x = \sum_1^\infty \varepsilon_k \theta^{-k}, \ \varepsilon_k = 0 \text{ or } 1. \tag{3.1}$$

2.2. *Construction of measures carried by* $E(\theta)$ *and* $E(\xi_1, \xi_2, \ldots)$

For each $k \geqslant 1$, let F_k be the finite subset of $E(\xi_1, \xi_2, \ldots)$ defined by $x \in F_k$ if $x = \sum_1^k \varepsilon_j t_j$, $\varepsilon_j = 0$ or 1, and let μ_k be the measure carried by F_k which assigns the mass 2^{-k} to each of the 2^k points of F_k. Then $\mu_k \geqslant 0$, and the total mass of μ_k is 1. Let σ_k, $k \geqslant 1$, be the measure carried by

$\{0, t_k\}$ which assigns the mass $\frac{1}{2}$ to each of these points. Then μ_k is the convolution product $\sigma_1 * \cdots * \sigma_k$ and the Fourier transform of μ_k is the product $(\frac{1}{2} + \frac{1}{2}e^{-ixt_1}) \cdots (\frac{1}{2} + \frac{1}{2}e^{-ixt_k}) = e^{-ixs_k} \cos t_1x/2 \cdots \cos t_kx/2$, where $s_k = t_1 + \cdots + t_k$.

Let μ be the weak star limit of the sequence of μ_k, $k \geqslant 1$; μ is carried by E and the Fourier transform $\hat{\mu}$ of μ is $e^{-ixs} \prod_1^\infty \cos t_kx/2$. It should be observed that $|\hat{\mu}(x)| = \prod_1^\infty |\cos t_kx/2|$.

3. Pisot numbers and uniqueness

3.1. Simple examples of sets of multiplicity and of uniqueness

A compact set E of positive Lebesgue measure is a set of multiplicity. The function which is 1 on E and 0 outside is an example of a pseudo-function carried by E.

On the other hand, a countable compact set E is a set of uniqueness. In fact each complex valued bounded continuous function φ whose spectrum is a countable compact set is almost periodic ([11, p. 167, th. 5.20) and an almost periodic function cannot vanish at infinity without being identically 0.

From these two examples, it might be conjectured that thin sets are sets of uniqueness and thick sets are sets of multiplicity. Testing this idea on the sets $E(\theta)$, one would think that $E(\theta)$ has more chance of being a set of uniqueness when θ is great, since it seems obvious that the greater θ is, the thinner $E(\theta)$ is.

(Using the Hausdorff dimension of a set, this is actually the case.)

The truth is quite different, as Salem and Zygmund showed in 1955.

3.2. Pisot numbers and the problem of uniqueness

THEOREM III: *The set $E(\theta)$ is a set of uniqueness if and only if θ is a Pisot number.*

3.3. The easy part of Theorem III

If θ is not a Pisot number, we shall prove that the Fourier transform of the measure defined in Section 2.2 tends to 0 at infinity. Let $\Pi(x)$ be the infinite product $\prod_1^\infty \cos \theta^{-k}x$. To prove that $\Pi(x)$ tends to 0 at infinity we shall, since Π is an even function of x, compute Π on the interval $[\theta^k,$

$\theta^{k+1}]$. Then, putting $x = \theta^k u$, $1 \leqslant u \leqslant \theta$, we have $|\Pi(x)| \leqslant P_k(u)$, where $P_k(u) = \prod_0^{k-1} |\cos \theta^j u|$.

LEMMA 7: *If* $\theta > 1$ *is not a Pisot number,* $P_k(u)$ *tends to* 0 *uniformly on each compact set not containing* 0.

Proof: For each real number x, let $\|x\|$ be the distance from x to the nearest integer. Then $1 - |\cos x|$ and $2^{-1}\pi^2 \|\pi^{-1}x\|^2$ are equivalent when $|\cos x| \to 1$. Therefore the convergence of the infinite product $\prod_{j\geqslant 0} |\cos \theta^j u|$ is equivalent to the convergence of the series $\sum_{j\geqslant 0} \|\pi^{-1}\theta^j u\|^2$; if $u \neq 0$, this convergence implies that θ is a Pisot number (Chapter I, Theorem XII).

Assume now that θ is not a Pisot number. On each compact set K of real numbers not containing 0, P_k is a decreasing sequence of continuous functions tending to 0. By Dini's theorem, the convergence is uniform. Put $\omega_k = \sup_{1\leqslant u\leqslant \theta} P_k(u)$. Then $\omega_k \to 0$ $(k \to +\infty)$ and so $|\Pi(x)| \leqslant \omega_k$ on $[\theta^k, \theta^{k+1}]$. Hence $\Pi(x)$ tends to 0 as x tends to infinity.

3.4. *The difficult part of Theorem III*

The main idea of the proof is to consider **R** as a dense subgroup of the compact group $\mathbf{T}^n$

Sketch of the proof: Let S be a distribution carried by E whose Fourier transform is equal to 1 at 0 and vanishes at infinity, let n be the degree of the algebraic number θ and $h:\mathbf{R} \to \mathbf{T}^n$ the continuous homomorphism from **R** to the n dimensional torus defined by $h(t) = (\exp 2\pi i t, \ldots, \exp 2\pi i \theta^{n-1} t)$.

(a) We define S_k by $S_k(t) = \theta^{-k} S(\theta^{-k} t)$; S_k is a distribution carried by $\theta^k E$, $|\hat{S}_k(x)| \leqslant A$, $\hat{S}_k(0) = \hat{S}(0) = 1$, while $\hat{S}_k(x) \to 0$ if $x \neq 0$ where ^ denotes the Fourier transform.

(b) We next define a distribution σ_k carried by $h(\theta^k E)$; σ_k is $h(S_k)$ and tends in the weak star topology to the limit $d\mathbf{x}$, where $d\mathbf{x}$ is the Haar measure on $\mathbf{T}^n$.

(c) Now if θ is a Pisot number, $h(\theta^k E)$ is contained in a given compact subset K of $\mathbf{T}^n$ of measure 0; σ_k is then carried by K and hence so is the weak star limit dx of the sequence (σ_k); we thus have a contradiction

If $\hat{S}(0) \neq 1$, we replace $S(t)$ by $c\, e^{-ix_0 t} S(t)$; $\hat{S}(t)$ becomes $c\hat{S}(t + x_0)$ and if $\hat{S}$ does not vanish identically, there is a complex number c and a real x_0 such that $c\hat{S}(x_0) = 1$ and $c\, e^{-ix_0 t} S(t)$ is always a pseudo-function. The proof is now complete.

*3.5. Details of the proof (*3.5, 3.6 *and* 3.7*)*

Verification of (a): Let φ be the Fourier transform of S and φ_k the Fourier transform of S_k. Then $\varphi_k(x) = \varphi(\theta^k x)$. Since φ vanishes at infinity, the complex valued continuous function φ is bounded. The three properties of φ_k are then clear.

Verification of (b): Since S_k is carried by the compact subset $\theta^k E$ of $\mathbf{R}$, the image $\sigma_k = h(S_k)$ is well defined by the rule $\langle \sigma_k, \alpha \rangle = \langle S_k, \alpha \circ h \rangle$; α is a test function on $\mathbf{T}^n$ and $\alpha \circ h$ is an indefinitely differentiable function on the real line. Clearly σ_k is carried by $h(\theta^k E)$.

Let $h^* : \mathbf{Z}^n \to \mathbf{R}$ be defined by $h^*(\mathbf{m}) = m_1 + m_2\theta + \cdots + m_n\theta^{n-1}$ for each $\mathbf{m} \in \mathbf{Z}^n$. The Fourier coefficients of σ_k are then

$$\hat{\sigma}_k(\mathbf{m}) = \hat{S}_k(h^*(\mathbf{m})) = \varphi_k(h^*(\mathbf{m})) = \varphi(m_1\theta^k + \cdots + m_n\theta^{k+n-1}).$$

Since h^* is injective, $\hat{\sigma}_k(\mathbf{m}) \to 0$ for $\mathbf{m} \neq 0$ while $\hat{\sigma}_k(0) = 1$; there is a constant A independent of k and $\mathbf{m}$ such that $|\hat{\sigma}_k(\mathbf{m})| \leqslant A$.

3.6. Pseudomeasures and weak-star topology (end of verification of (b))

Let $A(\mathbf{T}^n)$ be the Banach space of all continuous functions $\alpha : \mathbf{T}^n \to \mathbf{C}$, whose Fourier series are absolutely convergent; we put $\|\alpha\| = \sum_{\mathbf{m} \in \mathbf{Z}^n} |\hat{\alpha}(\mathbf{m})|$. It can be checked without difficulty that $A(\mathbf{T}^n)$ is a Banach algebra of functions on $\mathbf{T}^n$; each sufficiently regular function belongs to $A(\mathbf{T}^n)$ and $A(\mathbf{T}^n)$ will be used as a space of test functions replacing the space $D(\mathbf{T}^n)$ of indefinitely differentiable functions.

The dual of $A(\mathbf{T}^n)$ is the Banach space $PM(\mathbf{T}^n)$ of all distributions σ on $\mathbf{T}^n$ whose Fourier transforms are bounded; $\|\sigma\|_{PM} = \sup_{\mathbf{Z}^n} |\hat{\sigma}(\mathbf{m})|$. Each complex valued Radon measure σ belongs to $PM(\mathbf{T}^n)$; for this reason, we call the distributions of $PM(\mathbf{T}^n)$ pseudo-measures.

The weak-star topology of $PM(\mathbf{T}^n)$ is the topology $\sigma(PM(\mathbf{T}^n), A(\mathbf{T}^n))$ of simple convergence on $A(\mathbf{T}^n)$. A sequence σ_k, $k \geqslant 1$, of pseudo-measures tends to a weak-star limit σ if and only if $\sup \|\sigma_k\|_{PM} < +\infty$ and for each $\mathbf{m}$ in $\mathbf{Z}^n$, $\hat{\sigma}_k(\mathbf{m}) \to \hat{\sigma}(\mathbf{m})$.

If all the distributions σ_k are carried by a fixed compact subset K of $\mathbf{T}^n$, so is their weak-star limit σ.

In our case, $\hat{\sigma}_k(\mathbf{m}) \to 0$ if $\mathbf{m} \neq 0$ and $\hat{\sigma}_k(0) = 1$ while $\|\sigma_k\|_{PM} = \|\varphi\|_\infty$; hence $\sigma_k \to d\mathbf{x}$ $(k \to +\infty)$.

3.7. Verification of (c)

Let Λ be the set of all finite sums $\sum_{k \geqslant 0} \varepsilon_k \theta^k$, $\varepsilon_k = 0$ or 1. Then, by the form (3.1) of elements of E, $\theta^k E$ is contained in $\Lambda + E$.

Let H be the linear map from $\mathbf{R}$ into $\mathbf{R}^n$ defined by $H(t) = (t, \theta t, \ldots, \theta^{n-1} t)$ and $\Pi : \mathbf{R}^n \to \mathbf{T}^n$ the canonical projection defined by $\Pi(x_1, \ldots, x_n) = (\exp 2\pi i x_1, \ldots, \exp 2\pi i x_n)$.

Let U be the compact subset of $\mathbf{R}^n$ defined by the parametric representation $x_1 = u_1 + \cdots + u_n$, $x_2 = \theta_2 u_2 + \cdots + \theta_n u_n, \ldots, x_n = \theta_2^{n-1} u_2 + \cdots + \theta_n^{n-1} u_n$, where $u_2 = -\sum_0^\infty \varepsilon_k \theta_2^k, \ldots, u_n = -\sum_0^\infty \varepsilon_k \theta_n^k$ for all $\varepsilon_k = 0$ or 1 (ε_k is the same, for each $k \geqslant 0$ in $u_2, \ldots, u_n$). Then if θ_i and θ_j are complex conjugate, so are u_i and u_j $(1 < i < j \leqslant n)$: hence the numbers x_p, $1 \leqslant p \leqslant n$, are real and depend linearly on $n - 1$ real parameters and U is contained in an hyperplane L of $\mathbf{R}^n$. On the other hand U is compact.

Let $V = H(E)$ and $D = H(\mathbf{R})$; the straight line D is not contained in L since the determinant

$$\begin{vmatrix} 1 & \theta & \cdots & \theta^{n-1} \\ 1 & \theta_2 & \cdots & \theta_2^{n-1} \\ \cdots & \cdots & \cdots & \cdots \\ 1 & \theta_n & \cdots & \theta_n^{n-1} \end{vmatrix}$$

does not vanish. Hence $\mathbf{R}^n$ is equal to the direct sum $L \oplus D$; using this decomposition, we compute the Lebesgue measure of $U \oplus V$ by Fubini's theorem. Since E has measure zero, the measure of V with respect to the Lebesgue measure of D is 0. Hence the measure of $U + V = U \oplus V$ is 0.

Let $K = \Pi(U + V)$. The measure of $\Pi(C)$ in $\mathbf{T}^n$ does not exceed the measure of C in $\mathbf{R}^n$ (Chapter I, Section 2.1, Lemma 1). Hence the measure of the compact set K is 0.

We shall prove that $h(\Lambda + E)$ is contained in K or, equivalently, that $H(\Lambda + E)$ is contained in $U + V + \mathbf{Z}^n$. But $H(\Lambda + E) = H(\Lambda) + H(E) = H(\Lambda) + V$, and there is no more to prove if $H(\Lambda)$ is contained in $U + \mathbf{Z}^n$; this, however, is an obvious consequence of the congruences

$$\theta^k \equiv -\theta_2^k - \cdots - \theta_n^k \pmod 1. \tag{3.2}$$

Combining (3.2) with coefficients $\varepsilon_k = 0$ or 1 we get $H(\Lambda) \subset U + \mathbf{Z}^n$.

3.8. Some remarks on the proof

The inclusion $h(\Lambda) \subset \Pi(U)$, which is so crucial in the proof has already been proved in Proposition 8, Section 9, Chapter II, for models of real numbers. When θ is a Pisot number the set Λ of all finite sums $\sum_{k\geqslant 0} \varepsilon_k \theta^k$, $\varepsilon_k = 0$ or 1 is contained in a model.

The reason that we have here given a complete proof is to preserve the mutual independance of the different chapters of the book, while not forgetting the common features that unify them.

4. Results for symmetric sets

4.1. An analogue of Salem-Zygmund's theorem will be proved for sets constructed with variable dissection ratios: $E(\xi_1, \xi_2, \ldots)$ is a set of uniqueness if all the ξ_k, $k \geqslant 1$, are the inverses of Pisot numbers which lie in a given algebraic number field and which satisfy a very natural condition of uniformity.

Let $\mathfrak{K}$ be a real algebraic number field of degree $n \geqslant 2$. Let $\sigma_1, \ldots, \sigma_n$ be the n **Q**-isomorphisms from $\mathfrak{K}$ into **C** ordered in such a way that σ_1 is the identical map from $\mathfrak{K}$ to **R**.

The reader is referred to Section 2.1 for the definition of the compact set $E(\xi_1, \xi_k, \ldots)$ of all sums $\sum_{k\geqslant 1} \varepsilon_k t_k$, where $t_k = \xi_1 \cdots \xi_{k-1}(1 - \xi_k)$.

THEOREM IV: *Let $\mathfrak{K}$ be a real algebraic number field; let $n \geqslant 2$ be the degree of $\mathfrak{K}$. Let α and β be two real numbers such that $0 < \alpha < 1$ and $\beta > 2$. Let $(\omega_k)_{k\geqslant 1}$ be a sequence of algebraic integers in $\mathfrak{K}$ such that*

(a) $|\sigma_j(\omega_k)| \leqslant \alpha$ *for all* $2 \leqslant j \leqslant n$ *and* $k \geqslant 1$;

(b) $\omega_k \geqslant \beta$.

Then, putting $\xi_k = \omega_k^{-1}$, $E(\xi_1, \xi_2, \ldots)$ is a set of uniqueness

Before proving this theorem some comments should be made: we cannot have, in the case of symmetric sets, satisfactory necessary conditions. We shall prove in Chapter VII that if $\sum_1^\infty \xi_k^2 < +\infty$, $E(\xi_1, \xi_2, \ldots)$ is a set of uniqueness. Hence no necessary conditions of algebraic type exist.

An obvious corollary is the following: let $\mathfrak{K}$ be a real algebraic number field of degree n and F a finite set of Pisot numbers greater than 2, of degree n, lying in $\mathfrak{K}$. Then each symmetric set E constructed with dissection ratios whose inverses lie in F is a set of uniqueness.

If the degree of $\mathfrak{K}$ over $\mathbf{Q}$ is 1, $\mathfrak{K} = \mathbf{Q}$, Pisot numbers are rational integers and a weak version of Theorem IV says that if $\xi_k = 1/p_k$, $p_k \geqslant 3$, then $E(\xi_1, \xi_2, \ldots)$ is a set of uniqueness. The proof is easier than in the general case and is left to the reader as an exercise.

4.2. *Sketch of the proof*

Let S be a pseudofunction carried by E and such that $S(0) = 1$ (it has been shown in Section 3.4 how to reduce the problem to this case). Let $S_k = (\xi_1 \cdots \xi_k)^{-1} S(\xi_1 \cdots \xi_k t)$ and let φ_k be the Fourier transform of S_k.

Let θ be an algebraic integer of degree n of K and $h : \mathbf{R} \to \mathbf{T}^n$ the continuous injective homomorphism defined by $h(t) = (\exp 2\pi i t, \ldots, \exp 2\pi i \theta^{n-1} t)$.

(a) Let $T_k = h(S_k)$. Then the weak-star limit of T_k is the Lebesgue measure $d\mathbf{x}$ on $\mathbf{T}^n$ normalized by $\int_{\mathbf{T}^n} d\mathbf{x} = 1$;

(b) There exists a compact subset K of $\mathbf{T}^n$, $K \neq \mathbf{T}^n$, and a sequence $\mathbf{x}_k$ of elements of $\mathbf{T}^n$ such that T_k is supported by $\mathbf{x}_k + K$;

(c) Then, if T_k' denotes the translated distribution $T_k(\mathbf{x} - \mathbf{x}_k)$, (a) implies that the weak star limit of T_k' is $d\mathbf{x}$. But T_k' is carried by K. Hence $d\mathbf{x}$ is carried by K, and this is false.

4.3. *Verification of* (a)

The proof is essentially the same as that in Section 3.5.

4.4. *Verification of* (b)

A new definition is needed.

Definition 4: *We say that a sequence K_k, $k \geqslant 1$, of compact subsets of $\mathbf{T}^n$ admits a uniform room in $\mathbf{T}^n$ if there exist a compact subset K of $\mathbf{T}^n$, $K \neq \mathbf{T}^n$ and a sequence $\mathbf{x}_k$, $k \geqslant 1$, of points of $\mathbf{T}^n$ such that K_k is contained in $K + \mathbf{x}_k$ for each $k \geqslant 1$.*

Lemma 8: *Let K_k, $k \geqslant 1$, be a sequence of compact subsets of $\mathbf{T}^n$ and for each $k \geqslant 1$ and each $\varepsilon > 0$, let K_k^ε be the set of $\mathbf{x}$ in $\mathbf{T}^n$ whose distance from K_k does not exceed ε. If there exists a positive ε such that the Lebesgue measure of K_k^ε is less than that of $\mathbf{T}^n$, then K_k admits a uniform room.*

Proof: In fact, $K_k^\varepsilon \neq \mathbf{T}^n$; let $\mathbf{x}_k$ in $\mathbf{T}^n$ be such that $d(\mathbf{x}_k, K_k) > \varepsilon$ and let K be the set of all $\mathbf{x}$ in $\mathbf{T}^n$ such that $d(0, \mathbf{x}) \geqslant \varepsilon$. We may assume that ε is sufficiently small to ensure that $K \neq \mathbf{T}^n$; the distance d is translation-invariant and $K_k \subset x_k + K$.

LEMMA 9: *Let W be a compact subset of $\mathbf{T}^n$, m an integer $\geqslant 1$, and F_k, $k \geqslant 1$, a sequence of finite subsets of $\mathbf{T}^n$. Assume that* Card $F_k \leqslant m$ *for each $k \geqslant 1$ and $m\,|W| < 1$, where $|\ \ |$ denotes Lebesgue measure. Then $K_k = W + F_k$ admits a uniform room.*

Proof: Let W^ε (or K_k^ε) be the set of all points in $\mathbf{T}^n$ whose distance from W (or K_k) does not exceed ε. Then $W^\varepsilon + F_k = K_k^\varepsilon$; hence $|K_k^\varepsilon| \leqslant m\,|W^\varepsilon|$. We choose ε sufficiently small to ensure that $m\,|W^\varepsilon| < 1$, which is possible since $\lim_{\varepsilon \to 0} |W^\varepsilon| = |W|$, and we apply Lemma 8.

4.5. *Conclusion of the verification of (b)*

Let θ be an algebraic integer of degree n of $\mathfrak{K}$ and let $\theta_2 = \sigma_2(\theta), \ldots, \theta_n = \sigma_n(\theta)$ be its conjugates different from θ. The elements t of E are given by the formula

$$t = \varepsilon_1 (1 - \xi_1) + \varepsilon_2 \xi_1 (1 - \xi_2) + \cdots + \varepsilon_k \xi_1 \cdots \xi_{k-1} (1 - \xi_k) + \cdots$$

Hence $(\xi_1 \cdots \xi_k)^{-1}\, t = t' + t''$, where t' is an algebraic integer of $\mathfrak{K}$ such that $|\sigma_2(t')| \leqslant (1 + \alpha)(1 - \alpha)^{-1}, \ldots, |\sigma_n(t')| \leqslant (1 + \alpha)(1 - \alpha)^{-1}$ and t'' belongs to the compact set R_k of all sums

$$\varepsilon_{k+1} (1 - \xi_{k+1}) + \varepsilon_{k+2} \xi_{k+1} (1 - \xi_{k+2}) + \cdots + \varepsilon_{k+j} \xi_{k+1} \cdots \xi_{k+j-1} \times$$
$$\times (1 - \xi_{k+j}) + \cdots$$

For each j, $0 \leqslant j \leqslant n - 1$, $t'\theta^j$ is an algebraic integer of the field $\mathfrak{K}$; hence $\sigma_1 (t'\theta^j) + \cdots + \sigma_n (t'\theta^j) \in \mathbf{Z}$ and

$$t'\theta^j \equiv u_2 \theta_2^j + \cdots + u_n \theta_n^j \pmod 1 \tag{3.3}$$

where the complex numbers u_p corresponding to conjugate θ_p are also conjugate and where

$$|u_p| \leqslant (1 + \alpha)(1 - \alpha)^{-1}. \tag{3.4}$$

Let U be the compact set of all $(x_1, \ldots, x_n)$ in $\mathbf{R}^n$ defined by the parametric relations $x_j = u_2 \theta_2^{j-1} + \cdots + u_n \theta_n^{j-1}$, $1 \leqslant j \leqslant n$, for all sequences $(u_2, \ldots, u_n)$ satisfying (3.4). As in Section 3 it can be checked that U is a compact subset of a hyperplane L of $\mathbf{R}^n$. We have $H(t') \in U + \mathbf{Z}^n$.

We shall now compute $H(t'')$, Let N be a fixed positive integer whose exact value will be given shortly and let I be the closed interval of $\mathbf{R}^n$ defined by $\mathbf{x} = H(t)$, $0 \leqslant t \leqslant \beta^{-N}$ where β is given as in Theorem IV.

Then there exists a finite subset D_k of $D = H(\mathbf{R})$ containing 2^N points such that $H(R_k) \subset I + D_k$ as we observed in the construction of a symmetric set.

Consider $U + H(R_k)$; this set is contained in the direct sum $U \oplus (I + D_k) = U \oplus I + D_k$. For some constant C depending only on $\mathfrak{K}$, the Lebesgue measure of $U \oplus I$ is $C_1|U|\,|I| = C|U|\,\beta^{-N}$.

We choose N so large that $C|U|\,\beta^{-N}2^N < 1$. This is possible since $\beta > 2$. We put $W = \Pi\,(U + I)$, $F_k = \Pi(D_k)$ and $K_k = W + F_k$, where Π is the homomorphism from $\mathbf{R}^n$ to $\mathbf{T}^n$ whose kernel is $\mathbf{Z}^n$.

Lemma 9 shows that K_k admits a uniform room in $\mathbf{T}^n$ ($m = 2^N$ and $|W| \leqslant |U + I|$ since Π is measure-decreasing) and this ends the proof of Theorem IV.

5. Other results for symmetric sets

5.1. THEOREM V: *Let $\mathfrak{K}$ be a real algebraic number field, n the degree of $\mathfrak{K}$, and $\sigma_1, \ldots, \sigma_n$ the n $\mathbf{Q}$-isomorphisms from $\mathfrak{K}$ to $\mathbf{C}$, where σ_1 is the identity map from $\mathfrak{K}$ to $\mathbf{R}$.*

Let $\alpha, \alpha_2, \ldots, \alpha_n$ be n real numbers such that $0 < \alpha < \frac{1}{2}$, $1 < \alpha_2, \ldots, 1 < \alpha_n$ and $\alpha\alpha_2 \cdots \alpha_n = 1$. Let C be a positive real number.

Let $(t_k)_{k\geqslant 1}$ be a sequence of algebraic integers of $\mathfrak{K}$ such that $0 < t_k \leqslant \alpha^k$ and for $2 \leqslant j \leqslant n$, $|\sigma_j(t_k)| \leqslant C\alpha_j^k$.

Let E be the set of all sums $\sum_{k\geqslant 1} \varepsilon_k t_k$, $\varepsilon_k = 0$ or 1.

Then E is a set of uniqueness.

Before proving this theorem, some remarks should be made. Let $e_1, \ldots, e_n$ be a basis of the free $\mathbf{Z}$-module of algebraic integers of $\mathfrak{K}$. Let D be the determinant of the n vectors $\sigma_j(e_1), \ldots, \sigma_j(e_n)$, $1 \leqslant j \leqslant n$. Minkowski's theorem (Chapter I, Section 2) shows that there exists an algebraic integer t_k in $\mathfrak{K}$ such that $0 < t_k \leqslant \alpha^k$ and, for $2 \leqslant j \leqslant n$, $|\sigma_j(t_k)| \leqslant C\alpha_j^k$ provided $C \geqslant |D|$.

5.2. Now to the proof. We have to define the dilations used to transform the pseudo-function S carried by E into S_k. For each $k \geqslant 1$, let s_k be an algebraic integer of $\mathfrak{K}$ such that $0 < s_k < \alpha^{-k}$ and $|\sigma_j(s_k)| \leqslant C\alpha_j^{-k}$ for $2 \leqslant j \leqslant n$. By Minkowski's theorem, s_k can be found if $C \geqslant |D|$.

Then S_k is defined by $S_k(t) = s_k^{-1} S\,(s_k^{-1} t)$ and is carried by $s_k E$.

The continuous homomorphism $h : \mathbf{R} \to \mathbf{T}^n$ is defined by

$$h(t) = (\exp 2\pi i e_1 t, \ldots, \exp 2\pi i e_n t)$$

and the linear map $H : \mathbf{R} \to \mathbf{R}^n$ is given by $H(t) = (e_1 t, \ldots, e_n t)$.

We shall prove that $h(s_k t)$ admits a uniform room in $\mathbf{T}^n$. The proof can then be finished as in Section 4.

If $t = \sum_1^\infty \varepsilon_j t_j$, $s_k t = t' + t''$, where $t'' = s_k \sum_{k+1}^\infty \varepsilon_j t_j$. Let Λ_k be the set of the corresponding t'; each element of Λ_k is an algebraic integer λ of the field $\mathfrak{K}$ such that $|\sigma_j(\lambda)| \leqslant c^2\alpha_j/(\alpha_j - 1)$ for $2 \leqslant j \leqslant n$.

For each integer p, $1 \leqslant p \leqslant n$, and each λ in Λ_k, we put $u_j = -\sigma_j(\lambda)$ and we have

$$e_p\lambda \equiv \sigma_2(e_p)\,u_2 + \cdots + \sigma_n(e_p)\,u_n \pmod 1, \tag{3.5}$$

where $|u_j| \leqslant c^2\alpha_j/\alpha_j - 1$, $2 \leqslant j \leqslant n$ and where the u_j corresponding to complex conjugate σ_j are also complex-conjugate. Let U be the set of all the right-hand sides of (3.5) for all u_j satisfying the preceding conditions. Then U is compact and contained in a hyperplane L of $\mathbf{R}^n$. We have $H(t') \in U + \mathbf{Z}^n$.

Let R_k be the set of all sums $s_k \sum_{k+1}^\infty \varepsilon_j t_j$, $\varepsilon_j = 0$ or 1 and $V_k = H(R_k)$. Then as in the proof of Theorem IV, I is the image under H of $[0, \alpha^{N+1}/(1-\alpha)]$ and D_k is a set of 2^N elements of $D = H(\mathbf{R})$. We have $V_k \subset D_k + I$ and $H(s_k E) \subset U + I + D_k$. Let W be $\Pi(U + I)$, $F_k = \Pi(D_k)$. We have $|U + I| \leqslant C_1\alpha^N$, hence $|W| \leqslant C_1\alpha^N$. For N sufficiently large $\operatorname{Card} F_k|W| < 1$ and the sequence of compact sets $K_k = W + F_k$ admits a uniform room in $\mathbf{T}^n$. The proof is concluded by remarking that $h(s_k E) \subset K_k$ since $H(t') \in U + \mathbf{Z}^n$ and $H(t'') \in V_k$ for each $t' + t''$ in $s_k E$.

6. Sets of uniqueness in general l.c.a. groups

A natural space $PM(G)$ of distributions will be defined on each locally compact group. As usual we begin by defining the test functions. The reader is referred to [15] throughout this section.

6.1. *The spaces $L^1(G)$, $M(G)$, $L^\infty(G)$ and the Fourier transform*

In all the following, G will be a metrizable and separable locally compact abelian group. As usual $\mathscr{C}_0(G)$ will denote the Banach space of all complex valued continuous functions on G vanishing at infinity; the norm of an element of $\mathscr{C}_0(G)$ is $\|f\|_\infty = \sup_G |f|$.

The dual space of $\mathscr{C}_0(G)$ is the Banach space $M(G)$ of all complex valued bounded Radon measures μ on G such that $\|\mu\|$, the total variation of μ, is finite. Such a μ can also be defined as a bounded complex valued σ-additive set function on the Borel field of G.

If G is a metrizable and separable l.c.a. group, there exists a non-zero m in $M(G)$ (m is called a Haar measure on G) such that, for each Borel set E in G and each s in G, $m(E + s) = m(E)$; it is equivalent to state that for each f in $\mathscr{C}_0(G)$ and each s in G, $\int_G f(x)\, dm(x) = \int_G f(x + s)\, dm(x)$.

The Lebesgue space of all classes of complex valued m—integrable functions f on G is denoted $L^1(G)$—the classes are taken with respect to equality almost everywhere. The dual space of $L^1(G)$ is the Banach space $L^\infty(G)$ of all classes of complex valued Lebesgue measurable bounded functions on G. On $L^\infty(G)$ two topologies are of special interest. The first one is defined by the norm $\|f\|_\infty = \text{ess sup}\,|f(x)|$ and the second one, denoted by $\mathscr{T}$, is the topology $\sigma(L^\infty, L^1)$ of pointwise convergence on L^1 of the linear forms defined by elements of $L^\infty(G)$.

If f belongs to $L^\infty(G)$, the mapping from G to $L^\infty(G)$ defined by $s \to f_s$ ($f_s(x) = f(x - s)$) is continuous for $\mathscr{T}$; this is not true for the sup-norm topology.

Let $\Gamma = G^\wedge$ be the dual group of G. The Fourier transform of an f in $L^1(G)$ is an element of $\mathscr{C}_0(\Gamma)$ defined by $\hat{f}(\gamma) = \int_G \overline{\chi_\gamma(x)}\, f(x)\, dx$, where $^-$ denotes complex conjugation and, for each γ in Γ, $\chi_\gamma(x)$ is the value taken on x by the character γ in Γ.

For each μ in $M(G)$, the Fourier transform $\hat{\mu}$ of μ is a uniformly continuous bounded function defined on Γ by $\hat{\mu}(\gamma) = \int_G \overline{\chi_\gamma(x)}\, d\mu(x)$.

6.2. *The convolution product*

The convolution product $\mu * \nu$ of two elements μ and ν of $M(G)$ is given by $\langle \mu * \nu, f \rangle = \int_{G\times G} f(x + y)\, d\mu(x)\, d\nu(y)$ ($f \in \mathscr{C}_0(G)$).

The Fourier transform $(\mu * \nu)^\wedge$ of $\mu * \nu$ is the ordinary product $\hat{\mu}\hat{\nu}$. Hence we have a Banach algebra $B(\Gamma)$ of complex valued uniformly continuous and bounded functions on Γ given by the following rules:

(i) $\beta \in B(\Gamma)$ if there exists a μ in $M(G)$ such that $\beta = \hat{\mu}$;

(ii) the product of β_1 and β_2 in $B(\Gamma)$ is the ordinary (pointwise) product of functions;

(iii) $\|\beta\|_{B(\Gamma)} = \|\mu\|$.

The Lebesgue space $L^1(G)$ becomes a Banach algebra when the product is the convolution $f * g$—we recall that for almost all x in G, $t \to f(x - t)\, g(t)$ belongs to $L^1(G)$ and that $(f * g)(x) = \int_G f(x - t)\, g(t)\, dt$.

The homomorphisms from $L^1(G)$ to the complex field $\mathbf{C}$ are all of the form $f \to \hat{f}(\gamma)$ for a suitable γ in Γ and, under the Fourier transformation, $L^1(G)$ becomes a subalgebra $A(\Gamma)$ of $\mathscr{C}_0(\Gamma)$; if $\alpha = \hat{f}$, for some $f \in L^1(G)$ we put $\|\alpha\|_{A(\Gamma)} = \|f\|_1$; this makes $A(\Gamma)$ an isometric copy of the Banach algebra $L^1(G)$, in which convolutions are replaced by pointwise multiplications: this seems better, although the norm of an element α of $A(\Gamma)$ cannot be computed directly from α.

6.3. *Test functions*

On a l.c.a. group G, the test functions are the complex valued continuous functions α which belong both to $L^1(G)$ and to $A(G)$; by the inversion formula ([15], p. 22, th. 1.5.1) the Fourier transform $\hat{\alpha}$ of α also has these two properties. In [15] it is proved that we can construct partitions of unity by means of these test functions. The vector space of test-functions will be denoted by $\mathscr{A}(G)$.

6.4. *Pseudomeasures*

If φ belongs to $L^\infty(\Gamma)$ the Fourier transform $S = \hat{\varphi}$ of φ is a distribution S defined by $\langle S, \alpha\rangle = \int_\Gamma \varphi(\gamma)\, \hat{\alpha}(\gamma)\, d\gamma$ for each test function α in $\mathscr{A}(G)$. Since this integral is convergent for all $\hat{\alpha}$ in $L^1(\Gamma)$, it is sufficient to assume that $\alpha \in A(G)$. Such an S is called a pseudo-measure. We put $\|S\|_{PM} = \|\varphi\|_\infty$ and, with this norm, the pseudo-measures form a Banach space denoted by $PM\,(G)$.

In the preceding terminology, the spectrum of φ is the support of the pseudo-measure $\hat{\varphi}$.

Note that each μ in $M(G)$ is a pseudo-measure; in this case $\varphi(\gamma) = \int_G \chi_\gamma(x)\, d\mu\,(x)$ and μ is the Fourier transform of the complex valued bounded continuous function φ.

6.5. *Compactly supported pseudo-measures*

For a compactly supported pseudo-measure S, we can define a version of the Fourier transform $\hat{S}$ of S, which is a complex valued bounded continuous function on Γ: if $\hat{S} = \varphi \in L^\infty(\Gamma)$, let α be an element of $\mathscr{A}(G)$, which is equal to 1 on a neighborhood of the support of S and write $S = S\alpha$; hence $\varphi = \hat{S} = \hat{S} * \hat{\alpha} = \psi$ is the required continuous bounded function, where the equality is an almost everywhere equality and ψ is the convolution of an element of $L^1(\Gamma)$ and an element of $L^\infty(\Gamma)$ and is therefore continuous.

6.6. The scalar product of a compactly supported pseudo-measure S and a function g in $B(G)$ is defined using the same function α: αg belongs to $A(G)$ and $\langle S, g\rangle = \langle S, \alpha g\rangle$; this definition is independent of the particular α chosen.

6.7. Homomorphisms of groups

Let G_1 and G_2 be two l.c.a. groups and $h: G_1 \to G_2$ a continuous homomorphism. For each f in $A(G_2)$, $f \circ h = g$ belongs to $B(G_1)$ and $\|g\|_{B(G_1)} \leqslant \|f\|_{A(G_2)}$ ([15], 4.2.1, p. 79).

Hence for each compactly supported pseudomeasure S on G_1, we can define a pseudo-measure $h(S)$ on G_2 by the rule $\langle h(S), f\rangle = \langle S, f \circ h\rangle$. Clearly if E is the support of S, $h(S)$ is supported by $h(E)$.

6.8. Sets of multiplicity

A compact subset E of G is said to be a set of multiplicity if there exists a non zero pseudo-measure S carried by E whose Fourier transform vanishes at infinity. An equivalent condition is that there exists a continuous complex valued function $f: \Gamma \to \mathbf{C}$ vanishing at infinity whose spectrum lies in E.

7. The p-adic case

7.1. Let $\mathbf{Q}_p$ be the p-adic field, A_p be the subring of $\mathbf{Q}_p$ consisting of all mp^n, $m \in \mathbf{Z}$, $n \in \mathbf{Z}$, m and p relatively prime. The p-adic norm is defined on A_p by $|mp^n|_p = p^{-n}$ and for this norm A_p is dense in $\mathbf{Q}_p$. But A_p is also a subring of $\mathbf{R}$ and the norm $|mp^n|_\infty$ is induced on A_p by the Archimedean norm of $\mathbf{R}$.

Let $\mathbf{Z}_p$ be the subring of all x in $\mathbf{Q}_p$ such that $|x|_p \leqslant 1$: $\mathbf{Z}_p$ is the completion of $\mathbf{Z}$ with respect to the p-adic norm. Each x in $\mathbf{Q}_p$ can be written uniquely in the form $x = y + r$ where $y \in \mathbf{Z}_p$, $r \in A_p$ and $0 \leqslant r < 1$. We put $\exp 2\pi ix = \exp 2\pi ir$.

The dual group of $\mathbf{Q}_p$ is $\mathbf{Q}_p$ itself; each continuous character $\chi : \mathbf{Q}_p \to \mathbf{T}$ is of the form $\chi(x) = \exp 2\pi ixy$ for a suitable y in $\mathbf{Q}_p$.

Let θ be an element of $\mathbf{Q}_p$ such that $|\theta|_p > 2$. Let $E(\theta)$ be the compact set of all convergent sums $\sum_1^\infty \varepsilon_k \theta^{-k}$, $\varepsilon_k = 0$ or 1.

7.2. THEOREM VI: *The compact set $E(\theta)$ is a set of uniqueness if and only if θ is a Pisot-Chabauty number.*

The proof of the necessity of the condition is very similar to the proof given for the real case. We can construct as in Section 2.2 a probability measure μ carried by $E(\theta)$, whose Fourier transform is

$$\hat{\mu}(y) = \prod_1^\infty \frac{1 + \exp 2i\theta^{-k}y}{2}.$$

To study the behaviour of $\hat{\mu}$ at infinity, we divide $\mathbf{Q}_p$ into the cells $|\theta|_p^k \leqslant |y|_p \leqslant |\theta|_p^{k+1}$ which leads one to put $y = \theta^k u$ and

$$P_k(u) = \prod_0^{k-1} \left| \frac{1 + \exp 2i\theta^j u}{2} \right|.$$

We get, if $y = \theta^k u$ and $1 \leqslant |u|_p \leqslant |\theta|_p$, $|\hat{\mu}(y)| \leqslant P_k(u)$. If θ is not a Pisot-Chabauty number, $P_k(u)$ tends uniformly to 0 on each compact subset of $\mathbf{Q}_p$ not containing 0 by Pisot-Chabauty's theorem. Hence $\omega_k = \sup_{1 \leqslant |u|_p \leqslant |\theta|_p} |P_k(u)|$ tends to 0 as k tends to infinity and, for $|\theta|_p^k \leqslant |y|_p \leqslant |\theta|_p^{k+1}$, $|\hat{\mu}(y)| \leqslant \omega_k$.

7.3. *$E(\theta)$ is a set of uniqueness when θ is a Pisot-Chabauty number*

Let Λ be the set of all p-adic number $\sum_{k \geqslant 0} \varepsilon_k \theta^k$, where $\varepsilon_k = 0$ or 1 and where the sum is finite. The proof is implicitly contained in the statement that Λ is a part of a model.

Sketch of the proof: Let S be a pseudofunction carried by E such that $\hat{S}(0) = 1$. Let S_k be defined by $S_k(x) = \theta^{-k} S(\theta^{-k}x)$; S_k is then carried by $\theta^k E \subset \Lambda + E$.

(a) there is a compact group G and a continuous injective homomorphism $h : \mathbf{Q}_p \to G$ with the following properties:

(1) $h(\mathbf{Q}_p)$ is dense in G;

(2) there is a compact subset K of G of measure 0 such that $h(\Lambda + E) \subset K$;

(b) Putting $T_k = h(S_k)$, we get a bounded sequence of pseudo-measures carried by K and converging in the weak star topology of pseudo-measures to the limit $d\mathbf{x}$, where $d\mathbf{x}$ is the Haar measure of G. This is impossible.

Details of the proof: Part (b) of the proof is an easy modification of the corresponding argument used in the real case. We have to check (a).

Let n be the degree of θ, $\mathfrak{K}$ the field $\mathbf{Q}(\theta)$ and $\mathfrak{G}$ the product group $\mathbf{Q}_p \times \mathbf{R}^n$; we write $(x, y) \in \mathfrak{G}$ if $x \in \mathbf{Q}_p$ and $y \in \mathbf{R}^n$. Let $1, a_2, \ldots, a_n$ be a dual basis of the basis $(1, \theta, \ldots, \theta^{n-1})$ defined by $1 + a_2\theta_j + \cdots + a_n\theta_j^{n-1} = 0$ for $2 \leqslant j \leqslant n$, $\theta_2, \ldots, \theta_n$ being the conjugates of θ over $\mathbf{Q}$. Then $a_2, \ldots, a_n$ lie in $\mathfrak{K} \subset \mathbf{Q}_p$ and, provided $c > 0$ is large enough, Λ is contained in the set M of all elements λ of $\mathfrak{K}$ of the form $\lambda = t_1 + t_2\theta + \cdots + t_n\theta^{n-1}$ for all $t_1, \ldots, t_n$ in A_p satisfying the inequalities: $|t_2 - a_2t_1|_p \leqslant c, \ldots, |t_n - a_nt_1|_p \leqslant c$, $|t_1|_\infty \leqslant c, \ldots, |t_n|_\infty \leqslant c$ (Chapter II, Section 10).

Let D be the discrete and relatively dense subgroup of $\mathbf{Q}_p \times \mathbf{R}^n$ consisting of all (x, y) of the form $x = t_1 + t_2\theta + \cdots + t_n\theta^{n-1}$, $y = (t_1, t_2, \ldots, t_n)$ for $t_1, \ldots, t_n$ in A_p such that $|t_2 - a_2t_1|_p \leqslant c, \ldots, |t_n - a_nt_1|_p \leqslant c$. Let G be the quotient group $\mathfrak{G}/D$, Π the canonical projection of $\mathfrak{G}$ onto G and h the restriction of Π to $\mathbf{Q}_p$. Then the image of $\mathbf{Q}_p$ under h is a dense subgroup of G.

Definition of K: Let $B \subset \mathbf{R}^n$ be the set of all $y = (y_1, \ldots, y_n)$ in $\mathbf{R}^n$ such that $|y_1|_\infty \leqslant c, \ldots, |y_n|_\infty \leqslant c$. Then $K = \Pi(E \times B)$. Since D is discrete in $\mathfrak{G}$, Π is measure-decreasing. Hence, if $|\ \ |$ denotes the corresponding Haar measures with suitable normalizations, $|K| \leqslant |E \times B| = |E|\,|B| = 0$ since E has zero measure in $\mathbf{Q}_p$.

$h(\Lambda + E)$ is contained in K. Since h is the restriction of Π to $\mathbf{Q}_p$, it suffices to show that $\Lambda + E \subset E \times B + D$. But for each λ in Λ, we can find a y in B such that $(\lambda, 0) + (0, y) \in D$ or, since $-B = B$, a y in B such that $(\lambda, 0) \equiv (0, y) \pmod{D}$. If $t \in E$, then $(\lambda + t, 0) \equiv (t, y) \pmod{D}$ and the proof is complete.

8. Generalizations

As in the real case, the proof can be extended to symmetric sets constructed with variable dissection ratios whose inverses are 'uniform' Pisot numbers of a given algebraic number field contained in $\mathbf{Q}_p$.

THEOREM VII: *Let $\mathfrak{K}$ be an extension of the field $\mathbf{Q}$ of finite degree n contained in $\mathbf{Q}_p$ and let $\tau_1, \ldots, \tau_n$ be the n $\mathbf{Q}$-isomorphisms from $\mathfrak{K}$ into $\mathbf{C}$. Let $0 < \alpha < 1$, and let $(\omega_k)_{k \geqslant 1}$ be a sequence of Pisot-Chabauty numbers* *of degree n in* $\mathfrak{K}$ *satisfying the condition of uniformity:* $|\tau_j(\omega_k)|_\infty \leqslant \alpha$, *for all* $1 \leqslant j \leqslant n$ *and* $k \geqslant 1$, *and such that* $\|\omega_k\|_p > 2$ *for all* $k \geqslant 1$.

Then the symmetric set E constructed with the sequence (ω_k^{-1}) *of dissection ratios is a set of uniqueness.*

The definition of E is the same as that given in the real case: E is the set of all sums $\sum_1^\infty \varepsilon_k t_k$, $\varepsilon_k = 0$ or 1, $t_k = \xi_1 \cdots \xi_{k-1}(1 - \xi_k)$.

9. Other results on the problem of uniqueness

In Chapter VII ultrathin symmetric sets will be studied: when the dissection ratios tend to 0 sufficiently rapidly, no arithmetical properties are needed to get sets of uniqueness. The exact condition is $\sum_1^\infty \xi_k^2 < +\infty$ and it is the best possible in the sense that the exponent 2 cannot be replaced by a larger one.

10. Notes

A remarkable survey on the problem of uniqueness and multiplicity is given in [2], Ch. XIV. A first version of Theorem IV may be found in [10], where there is a requirement of periodicity on the sequence of dissection ratios. For other results, the reader is referred to [18], [9], [58] and [92].

CHAPTER IV

PROBLEMS ON A-PERIODIC TRIGONOMETRIC SUMS

Almost periodicity is a useful and natural generalization of periodicity: a bounded continuous $f: \mathbf{R} \to \mathbf{C}$ is almost periodic if for each positive ε, there exists a $T(\varepsilon) > 0$ such that each interval of length $T(\varepsilon)$ contains an ε-almost period τ of f, i.e. $\sup_{\mathbf{R}} |f(t + \tau) - f(t)| \leqslant \varepsilon \sup_{\mathbf{R}} |f|$.

The theory shows that $T(\varepsilon)$ depends on f in a very intricate way: to make any estimate of $T(\varepsilon)$ one needs to have knowledge of f right up to the end of time and $T(\varepsilon)$ cannot be used to predict what f will be. The general theory only gives a posteriori estimates of ε-almost periods.

This is quite restricting and our first problem is to get a priori estimates of ε-almost periods for reasonable spaces $\mathscr{C}$ of almost periodic functions. By reasonable spaces $\mathscr{C}$, we mean spaces that are translation invariant and linear, and in this case the answer to our first problem is given by the following rule: the frequencies of any f in $\mathscr{C}$ belong to a given harmonious set.

A second problem is to give a priori estimates of the size of almost periodic functions f belonging to such reasonable spaces, where the size of f is taken to be $\sup_{\mathbf{R}} |f|$. We are led to the following problem: let Λ be a set of real numbers; do a compact set K and a constant C exist such that for any trigonometric sum whose frequencies belong to Λ, $\sup_{\mathbf{R}} |f| \leqslant C \sup_{\mathbf{K}} |f|$? If they do exist, Λ is said to be a coherent set of frequencies.

A third problem is to study the effect of slow perturbations on trigonometric sums whose frequencies belong to a given set Λ of frequencies. Perturbations are said to be slow when constant coefficients are replaced by continuous functions whose spectra are contained in a small neighborhood of 0: in the real case, we get small frequencies and great periods as in the secular perturbations of astronomy. If $f(t) = \sum_{\lambda \in \Lambda} a_\lambda(t) \exp 2\pi i\lambda t$ is such a slowly-perturbed trigonometric sum and if $f(t_0, t) = \sum_{\lambda \in \Lambda} a_\lambda(t_0) \exp 2\pi i\lambda t$ is the corresponding family of pure trigonometric sums, it is natural to expect $f(t_0, t)$ to be a good approximation of $f(t)$ in a suitable neighborhood of t_0.

Surprisingly, two extra conditions are needed to obtain a good result: an a priori estimate of the behaviour of f at infinity ($|f(t)| \leqslant C \exp |t^\alpha|$, $0 < \alpha < 1$, is sufficient) and a remarkable arithmetical structure for Λ (it is sufficient to assume that Λ is a coherent set of frequencies).

1. Elementary properties of almost-periodic functions

We are mainly interested in spaces of trigonometric sums defined on a l.c.a. group G; a trigonometric sum P is a sum $\sum_{\gamma \in \Gamma} a_\gamma \chi_\gamma$ of continuous characters with complex coefficients indexed by the dual group Γ such that the complex numbers a_γ are equal to 0 for all γ in Γ with the exception of at most a finite number of γ; the set of all $\gamma \in \Gamma$ such that $a_\gamma \neq 0$ is called the spectrum of P or the set of frequencies of P. In the real case, $P(t) = \sum_{\gamma \in \mathbf{R}} a_\nu \exp 2\pi i \nu t$ and the real numbers ν for which $a_\nu \neq 0$ are the frequencies of P.

A great number of properties of such sums hold for uniform limits on G of trigonometric sums. We have just defined the space of almost periodic functions on G.

We recall without proofs the elementary properties of complex valued almost periodic functions on G. The reader is referred to [11] for the proofs.

Let $\mathscr{C}_b(G)$ be the Banach space of all complex-valued bounded continuous functions on G. The norm of $f \in \mathscr{C}_b(G)$ is $\|f\|_\infty = \sup_{x \in G} |f(x)|$ and is also written $\sup_G |f|$.

For each $f \in \mathscr{C}_b(G)$ and each $s \in G$, the translated function f_s is defined by $f_s(x) = f(x - s)$ for all x in G; for each positive ε, the set $M(\varepsilon, f)$ of ε-almost periods of f is defined by $s \in M(\varepsilon, f)$ if $\|f_s - f\|_\infty \leqslant \varepsilon \|f\|_\infty$. This is a slight modification of the usual definition: the normalization by $\|f\|_\infty$ is fundamental in the following.

Definition 1: *A function $f \in \mathscr{C}_b(G)$ is said to be almost-periodic if one of the following two equivalent conditions is satisfied.*

(1) *for each positive ε, the set $M(\varepsilon, f)$ of ε-almost periods of f is relatively dense;*

(2) *for each positive ε, there exist a finite subset F of Γ and a trigonometric sum $P(x) = \sum_{\gamma \in F} a_\gamma \chi_\gamma(x)$ such that $\|f - P\|_\infty \leqslant \varepsilon$.*

If f is almost periodic, some precise information can be given about the sets F used in (2). The closed linear subspace of $\mathscr{C}_b(G)$ spanned by all the f_s, $s \in G$, is denoted by $E(f)$.

DEFINITION 2: *The spectrum, denoted by* $Sp(f)$, *of an almost periodic function* f *on* G *is the set of all* γ *in the dual group* Γ *of* G *such that the corresponding character* χ_γ *belongs to* $E(f)$.

The theorem of spectral synthesis for almost periodic functions is an improvement of property 2 in Definition 1.

THEOREM I: *Let* f *be a complex-valued almost periodic function defined on an l.c.a. group* G *and* Λ *be the spectrum of* f. *Then there exists a sequence* $(P_k)_{k \geqslant 1}$ *of trigonometric sums whose frequencies belong to* Λ *such that* $\sup_{G} |f - P_k| \to 0$.

At this point, it seems that the theory of almost periodic functions is going to be quite easy-going: we can define the Fourier series of almost-periodic functions, the Parseval relation is valid, and so on. How do new problems arise?

2. Classes of almost-periodic functions

2.1. We come back to the real line. The space, denoted by $AP(\mathbf{R})$, of all complex-valued almost-periodic functions on the real line is a huge space containing all complex-valued continuous periodic functions; indeed, $AP(\mathbf{R})$ is not separable. The properties of an individual almost-periodic function on the line are very similar to the properties of an individual periodic function, but the common properties of all complex-valued continuous 2π-periodic functions are very different from the common properties of all almost periodic functions. We shall list these properties:

Each $\tau \in 2\pi\mathbf{Z}$ is a period for each 2π-periodic function. (4.1)

The 2π-periodic functions are given by their values on the compact subset $[0, 2\pi]$ of $\mathbf{R}$. (4.2)

The space of all continuous 2π-periodic functions is closed in $\mathscr{C}_b(\mathbf{R})$ and is translation invariant. (4.3)

If K is a compact subset of $[0, 2\pi]$, different from $[0, 2\pi]$, each complex-valued continuous function on K can be extended to a 2π-periodic function. (4.4)

2.2. In an attempt to generalize this situation, we are led to the following definition:

DEFINITION 3: *Let $\mathscr{C}$ be a linear subspace of $AP(G)$. We say that $\mathscr{C}$ is a class of almost-periodic functions if $\mathscr{C}$ has the following three properties:*

(a) *$\mathscr{C}$ is closed in $AP(G)$;*

(b) *$\mathscr{C}$ is translation-invariant;*

(c) *for all positive ε, there exists a relatively dense set $M(\varepsilon)$ in G such that for each s in $M(\varepsilon)$ and each f in $\mathscr{C}$, $\|f_s - f\|_\infty \leqslant \varepsilon \|f\|_\infty$.*

Example: the space of all complex valued continuous 2π-periodic functions on the line is a class of almost-periodic functions.

THEOREM II: *Let Λ be a harmonious subset of the dual group Γ of a l.c.a. group G. Let $\mathscr{C}(\Lambda)$ be the linear subspace of all almost-periodic functions on G whose spectra lie in Λ. Then $\mathscr{C}(\Lambda)$ is a class of almost periodic functions and each class of almost-periodic functions is a $\mathscr{C}(\Lambda)$ for a suitable harmonious set Λ.*

It will be shown in the interlude that (4.4) holds for $\mathscr{C}(\Lambda)$ if Λ is relatively dense in Γ and if K is a suitable compact neighbourhood of 0.

2.3. Proof of the first half of Theorem II

LEMMA 1: *Let l be a real number in the open interval $]0, \pi[$ and $a: \mathbf{R} \to [-1, 1]$ the continuous $4l$-periodic function defined by $a(t) = a(-t)$ for all real t and $a(t) = \sin (l - t)/2$ if $0 \leqslant t \leqslant 2l$. The Fourier expansion of $a(t)$ is $\sum_0^\infty c_k \cos (k + \frac{1}{2}) \pi l^{-1} t$, where $c_k \geqslant 0$ and $\sum_0^\infty c_k = \sin l/2$.*

Proof: An easy computation gives

$$c_k = \frac{2l \cos l/2}{(2k + 1)^2 \pi^2 - l^2} > 0.$$

Hence $\sum_0^\infty c_k < +\infty$, the function is the sum of its Fourier series and $\sum_0^\infty c_k = a(0) = \sin l/2$.

LEMMA 2: *Let $p = 2q + 1$ be an odd integer and $l = \pi/p$. There exists a sequence $(d_n)_{-\infty < n < +\infty}$ of complex numbers such that*

(a) $\sum_{-\infty}^{+\infty} |d_n| = |e^{il} - 1| = 2 \sin l/2$;

(b) *if* $-l \leqslant x \leqslant l$, $e^{ix} - 1 = \sum_{-\infty}^{+\infty} d_n \, e^{inx}$.

Proof: We write $e^{ix} - 1 = e^{ix/2}(e^{ix/2} - e^{-ix/2}) = 2i\, e^{ix/2} \sin x/2$. Put $t = l - x$ and apply Lemma 1. The hypothesis that p is odd and $l = \pi/p$ ensures that each $e^{-it/2} \cos (k + \frac{1}{2})\, \pi l^{-1}\, t$ can be written e^{int}, for some $n \in \mathbf{Z}$. It can be shown that the hypothesis $l = \pi/p$, p odd is necessary for (a) and (b) to hold.

LEMMA 3: *For each real number* δ *in* $[0, 1]$, *there exists a sequence* $(d_n)_{-\infty<n<+\infty}$ *of complex numbers such that*

(a) $\sum_{-\infty}^{+\infty} |d_n| \leqslant 2\delta$;

(b) $|z| = 1$ *and* $|z - 1| \leqslant \delta$ *imply* $z - 1 = \sum_{-\infty}^{+\infty} d_n z^n$.

Proof: l is defined by $l \in [0, \pi/3]$ and $\sin l/2 = \delta/2$, and p is the greatest odd integer less than π/l. Then $p \geqslant 3$ and Lemma 2 yields the sequence $(d_n)_{-\infty<n<+\infty}$ such that $\sum_{-\infty}^{+\infty} |d_n| = 2 \sin \pi/2p$. Moreover, $p \leqslant \pi/l < p + 2 < 2p$ implies $\sin \pi/2p < \sin l < 2 \sin l/2 = \delta$.

LEMMA 4: *Let G be a l.c.a. group,* Γ *the dual group and* Λ *a subset of* Γ. *Each element s of G such that* $\sup_{\Lambda} |\chi_s(\lambda) - 1| = \delta \leqslant 1$ *is a* 2δ*-almost period of each almost-periodic function f whose spectrum lies in* Λ.

Since the subset of $AP(G)$ defined by $\|f_s - f\|_\infty \leqslant 2\delta \|f\|_\infty$ is closed, Theorem I shows that we can restrict our attention to trigonometric sums whose frequencies belong to Λ.

We define the sequence $(d_n)_{n\in\mathbf{Z}}$ by Lemma 3 and a discrete measure σ on G by $\sigma = \sum_{-\infty}^{+\infty} d_n \varepsilon_{ns}$, where ε_x is the unit point-mass at x.

Then for each λ in Λ, $\int_G \chi_\lambda\, d\sigma = \sum_{-\infty}^{+\infty} d_n \chi_\lambda(ns) = \sum_{-\infty}^{+\infty} d_n (\chi_\lambda(s))^n = \chi_\lambda(s) - 1$.

Assume that f is a finite trigonometric sum $\sum_{\lambda\in\Lambda} c_\lambda \chi_\lambda$; then $f(s) - f(0) = \sum_{\lambda\in\Lambda} c_\lambda (\chi_\lambda(s) - 1) = \sum_{\lambda\in\Lambda} c_\lambda \int_G \chi_\lambda\, d\sigma = \int_G f\, d\sigma$ implies $|f(s) - f(0)| \leqslant 2\delta \|f\|_\infty$. To get the inequality $|f(s + x) - f(x)| \leqslant 2\delta \|f\|_\infty$ for each x in G, we apply the preceding proof to the function g defined on G by $t \to f(t+x)$.

We now return to Theorem II. If Λ is harmonious, there is, for each positive δ, a relatively dense subset $N(\delta)$ of G such that $\sup_{\Lambda} |\chi_\lambda(s) - 1| \leqslant \delta$ for each $s \in N(\delta)$. Lemma 4 gives $\|f_s - f\|_\infty \leqslant 2\delta\, \|f\|_\infty$ and the first part of Theorem II is proved [91].

2.4. The proof of the second part depends on the following lemma:

LEMMA 5: *Let* $\mathscr{C} \subset AP(G)$ *be a closed linear subspace which is translation-invariant. Then there exists a subset* Λ *of the dual group* Γ *such that* $\mathscr{C}$ *is the set of all f in* $AP(G)$ *whose spectra lie in* Λ.

Proof: Let Λ be the union of all $Sp(f)$ for all f in $\mathscr{C}$. Definition 2 shows that χ_λ belongs to $\mathscr{C}$ for each λ in Λ, and Theorem I shows that each f in $AP(G)$ whose spectra lie in Λ belongs to $\mathscr{C}$

If $\mathscr{C}$ is a class of almost-periodic functions, we get $|\chi_\lambda(x) - 1| \leqslant \varepsilon$ for each x in $M(\varepsilon)$ and each λ in Λ; since $M(\varepsilon)$ is relatively dense, Λ is harmonious

3. Dilations and classes of almost-periodic functions on the line

Let $\theta > 0$ be a positive real number. The dilation with ratio θ is defined on $\mathbf{R}$ by $x \to \theta x$. If f is any complex-valued function on $\mathbf{R}$, $T_\theta f$ will be the complex-valued function g defined by $g(x) = f(\theta x)$, rather than the expected $f(\theta^{-1}x)$.

The product of two dilations is clearly another dilation and if $\mathscr{C}$ is a set of complex-valued functions on the line, we shall study the set S of all $\theta > 0$ such that $T_\theta(\mathscr{C}) \subset \mathscr{C}$.

THEOREM III: *Let $\mathscr{C}$ be a vector space of almost-periodic functions on* $\mathbf{R}$. *Assume that*

(a) *$\mathscr{C}$ is translation-invariant;*

(b) *for each positive ε there is a relatively dense subset $M(\varepsilon)$ of* $\mathbf{R}$ *such that each $s \in M(\varepsilon)$ is an ε-almost period for all f in $\mathscr{C}$;*

(c) *there is a function f in $\mathscr{C}$ which is not a constant.*

Let $(T_\theta)_{\theta \in S}$ be the semigroup of all dilations such that $T_\theta(\mathscr{C}) \subset \mathscr{C}$. Then there exists a real algebraic number field K, of degree n over $\mathbf{Q}$ *say, such that S is contained in $\{1\} \cap \Sigma$, where Σ is the set of all Pisot or Salem numbers of degree n lying in K.*

Before proving this theorem, a remark should be made. It may happen that the only θ for which $T_\theta(\mathscr{C}) \subset \mathscr{C}$ is $\theta = 1$. All the other such θ are Pisot or Salem numbers.

Proof: Let $\overline{\mathscr{C}}$ be the closure of $\mathscr{C}$ with respect to the uniform norm. Then $\overline{\mathscr{C}}$ has the properties (a) and (b) and, using Theorem II of Chapter IV, $\overline{\mathscr{C}}$ is the space of all almost-periodic functions f on the line whose spectra lie in an harmonious set Λ of real numbers. If $\theta \in S$ and $\theta \neq 1$, we apply the inclusion $T_\theta(\overline{\mathscr{C}}) \subset \overline{\mathscr{C}}$ to each element $\exp 2\pi i\lambda x$, $\lambda \in \Lambda$, of $\overline{\mathscr{C}}$; we get $\theta\Lambda \subset \Lambda$. Since there is a function f in $\mathscr{C}$ which is not a constant, there is a λ in Λ which is not zero. Hence $S\lambda \subset \Lambda$ is harmonious and so is S. We apply Theorem VI of Chapter II (Section 8) to S.

4. Classes of almost-periodic functions and a priori estimates on the size

We shall see that each harmonious subset Λ of the dual group Γ of G is a coherent set of frequencies; the converse, however, is not true.

PROPOSITION 1: *Let $\mathscr{C}$ be a class of almost periodic functions, $M(\varepsilon)$ a relatively dense set of ε-almost periods for all f in $\mathscr{C}$, and $K(\varepsilon)$ a compact subset of G such that $K(\varepsilon) + M(\varepsilon) = G$. For each f in $\mathscr{C}$,* $\sup_{K_\varepsilon} |f| \geqslant (1 - \varepsilon) \|f\|_\infty$.

This means that, within a relative error not exceeding ε, the functions in $\mathscr{C}$ are 'known' on the whole group as soon as they are known on K_ε, just as the 2π-periodic functions on the line are fully determined by their values on $[0, 2\pi]$.

The proof of Proposition 1 is very simple. Each y in G can be written $y = s + x$, where $s \in M(\varepsilon)$, $x \in K_\varepsilon$. Since each $s \in M(\varepsilon)$ is an ε-almost period for each f in $\mathscr{C}$, we get $|f(y) - f(x)| \leqslant \varepsilon \|f\|_\infty$; hence $|f(y)| \leqslant \sup_{K_\varepsilon} |f| + \varepsilon \|f\|_\infty$ and taking the supremum over $y \in G$, we get Proposition 1.

Let $G = \mathbf{R}$ and let $(t_k)_{k \geqslant 0}$ be an increasing sequence of positive real numbers such that $\sum_{k \geqslant 0} (t_k/t_{k+1})^2 < +\infty$. Let Λ be the set of all finite sums $\sum_{k \geqslant 0} \varepsilon_k t_k$, where $\varepsilon_k = 0$ or 1. We shall prove (Chapter VIII, Section 5, Theorem IV) that, for each positive ε, there exists a compact subset K_ε of $\mathbf{R}$ such that $\sup_{K} |f| \geqslant (1 - \varepsilon) \|f\|_\infty$ for each almost periodic function f whose spectrum lies in Λ. However, the sequence $(t_k)_{k \geqslant 0}$ can be chosen in such a way that Λ is not harmonious (Chapter VIII, Theorem XI).

5. Pisot numbers and coherent sets of frequencies: the real case

THEOREM IV: *Let θ be a real number greater than one and Λ_θ the set of all finite sums $\sum_{k \geqslant 0} \varepsilon_k \theta^k$, where $\varepsilon_k = 0$ or 1. Then the following two properties of θ are equivalent:*

(a) *θ is a Pisot number;*

(b) *Λ_θ is a coherent set of frequencies.*

Before proving Theorem IV, some remarks should be made: if $\theta = 1$, Λ_θ is the set $\mathbf{N}$ of all positive integers and Λ_θ is a coherent set of frequencies; if $\theta = -1$, $\Lambda_\theta = \mathbf{Z}$ and is a coherent set of frequencies. If $\theta < -1$ the proof given below will show that Λ_θ is a coherent set of frequencies if and only if $-\theta$ is a Pisot number. If $-1 < \theta < 1$, Proposition 3 below shows that Λ_θ cannot be a coherent set of frequencies.

Now to the proof.

(a) Assume θ is a Pisot number: then Λ_θ is harmonious and hence Λ_θ is a coherent set of frequencies;

(b) Now assume θ is not a Pisot number. The following lemma will be needed.

LEMMA 6: *If θ is not a Pisot number there exists a sequence $(P_k)_{k\geqslant 0}$ of finite trigonometric sums whose frequencies belong to Λ_θ such that*

(1) $\sup_{\mathbf{R}} |P_k| = P_k(0) = 1.$

(2) *on each compact subset K of $\mathbf{R}$ not containing* 0, *P_k tends uniformly to* 0 *as k tends to infinity.*

Proof of the lemma: Let $P_k(t) = \exp \pi it\,(1 + \cdots + \theta^k) \cos \pi t \cdots \cos \pi\theta^k t$; $P_k(t)$ is a finite trigonometric sum of the type $\sum_{\lambda\in\Lambda} a_\lambda \exp 2\pi i\lambda t$. If θ is not a Pisot number, $|P_k(t)| = |\cos \pi t| \cdots |\cos \pi\theta^k t|$ is a decreasing sequence tending to 0 for each $t \neq 0$: if we write $|\cos \pi\theta^k t|^2 = 1 - \sin^2 \pi\theta^2 t$, this fact is a consequence of Pisot's theorem.

Dini's theorem shows that $(P_k)_{k\geqslant 0}$ tends uniformly to 0 on each compact set of real numbers not containing 0.

5.1. *Proof of Theorem IV*

Assume that there exist a compact set K of real numbers and a positive C such that, for each trigonometric sum whose frequencies belong to Λ_θ, $\sup_{\mathbf{R}} |P| \leqslant C \sup_{K} |P|$. Let s be a real number such that $s + K$ does not contain 0 and set $Q_k(t) = P_k(s + t)$. Then Q_k is a finite trigonometric sum whose frequencies belong to Λ, $\sup_{K} |Q_k| \to 0$ as $k \to +\infty$ and $\|Q_k\|_\infty = 1$. Hence Λ_θ is not a coherent set of frequencies.

5.2. *Improvement on the preceding result*

Let θ be a Pisot number, η a real number in $]0, 1[$ and $T(\eta)$ the infimum of those positive T for which $\sup_{|t|\leqslant T} |P(t)| \geqslant (1 - \eta)\, \|P\|_\infty$ for each trigonometric sum P whose frequencies belong to Λ_θ.

Two problems can be stated: what is the behaviour of $T(\eta)$ as η tends to 0? What is the behaviour of $T(\eta)$ as η tends to 1?

PROPOSITION 2: *Let n be the degree of the Pisot number θ. There exist two positive constants a and b depending on θ, such that, for each η in* $]0, 1[$, $a\eta^{(-n+1)/2} \leqslant T(\eta) \leqslant b\eta^{(-n+1)/2}$.

This is the best possible result as Proposition 3 of Chapter I shows. This remarkable fact can be proved by the methods used in Chapter I, Section 4, and with the definition of models given in Chapter II, Section 9.

Observe that when η tends to 0, the quality of the prediction of the size of trigonometric sums whose frequencies belong to Λ increases at a corresponding 'cost', here the size of $T(\eta)$, which is of the order of magnitude of $\eta^{-(n-1)/2}$.

On the other hand when η tends to 1, the quality of the prediction decreases while the 'cost' of the prediction also decreases. What is the lower bound of this 'cost'? This is an open question. When η tends to 1, a natural conjecture is $\lim_{\eta \to 1} T(\eta) = \text{Dens}\,\Lambda_\theta$. The density of Λ_θ is defined to be the uniform limit of $T^{-1}\,\text{Card}\,(\Lambda \cap [t, t + T])$ as T tends to infinity. If $\theta > 2$ this density is 0, if $\theta = 2$ it is 1 and if $\theta = (1 + \sqrt{5})/2$ (the golden number of Greek mathematicians) it is $\frac{1}{2} + 3\sqrt{5}/10$. This conjecture will be proved in the case where θ is the quadratic Pisot number $(1 + \sqrt{5})/2$ (Chapter V, Theorem II).

6. A set of powers

Let θ be a real number greater than one and Λ the set of all powers θ^k, $k \geqslant 0$. It will be proved that Λ is always a coherent set of frequencies (Chapter VI, Theorem II). However, some problems remain open.

Conjecture: Let θ and Λ be as above. Then the following two properties are equivalent:

(a) θ is a Pisot or a Salem number;

(b) for each positive η one can find a $T(\eta) > 0$ such that

$$\sup_{|t| \leqslant T(\eta)} |P(t)| \geqslant (1 - \eta) \sup_{\mathbf{R}} |P(t)|$$

for each trigonometric sum P whose frequencies belong to Λ.

7. Pisot numbers and coherent sets of frequencies: the p-adic case

7.1. THEOREM V: *Let θ be a p-adic number and Λ_θ the set of all finite sums $\sum_{k \geqslant 0} \varepsilon_k \theta^k$, where $\varepsilon_k = 0$ or 1. Then the following two properties of θ are equivalent:*

(a) *θ is a Pisot-Chabauty number;*

(b) *Λ is a coherent set of frequencies in $\mathbf{Q}_p$.*

Before proving Theorem IV, some remarks should be made: if $|\theta|_p \leqslant 1$, Λ_θ is an infinite subset of the unit ball of $\mathbf{Q}_p$. By Proposition 3 below, Λ_θ cannot be a coherent set of frequencies. We can therefore restrict our attention to those θ in $\mathbf{Q}_p$ such that $|\theta|_p > 1$.

Now to the proof.

(a) Assume θ is a Pisot-Chabauty number: Λ_θ is harmonious and hence Λ_θ is a coherent set of frequencies;

(b) Now assume θ is not a Pisot-Chabauty number. The following lemma will be needed:

7.2. LEMMA 7: *If θ is not a Pisot-Chabauty number, there exists a sequence $(P_k)_{k\geqslant 0}$ of finite trigonometric sums whose frequencies belong to Λ_θ such that*

(1) $\sup\limits_{\mathbf{Q}_p} |P_k| = P_k(0) = 1.$

(2) *on each compact subset K of $\mathbf{Q}_p$ not containing* 0, *P_k tends uniformly to* 0 *as k tends to infinity.*

Proof of the lemma: We recall that each p-adic number can be written uniquely as a sum $x = \xi + r$, where $\xi \in \mathbf{Z}_p$ and $-\frac{1}{2} \leqslant r < \frac{1}{2}$, $r \in A_p$, where A_p is the subring of $\mathbf{Q}$ generated by p^{-1}. We put $r = \varrho_p(x)$ and the characters on $\mathbf{Q}_p$ are of the form $x \to \exp 2\pi i \varrho_p(x\lambda)$ for a suitable λ in $\mathbf{Q}_p$. A finite trigonometric sum on $\mathbf{Q}_p$ can be written $\sum_{\lambda\in\mathbf{Q}_p} a_\lambda \exp 2\pi i\varrho_p(\lambda x)$.

For each $k \geqslant 1$, let F_k be the subset of all sums $\sum_0^{k-1} \varepsilon_j\theta^j$, $\varepsilon_j = 0$ or 1 in Λ_θ and let P_k be the trigonometric sum $2^{-k}\sum_{\lambda\in F_K} \exp 2\pi i\varrho_p(\lambda x)$. Then $P_k(x)$ can be written $2^{-k}\prod_0^{k-1} (1 + \exp 2\pi i\varrho_p(\theta^j x))$ and $|P_k| = \prod_0^{k-1} |\cos \pi\varrho_p(\theta^j x)|$. Chabauty's theorem (Proposition 12, Section 12, Chapter II) shows that if θ is not a Pisot-Chaubauty number, $P_k(x)$ tends to zero as k tends to infinity for each $x \neq 0$.

The proof is concluded as in the real case.

8. The groups C and $\mathbf{R}^2$

Complex Pisot numbers can be defined in the following way: a complex number θ is a complex Pisot number if there exists an irreducible element $P(X) = X^n + a_1X^{n-1} + \cdots + a_n$ of $\mathbf{Z}[X]$ with the following properties:

(a) $P(\theta) = P(\bar{\theta}) = 0$ and $\theta \neq \bar{\theta}$;

(b) the other $n - 2$ roots z of $P(z) = 0$ satisfy $|z| < 1$.

It can be proved [43] that for a complex number θ, which is not real, the following two properties are equivalent:

(a) θ is a complex Pisot number;

(b) θ is neither purely real nor purely imaginary and the set Λ_θ of all finite sums $\sum_{k\geqslant 0} \varepsilon_k \theta^k$, $\varepsilon_k = 0$ or 1, is a coherent set of frequencies.

Pisot pairs (α, β) in $\mathbf{R}^2$ are defined in the following way: an element (α, β) in $\mathbf{R}^2$ is a Pisot pair if there is an irreducible element $P(X) = X^n + a_1 X^{n-1} + \cdots + a_n$ of $\mathbf{Z}[X]$ with the following properties:

(a) $P(\alpha) = P(\beta) = 0$;

(b) the other $n - 2$ roots z of $P(z) = 0$ satisfy $|z| < 1$.

It can be proved [43] that for an element (α, β) of $\mathbf{R}^2$, the following two properties are equivalent:

(a) (α, β) is a Pisot pair or $|\alpha|$ and $|\beta|$ are two Pisot numbers;

(b) the set $\Lambda_{\alpha\,\beta}$ of all finite sums $(\sum_{k\geqslant 0} \varepsilon_k \alpha^k, \sum_{k\geqslant 0} \varepsilon_k \beta^k)$ $\varepsilon_k = 0, 1$ is a coherent set of frequencies.

9. Coherent sets of frequencies in l.c.a. groups

The definition of a coherent set of frequencies will be given for the case of a general l.c.a. group.

DEFINITION 4: *Let G be a l.c.a. group, Γ the dual group of G and Λ a subset of Γ.*

Λ is a coherent set of frequencies if there exist a positive real number C and a compact subset K of G such that, for each trigonometric sum $P(x) = \sum_{\lambda \in \Lambda} c_\lambda \chi_\lambda(x)$ we have

$$(1) \quad \sup_G |P| \leqslant C \sup_K |P|.$$

If (1) *holds we say that the pair (K, C) is suitable for Λ.*

Example: If G is compact, each subset Λ of Γ has this property. In the general case, each harmonious subset Λ of Γ has this property. But there seems to be no hope, even if Λ is relatively dense in Γ, of obtaining a characterization of Λ as we did for harmonious sets.

PROPOSITION 3: *Let $\Lambda \subset \Gamma$ be a coherent set of frequencies. There exists a neighborhood V of 0 in Γ such that, if $\lambda \in \Lambda$, $\lambda' \in \Lambda$ and $\lambda \neq \lambda'$, then $\lambda' - \lambda \notin V$.*

The proof depends on the following lemma:

LEMMA 8: *If* $\gamma \in \Gamma$, $\gamma' \in \Gamma$ *and* $\gamma \neq \gamma'$, *then* $\sup_{x \in G} |\chi_\gamma(x) - \chi_{\gamma'}(x)| \geqslant \sqrt{3}$.

Proof: Let $\delta = \gamma - \gamma'$; then, for each x in G, $|\chi_\gamma(x) - \chi_{\gamma'}(x)| = |\chi_\delta(x) - 1|$. Since $\delta \neq 0$, the subgroup $\chi_\delta(G)$ of $\mathbf{T}$ is not the trivial subgroup $\{1\}$. Hence there is an x in G such that $|\chi_\delta(x) - 1| \geqslant \sqrt{3}$.

Now to the proof of Proposition 3. Let V be a neighborhood of 0 in Γ such that $\sup_{K \times V} |\chi_\delta(x) - 1| < 1/(C\sqrt{3})$. If $\delta = \gamma - \gamma' \in V$, we have $\sup_K |\chi_\delta(x) - 1| = \sup_K |\chi_{\gamma'}(x) - \chi_\gamma(x)| \geqslant 1/(\sqrt{3}\, C)$, which contradicts the definition of V.

We shall say that Λ is uniformly discrete to express the conclusion of Proposition 3. If G is discrete, Γ is compact and each coherent set of frequencies Λ contained in Γ is finite. Hence coherent sets of frequencies are interesting when G is neither compact nor discrete.

10. Coherent sets of frequencies and restriction algebras

10.1. For each closed subset Λ of Γ, two Banach algebras $A(\Lambda)$ and $B(\Lambda)$ can be constructed: $A(\Lambda)$ is the quotient algebra of the group algebra $L^1(G)$ by the closed ideal $I(\Lambda)$ of $L^1(G)$ defined by $f \in I(\Lambda)$ if its Fourier transform vanishes on Λ; $B(\Lambda)$ is defined in the same way with the measure algebra $M(G)$ in place of $L^1(G)$.

We have the following surprising characterization of coherent sets of frequencies: Λ is a coherent set of frequencies if and only if Λ is discrete and $B(\Lambda)$ is the multiplier algebra of $A(\Lambda)$. We first give some new definitions.

DEFINITION 5: *Let E be a closed subset of Γ, the dual group of G. The Banach algebra $A(E)$ is defined by the following two conditions:*

(1) *φ belongs to $A(E)$ if there exists an $f \in L^1(G)$ such that $\varphi = \hat{f}$ on E;*

(2) $\|\varphi\|_{A(E)} = \inf \|f\|_1$, *where the infimum is taken with respect to those $f \in L^1(G)$, such that $\varphi = \hat{f}$ on E.*

DEFINITION 6: *Let E be a closed subset of Γ. The Banach algebra $B(E)$ is defined by the following two conditions:*

(3) *φ belongs to $B(E)$ if there exists a bounded complex-Radon measure $\mu \in M(G)$ such that $\varphi = \hat{\mu}$ on E;*

(4) $\|\varphi\|_{B(E)} = \inf \|\mu\|$, *where $\|\mu\|$ is the 'total variation' of μ.*

DEFINITION 7: *Let E be a closed subset of Γ. The subspace $AP_E(G)$ of $AP(G)$ is defined by*

(5) $f \in AP_E(G)$ *if the spectrum of the almost-periodic function* f *is contained in* E *and the subspace* $L_E^\infty(G)$ *is defined by*

(6) $f \in L_E^\infty(G)$ *if the spectrum of the bounded Lebesgue-measurable function* φ *is contained in* E.

Remarks: The Banach spaces $AP_E(G)$ and $B(E)$ can be put in duality. If $f \in AP_E(G)$ is a finite trigonometric sum $f(x) = \sum_{\gamma \in E} a_\gamma \chi_\gamma(x)$ and if $\varphi = \hat{\mu}$ on E, we define

$$\langle f, \varphi \rangle = \int_G f(-x)\, d\mu(x) = \sum_{\gamma \in E} a_\gamma \varphi(\gamma).$$

We get $|\langle f, \varphi \rangle| \leq \|f\|_\infty \|\varphi\|_{B(E)}$ and, by Theorem I, the linear form $f \to \langle f, \varphi \rangle$ can be extended to the whole of $AP_E(G)$.

10.2. Duality and coherent sets of frequencies

It may be asked whether $B(E)$ is the dual space of $AP_E(G)$.

The answer is surprising:

THEOREM VI: *Let* G *be a metrizable and separable l.c.a. group and* Λ *a closed subset of the dual group* Γ. *The following two properties of* Λ *are equivalent:*

(a) Λ *is a coherent set of frequencies;*

(b) *each continuous linear form* L *on the space of all almost periodic functions* P *whose spectra lie in* Λ *can be written* $L(P) = \int_G P(x)\, d\mu(x)$ *for a suitable complex valued bounded Radon measure* μ *on* G.

Before proving Theorem VI some remarks should be made. If for each positive η in $]0, 1[$ there exists a compact subset K of G such that the pair (K, η) is suitable for Λ, then the dual space of $AP_\Lambda(G)$ is isometric to $B(\Lambda)$, as the following proof shows. In all other cases, the isomorphism between $B(\Lambda)$ and the dual space of $AP_\Lambda(G)$ is not an isometry.

10.3. Proof of Theorem VI

A coherent set of frequencies Λ is discrete in Γ; hence Λ is closed.

(a) $\Rightarrow$ (b): Let (K, C) be suitable for Λ and let $\mathscr{E}$ be the closed subspace of $\mathscr{C}(K)$, whose elements are the restrictions g to K of all f in $AP_\Lambda(G)$. A continuous linear form l on $AP_\Lambda(G)$ defines a continuous linear form L on $\mathscr{E}$. The Hahn–Banach theorem shows that there is a measure μ carried by K such that, for each $f \in AP_\Lambda(G)$, $l(f) = \int_G f\, d\mu$ and (b) follows. The norm of μ satisfies the inequality $\|\mu\| \leq C\|l\|$.

(b) ⇒ (a): The proof depends on the following construction of linear forms over the space $AP_\Lambda(G)$. Let $\chi \in \mathrm{Hom}(\Gamma_d, \mathbf{T})$ be any homomorphism from the additive group Γ (with the discrete topology) to $\mathbf{T}$; for each $f \in AP(G)$, $f(\chi)$ is defined by the following rule: if f is a finite trigonometric sum $f(x) = \sum_{\gamma\in\Gamma} a_\gamma \chi_\gamma(x)$, $f(\chi)$ is the corresponding sum $\sum_{\gamma\in\Gamma} a_\gamma \chi(\gamma)$. Kronecker's theorem shows that, on the finite spectrum Λ of f, the complex numbers $\chi(\gamma)$, $\gamma \in \Lambda$, can be approximated by restrictions to Λ of continuous characters on Γ; hence $|f(\chi)| \leqslant \sup_G |f|$ and the linear form $f \to f(\chi)$ can be extended by continuity to the whole of $AP(G)$.

The following lemma will now be used:

LEMMA 9: *Let G be a metrizable and separable l.c.a. group, Γ the dual group and Λ a subset of Γ. Assume that Λ is not a coherent set of frequencies. Then for each positive ε, there exist a sequence $(P_k)_{k\geqslant 0}$ of trigonometric sums whose frequencies belong to Λ and a χ in* $\mathrm{Hom}(\Gamma_d, \mathbf{T})$ *such that*

(a) $\sup_G |P_k| = 1$;

(b) *on each compact subset K of G, $\sup_K |P_k| \to 0$, as $k \to +\infty$;*

(c) $|P_k(\chi)| \geqslant 1 - \varepsilon$ *for each* $k \geqslant 0$.

Before proving the lemma, we show why it implies Theorem VI. Assume that the linear form $f \to f(\chi)$, defined on $AP_\Lambda(G)$ by the element χ of $\mathrm{Hom}(\Gamma_d, \mathbf{T})$ found in Lemma 9, can be written $f(\chi) = \int_G f(-x)\, d\mu(x)$ with $\mu \in M(G)$. Replacing f by P_k and using Lebesgue's dominated convergence theorem, we get $\int_G P_k(-x)\, d\mu(x) \to 0$, while $|P_k(\chi)| \geqslant 1 - \varepsilon$. This is impossible and Theorem VI is proved.

10.4. We return to the proof of Lemma 9. Let $(K_k)_{k\geqslant 0}$ be an increasing sequence of compact subsets of G whose union is the whole group G. A sequence $(P_k)_{k\geqslant 0}$ of trigonometric sums on G whose frequencies belong to Λ and a sequence $(x_k)_{k\geqslant 0}$ of elements of G will be constructed by induction in such a way that

$$\sup_G |P_k| = 1 \tag{4.5}$$

$$\sup_{K_k} |P_k| \leqslant k^{-1} \tag{4.6}$$

$$\mathrm{Re}\,[P_1(x_k) + \cdots + P_k(x_k)] \geqslant k - \varepsilon(2^{-1} + \cdots + 2^{-k}) \tag{4.7}$$

We assume that $x_1, \ldots, x_k$, $P_1, \ldots, P_k$ have been chosen and we now define x_{k+1} and P_{k+1}, noting that the choice of x_1 and P_1 is easy.

Since Λ is not a coherent set of frequencies, there exists a sequence $(Q_k)_{k \geqslant 0}$ of trigonometric sums whose frequencies belong to Λ such that $\sup_G |Q_k| = 1$ and $\sup_{K_k} |Q_k| \to 0$ as $k \to +\infty$. The trigonometric sum $S_k = P_1 + \cdots + P_k$ is an almost periodic function, as is every trigonometric sum; so is Re $[S_k]$ and there exists a compact subset L_k of G such that each translate $s + L_k$ contains an x satisfying
$\operatorname{Re}[S_k(x)] \geqslant \sup_G \operatorname{Re}[S_k] - \varepsilon 2^{-k-2} \geqslant k - \varepsilon(2^{-1} + \cdots + 2^{-k} + 2^{-k-2})$.

Let $m = m_k$ be a sufficiently large integer that $|Q_m(x)| \leqslant 1/(k+1)$ on $K_{k+1} - L_k$. Now regarding m as fixed, there exists an element $s = s_k$ in G such that $|Q_m(s_k)| \geqslant 1 - \varepsilon 2^{-k-2}$. We can multiply Q_m by a constant complex number of absolute value 1, which does not affect the other properties of Q_m, to get $\operatorname{Re}[Q_m(s_k)] \geqslant 1 - \varepsilon 2^{-k-2}$. Now $s_k + L_k$ contains an element x_{k+1} of G such that

$$\operatorname{Re}[S_k(x_{k+1})] \geqslant k - \varepsilon(2^{-1} + \cdots + 2^{-k} + 2^{-k-2}).$$

Putting $P_{k+1}(x) = Q_m(x - x_{k+1} + s_k)$, we have

$$\sup_{K_{k+1}} |P_{k+1}| \leqslant \sup_{K_{k+1} - L_k} |Q_m| \leqslant 1/(k+1), \quad \sup_G |P_{k+1}| = \sup_G |Q_m| = 1$$

and

$$\operatorname{Re}[P_1(x_{k+1}) + \cdots + P_{k+1}(x_{k+1})] = \operatorname{Re}[S_k(x_{k+1})] + \operatorname{Re}[Q_m(s_k)]$$
$$\geqslant k + 1 - \varepsilon(2^{-1} + \cdots + 2^{-k-1}).$$

The induction is thus complete.

The inequalities (4.7) imply $|P_j(x_k)| \geqslant 1 - \varepsilon$ for each $k \geqslant 0$ and each j less than k. Following Section 1 of Chapter II, we regard G as a part of $\tilde{G} = \operatorname{Hom}(\Gamma_d, \mathbf{T})$; let C_k be the closure in $\tilde{G}$ of the tail-set $\{x_k, x_{k+1}, \ldots\}$. Since $\tilde{G}$ is compact, there exists a χ in $\tilde{G}$ belonging to all the C_k, $k \geqslant 0$. Using the continuity of P_j for each fixed j, we get $|P_j(\chi)| \geqslant 1 - \varepsilon$; since χ is independent of j, Lemma 9 is proved.

11. Other definitions of coherent sets of frequencies

11.1. We recall that Hom $(\Lambda_d, \mathbf{T})$ is the group of restrictions to Λ of all homomorphisms of the discretized group Γ into the circle group $\mathbf{T}$.

Then the following two properties of a closed subset Λ of Γ are equivalent:

(a) *Λ is a coherent set of frequencies;*

(b) Hom $(\Lambda_d, \mathbf{T})$ *is contained in* $B(\Lambda)$.

Let Λ be a coherent set of frequencies. Each $\chi \in$ Hom $(\Lambda_d, \mathbf{T})$ defines a linear form $f \to f(\chi)$ on $AP_\Lambda(G)$ and there is a $\mu \in M(G)$ such that $f(\chi) = \int_G f(-x)\, d\mu(x)$; this implies $\hat{\mu}(\lambda) = \chi(\lambda)$ for each λ in Λ.

If Λ is not a coherent set of frequencies, Lemma 9 gives a $\chi \in$ Hom $(\Lambda_d, \mathbf{T})$, which is not in $B(\Lambda)$.

11.2. Let Λ be a closed subset of Γ. We say that a function $\varphi : \Lambda \to \mathbf{C}$ is a multiplier of $A(\Lambda)$ if for each $f \in A(\Lambda)$, φf (the usual product) also lies in $A(\Lambda)$.

THEOREM VII: *If Λ is a discrete subset of Γ, the following two properties of Λ are equivalent:*

(a) *Λ is a coherent set of frequencies;*

(b) *the multiplier space of $A(\Lambda)$ is $B(\Lambda)$.*

The proof of this result depends on the following interesting lemma:

LEMMA 10: *Let Λ be a discrete subset of a l.c.a. group Γ and let $f : \Lambda \to \mathbf{C}$ be a function such that $f(\lambda) = 0$ for all λ in Λ with the exception of at most a finite number of $\lambda \in \Lambda$. Then $\|f\|_{A(\Lambda)} = \sup |\sum_{\lambda \in \Lambda} c_\lambda f(\lambda)|$, this supremum being taken over all finite complex sequences $(c_\lambda)_{\lambda \in \Lambda}$ such that $P(x) = \sum c_\lambda \chi_\lambda(x)$ satisfies $\|P\|_\infty \leqslant 1$.*

Proof of Lemma 10: Let $a = \|f\|_{A(\Lambda)}$ and let b the right-hand side of the hoped-for equality. The inequality $b \leqslant a$ is obvious. For each $\varepsilon > 0$, there is an F in $L^1(G)$ such that $\|F\|_1 \leqslant a + \varepsilon$ and $\hat{F} = f$ on Λ. Hence $\sum_{\lambda \in \Lambda} c_\lambda f(\lambda) = \int_G P(-x) F(x)\, dx$ implies $|\sum_{\lambda \in \Lambda} c_\lambda f(\lambda) \leqslant \|P\|_\infty \|F\|_1 \leqslant a + \varepsilon$. Thus $b \leqslant a + \varepsilon$ for all positive ε, which gives $b \leqslant a$.

We want to get the reverse inequality. Let ε be any positive real number and g an element of $L^1(G)$ with the following three properties: $\hat{g}(\lambda) = 1$ if $\lambda \in \Lambda$ and $f(\lambda) \neq 0$, $\hat{g}$ has compact support and $\|g\|_1 \leqslant 1 + \varepsilon$ ([15], p. 53, Theorem 2.6.8).

Let E be the finite set of all λ in Λ for which $\hat{g}(\lambda) \neq 0$. Theorem VI applied to the harmonious set E gives $\|f\|_{B(E)} \leqslant b$. This means that, for each positive ε, there is an element μ of $M(G)$ such that $\|\mu\| \leqslant b + \varepsilon$ and

$\hat{\mu} = f$ on E. The convolution $F = \mu * g$ belongs to $L^1(G)$; the norm of F in $L^1(G)$ is less than $(b + \varepsilon)(1 + \varepsilon)$ and $\hat{F} = f$ on Λ. Hence $a \leqslant (b + \varepsilon)(1 + \varepsilon)$ for all positive ε and Lemma 10 is proved.

Now Theorem VII can be proved. The space of multipliers of $A(\Lambda)$, for a discrete Λ, is exactly the dual space of $AP_\Lambda(G)$ and our theorem is a corollary of Theorem VI.

12. Other relations between $A(E)$, $B(E)$ and L_E^∞

12.1. Theorem VI shows that $B(E)$ is only exceptionally the dual space of $AP_E(G)$. However, it is interesting to seek other relations of duality between the spaces $A(E)$, $B(E)$ and $L_E^\infty(G)$.

PROPOSITION 4: *Let G be a l.c.a. group, Γ the dual group and E, F two closed subsets of Γ such that E is contained in the interior of F. For each f in $L_E^\infty(G)$ and each φ in the restriction algebra $B(F)$ a g in L_E^∞ can be defined by $\hat{g} = \varphi\hat{f}$, which furthermore satisfies $\|g\|_\infty \leqslant \|\varphi\|_{B(F)} \|f\|_\infty$.*

Proof: Let ε be any positive real number. There exists a complex valued bounded Radon measure μ on G such that $\hat{\mu} = \varphi$ on F and the total variation of μ, $\|\mu\| \leqslant (1 + \varepsilon) \|\varphi\|_{B(F)}$. Let $g = f * \mu$. This definition is independent of the choice of μ: if $\hat{\mu} = \hat{\nu}$ on F, then $\mu * f = \nu * f$ since E is contained in the interior of F. We get $\|g\|_\infty \leqslant \|\mu\| \|f\|_\infty \leqslant (1 + \varepsilon) \|\varphi\|_{B(F)} \|f\|_\infty$. Since g is independent of $\varepsilon > 0$, Proposition 4 is proved.

COROLLARY 1: *The notation is that of Proposition 4. Assume that E is compact. Then, for each $f \in L_E^\infty(G)$, there exists a complex valued bounded continuous function g on G such that $f = g$ almost everywhere with respect to the Haar measure of G.*

Proof: There is an element k of $L^1(G)$ such that $\hat{k} = 1$ on a neighborhood of E. Take $g = f * k$.

COROLLARY 2: *The notation is that of corollary 1 and E is compact. For each $f \in L_E^\infty(G)$ and each φ in $B(F)$, a scalar product of f and φ is given by $\langle f, \varphi \rangle = \int_G f(-x)\, d\mu(x)$, where $\hat{\mu} = \varphi$ on F.*

COROLLARY 3: *Let f be a complex-valued bounded continuous function on G whose spectrum is the union $E_1 \cup \cdots \cup E_n$ of n disjoint compact subsets of Γ. Then f can be written $f = f_1 + \cdots + f_n$, where the f_j are complex-valued bounded continuous functions on G and where the spectrum of each f_j is contained in E_j. Moreover $\|f_j\|_\infty \leqslant C \|f\|_\infty$, where C depends only on $E_1, \ldots, E_n$.*

Proof: Let k_j be an element of $L^1(G)$ such that $\hat{k}_j = 1$ on a neighbourhood of E_j and $\hat{k}_j = 0$ on a neighbourhood of the union of the remaining E_k, $k \neq j$, $1 \leqslant k \leqslant n$, and take f_j to be $f * k_j$.

12.2. We have the following results on the algebras $A(E)$ and $B(E)$.

PROPOSITION 5: *If φ belongs to $B(E)$ and has compact support, then φ belongs to $A(E)$ and $\|\varphi\|_{A(E)} = \|\varphi\|_{B(E)}$.*

Proof: The inequality $\|\varphi\|_{B(E)} \leqslant \|\varphi\|_{A(E)}$ is immediate from the definition of these norms. To get the reverse inequality, let μ, for each positive ε, be a complex-valued bounded Radon measure on G such that $\hat{\mu} = \varphi$ on E and $\|\mu\| \leqslant (1 + \varepsilon) \|\varphi\|_{B(E)}$. Let k be an element of $L^1(G)$ such that $\|k\|_1 \leqslant 1 + \varepsilon$ and $\hat{k} = 1$ on the support of φ. Then $h = \mu * k$ belongs to $L^1(G)$, $\hat{h} = \varphi$ on E and $\|h\|_1 \leqslant (1 + \varepsilon)^2 \|\varphi\|_{B(E)}$. Hence $\|\varphi\|_{A(E)} \leqslant \|\varphi\|_{B(E)}$.

PROPOSITION 6: *The subalgebra $A(E)$ of $B(E)$ can be defined as the closure in $B(E)$ of the subspace of compactly supported elements. If $\varphi \in A(E)$, then $\|\varphi\|_{A(E)} = \|\varphi\|_{B(E)}$.*

Proof: The subset of all those f in $L^1(G)$ whose Fourier transforms have compact support is dense in $L^1(G)$ ([15], th 2.6.6, p. 51). Proposition 6 now follows immediately from Proposition 5.

13. Slowly-perturbed trigonometric sums

13.1. Intuitive motivation

Consider the motion of the earth around the sun; the position of the centre of gravity of the earth at time t is given by a continuous function $\varphi: \mathbf{R} \to \mathbf{R}^3$. To a first approximation, φ is a periodic function whose period T is one year and whose Fourier series is $\sum_{-\infty}^{+\infty} a_k \exp 2k\pi i\,(t/T)$. More precise observations show that the a_k are not constants but functions of time whose periods are very great compared with T (secular perturbation).

What is the quality of the approximation of these perturbed trigonometric sums $\varphi(t) = \sum_{-\infty}^{+\infty} a_k(t) \exp 2k\pi i\,(t/T)$ by the corresponding periodic functions $\varphi\,(t_0, t) = \sum_{-\infty}^{+\infty} a_k(t_0) \exp 2k\pi i\,(t/T)$; it is to be expected that the error term will be quite small in a neighbourhood of t_0.

To state this problem in the context of general groups, the set $\mathbf{Z}T^{-1}$ of frequencies used in the periodic case must be replaced by a coherent set Λ of frequencies.

Surprisingly, in order to solve our problem, it is necessary to know the behaviour of φ at infinity; in the real case, an apriori estimate of the type $|\varphi(t)| \leqslant C \exp |t|^{\alpha}$, $C > 0$, $0 < \alpha < 1$, is sufficient (Section 17) and, in the general case of l.c.a. groups, it will be assumed that φ is bounded.

If Λ is a coherent set of (real) frequencies and if $\varphi(t) = \sum_{\Lambda \in \lambda} a_\lambda(t) \exp 2\pi i \lambda t$ is a slowly perturbed trigonometric sum by which we mean that the spectrum of each a_λ lies in a small interval $[-\varepsilon, \varepsilon]$, where ε depends only on Λ, it will be proved that $\varphi(t_0, t) = \sum_{\lambda \in \Lambda} a_\lambda(t_0) \exp 2\pi i \lambda t$ is a very good approximation to $\varphi(t)$. In fact $|\varphi(t) - \varphi(t_0, t)| \leqslant A\, |t - t_0| \sup_{\mathbf{R}} (|\varphi(x)| \exp -|x - t_0|^{\alpha})$ and $\sup_{t \in \mathbf{R}} |\varphi(t_0, t)| \leqslant A \sup_{\mathbf{R}} (|\varphi(x)| \exp -|x - t_0|^{\alpha})$, where A depends only on Λ, ε and α.

13.2. Slowly-perturbed coefficients

Let G be a l.c.a. group, Γ the dual group and Λ a subset of Γ. We are interested in slowly-perturbed trigonometric sums of the type $\sum_{\lambda \in \Lambda} f_\lambda(x) \chi_\lambda(x)$, where the sum is finite and the f_λ are functions with slow variations. To be more precise, the following definition will be needed:

DEFINITION 8: *Let G be a l.c.a. group, K a compact subset of G and ε a positive real number. A complex valued bounded continuous function f on G is (K, ε)-slow if for each x in G and each s in K, $|f(x + s) - f(x)| \leqslant \varepsilon \|f\|_\infty$.*

In other words, each s in K is an ε-almost period of f. The following proposition provides examples of (K, ε)-slow functions:

PROPOSITION 7: *Let G be a l.c.a, group, Γ the dual group, E a closed subset of Γ and δ a real number in* $]0, 1[$. *Then each s in G such that $\sup_{\gamma \in E} |\chi_\gamma(s) - 1| \leqslant \delta$ is a 2δ-almost-period of each compex-valued bounded continuous function f on G whose spectrum lies in E.*

Before giving the proof, we show why Proposition 7 gives examples of (K, ε)-slow functions. Let U be a compact neighbourhood of 0 in Γ such that $\sup_{K \times U} |\chi_\gamma(x) - 1| \leqslant \varepsilon/2$. Then each complex-valued bounded continuous function f on G whose spectrum lies in U is (K, ε)-slow.

Now to the proof. Let $\alpha > 0$ be an arbitrary positive real number sufficiently small that $\delta + \alpha < 1$ and let U be a compact neighbourhood of 0 in Γ, small enough to satisfy $\sup_{\gamma \in U} |\chi_\gamma(s) - 1| \leqslant \alpha$, where s is fixed throughout the proof. Let $F = E + U$.

As in Lemma 4, we construct a discrete measure $\sigma = \sum_{-\infty}^{+\infty} d_n \varepsilon_{-ns}$ on G such that $\|\sigma\| \leqslant 2(\delta + \alpha)$ and $\hat{\sigma}(\gamma) = \sum_{-\infty}^{+\infty} d_n \chi_\gamma(ns) = \chi_\gamma(s) - 1$ for each γ in F. At this point, the following lemma is useful:

LEMMA 11: *The function $\Delta_s f$ defined on G by $\Delta_s f(x) = f(x+s) - f(x)$ is equal to $f * \sigma$.*

Proof: The Fourier transforms of $\Delta_s f$ and $f * \sigma$ are respectively $(\chi_\gamma(s) - 1)\hat{f}$ and $\hat{\sigma}\hat{f}$. Since $\hat{\sigma} = \chi_s - 1$ on a neighbourhood of the support of the distribution $\hat{f}$, we have the required equality.

Returning to Proposition 7, we have $\|\Delta_s f\|_\infty \leqslant \|\sigma\| \, \|f\|_\infty \leqslant 2(\delta + \alpha)\|f\|_\infty$. Since this inequality holds for each $\alpha > 0$, we obtain $\|\Delta_s f\|_\infty \leqslant 2\delta\|f\|_\infty$. If f_s is defined on G by $f_s(x) = f(x - s)$, the spectrum of f_s lies in E and we can apply the preceding proof to f_s instead of f. We get $\|f_s - f\|_\infty \leqslant 2\delta\|f\|_\infty$ and Proposition 7 is proved.

Slow families: DEFINITION 9: *Let G be a l.c.a. group, K a compact subset of G, and ε a positive real number. We say that a family $(f_\lambda)_{\lambda \in \Lambda}$ of complex-valued bounded continuous functions on G is a (K, ε)-slow family if, for every family $(c_\lambda)_{\lambda \in \Lambda}$ of complex numbers of which all but a finite number are* 0, *the finite sum $\sum_{\lambda \in \Lambda} c_\lambda f_\lambda$ is (K, ε)-slow.*

Example: Let U be a compact neighbourhood of 0 in the dual group Γ such that $\sup_{K \times U} |\chi_\gamma(x) - 1| \leqslant \varepsilon/2$. Then each family $(f_\lambda)_{\lambda \in \Lambda}$ of complex-valued bounded continuous functions whose spectra lie in U is a (K, ε)-slow family.

13.3. Approximation of slowly-perturbed trigonometric sums

A fundamental result can now be stated:

THEOREM VIII: *Let G be a l.c.a. group, Γ the dual group and Λ a subset of Γ. The following two properties of Λ are equivalent:*

(a) *Λ is a coherent set of frequencies;*

(b) *There exist a compact subset K of G, a positive real number ε and a constant C such that, for each finite (K, ε)-slow family $(f_\lambda)_{\lambda \in \Lambda}$ and each x_0 in G, the pure trigonometric sum $f(x_0, x) = \sum_{\lambda \in \Lambda} f_\lambda(x_0)\chi_\lambda(x)$ is a good approximation to the slowly-perturbed trigonometric sum $f(x) = \sum_{\lambda \in \Lambda} f_\lambda(x)\chi_\lambda(x)$ in the following sense:*

$$|f(x_0, x) - f(x)| \leqslant C\varepsilon \sup_G |f| \quad \textit{for each } x \textit{ in } x_0 + K, \tag{4.8}$$

$$\sup_{x \in G} |f(x_0, x)| \leqslant C \sup_G |f|. \tag{4.9}$$

Before proving this result, some remarks should be made. A finite family $(f_\lambda)_{\lambda\in\Lambda}$ is a family all of whose terms are the zero-function with at most a finite number of exceptions; (4.8) means that the approximation of f by $f(x_0, x)$ is good in the neighbourhood $x_0 + K$ of x_0 and the estimates given by (4.9) will play a fundamental role in the problem of spectral synthesis.

Proof (a) $\Rightarrow$ (b): Assume that the pair (K, c) is suitable for Λ, and let ε so small that $\varepsilon c < 1$. Let $M = \sup |f(x, x_0)|$ over all $x \in G$ and $x_0 \in G$. For each fixed u, $x \to f(x, u)$ is (K, ε)-slow since the family $(f_\lambda)_{\lambda\in\Lambda}$ is (K, ε)-slow; hence if $x \in y + K$, $|f(y, u) - f(x, u)| \leqslant \varepsilon M$. Applying this remark with $y = x_0$ and $u = x$, we get $|f(x_0, x) - f(x)| \leqslant \varepsilon M$ whenever $x \in x_0 + K$.

Since the pair (K, c) is suitable for Λ, so is the pair $(K + x_0, c)$ and, for each x_0 fixed in G, we have $\sup\limits_{x\in G} |f(x_0, x)| \leqslant c \sup\limits_{x\in x_0+K} |f(x_0, x)| \leqslant c \sup\limits_{G} |f(x)| + c\varepsilon M$ since $|f(x_0, x)| \leqslant |f(x)| + |f(x_0, x) - f(x)|$. Taking the sup over $x_0 \in G$, we get $M \leqslant c\,\|f\|_\infty + c\varepsilon M$ and $M \leqslant c(1 - \varepsilon c)^{-1}\,\|f\|_\infty$; (4.9) is satisfied if we take $C = c/1 - \varepsilon c$. Substituting this estimate of M in $|f(x_0, x) - f(x)| \leqslant \varepsilon M$, we get (4.8).

(b) $\Rightarrow$ (a): If Λ is not coherent, there exists a sequence $(P_k)_{k\geqslant 0}$ of trigonometric sums whose frequencies belong to Λ such that $\|P_k\|_\infty = 1$ and $P_k \to 0$ uniformly on compact subsets of G.

Let f be any (K, ε)-slow function vanishing at infinity and such that $f(0) = 1$. For example, let U be a compact neighbourhood of 0 in Γ such that $\sup\limits_{K\times U} |\chi_\gamma(x) - 1| \leqslant \varepsilon/2$ and put $f = \hat{g}$, where g belongs to $L^1(\Gamma)$, g is carried by U and $\int_\Gamma g(\gamma)\, d\gamma = 1$. By Riemann–Lebesgue's theorem, f vanishes at infinity and Proposition 7 shows that f is (K, ε)-slow.

Applying (4.9) with $x_0 = 0$ to the perturbed trigonometric sums fP_k, we get $1 = \|P_k\|_\infty \leqslant C\,\|fP_k\|_\infty \to 0$, as $k \to +\infty$. We thus have a contradiction and so Λ is a coherent set of frequencies.

13.4. *Better results on the approximation of slowly-perturbed trigonometric sums*

Our next theorem is an improvement of Theorem VIII.

The terminology of Theorem IX is the following:

(a) G is a metrizable and separable l.c.a. group;

(b) Γ is the dual group;

(c) Λ is a subset of Γ;

(d) U is a compact neighbourhood of 0 in Γ and for distinct λ in Λ, the corresponding $\lambda + U$ are disjoint;

(e) for each x in G, $r(x) = \sup_{\gamma \in U} |\chi_\gamma(x) - 1|$;

(f) by a suitable normalization of the metric on G, we may assume that $r(x) \leqslant |x|$, where $|x|$ is the distance from 0 to x.

With this notation, we can state

THEOREM IX: *The following two statements are equivalent:*

(a) *Λ is a coherent set of frequencies;*

(b) *There exist a compact neighbourhood U of 0 in Γ and a positive constant C such that, for each x_0 in G and each perturbed trigonometric sum $f = \sum_{\lambda \in \Lambda} f_\lambda \chi_\lambda$ for which the spectrum of every f_λ is contained in U, the corresponding pure trigonometric sum $f(x_0, x) = \sum_{\lambda \in \Lambda} f_\lambda(x_0)\, \chi_\lambda(x)$ satisfies the two relations*

$$|f(x_0, x) - f(x)| \leqslant Cr\,(x - x_0)\, \|f\|_\infty \leqslant C\,|x - x_0|\, \|f\|_\infty \quad (4.10)$$

$$\sup_{G^2} |f(x_0, x)| \leqslant C \sup_{G} |f(x)|. \quad (4.11)$$

Proof: The implication (b) ⇒ (a) can be proved just as in Theorem VIII. If (a) is satisfied, let (K, c) be suitable for Λ and let ε be a positive real number so small that $\varepsilon c < 1$. Let U be a compact neighbourhood of 0 in Γ such that $\sup_{U \times K} |\chi_\gamma(x) - 1| \leqslant \varepsilon/2$. Proposition 7 now shows that $(f_\lambda)_{\lambda \in \Lambda}$ is a (K, ε)-slow family and Theorem VIII gives (4.11) with $C = (1 - \varepsilon c)^{-1}$.

We shall prove (4.10) using a constant D instead of C. We put $\alpha = r\,(x - x_0)$. If $\alpha \geqslant 1$, there is nothing to prove: (4.11) implies $|f_{x_0}(x) - f(x)| \leqslant (C + 1)\, \|f\|_\infty \leqslant D\alpha\, \|f\|_\infty$, where $D = C + 1$. If $\alpha < 1$, we define, for each fixed x in G, a function g on G by $g(u) = \sum_{\lambda \in \Lambda} f_\lambda(u)\, \chi_\lambda(x)$. Proposition 7, applied with $s = x - x_0$, gives $|g(x) - g(x_0)| \leqslant 2\alpha \|g\|_\infty \leqslant 2C\alpha\, \|f\|_\infty$; (4.10) is satisfied with $C = \max\,((1 - \varepsilon c)^{-1} + 1, 2\,(1 - \varepsilon c)^{-1})$.

DEFINITION 10: *Let G be a l.c.a. group, Λ a subset of the dual group Γ and U a compact subset of Γ. The pair (Λ, U) is a harmonic discretization of Λ if, for distinct λ in Λ, the corresponding $\lambda + U$ are disjoint and if the conclusion of Theorem IX holds.*

It will be now shown that each harmonic discretization of Λ gives rise to an isomorphism between $A\,(\Lambda + U)$ and $A\,(\Lambda \times U)$.

14. Coherent sets of frequencies and isomorphisms

14.1. Let G be a l.c.a. group, Γ the dual group and Λ a uniformly discrete subset of Γ. This means that there is a compact neighbourhood U of 0 in Γ such that for different λ in Λ, the corresponding $\lambda + U$, $\lambda \in \Lambda$ are pairwise disjoint.

There is thus a homeomorphism H from the subset $\Lambda + U$ of Γ onto the subset $\Lambda \times U$ of Γ^2 given by the rule $H(\lambda + t) = (\lambda, t)$ if $\lambda \in \Lambda$, $t \in U$. This homeomorphism defines an isometry $\mathscr{H}$ between $\mathscr{C}_0(\Lambda \times U)$ and $\mathscr{C}_0(\Lambda + U)$: for each f in $\mathscr{C}_0(\Lambda \times U)$, $\mathscr{H}(f) = f \circ H$.

We recall that $A(\Lambda + U)$ is the restriction algebra to $\Lambda + U$ of all Fourier transforms of elements of $L^1(G)$, that $B(\Lambda + U)$ is the restriction algebra to $\Lambda + U$ of all Fourier transforms of all complex-valued bounded Radon measures μ in $M(G)$ and that the corresponding notations are used for $\Lambda \times U$, where G is replaced by G^2.

THEOREM X: *Let G be a l.c.a, group, Γ the dual group and Λ a subset of Γ. Then the following three properties of Λ are equivalent:*

(a) *Λ is a coherent set of frequencies;*

(b) *Λ is uniformly discrete and there exists a compact neighbourhood U of 0 in Γ with the following property: for each element f of $A(\Lambda \times U)$, $f \circ H$ belongs to $A(\Lambda + U)$, where H is the homeomorphism from $\Lambda + U$ onto $\Lambda \times U$ defined above;*

(c) *there is a constant C such that, for each f in $A(\Lambda \times U)$, $\|f \circ H\|_{A(\Lambda+U)} \leqslant C\|f\|_{A(\Lambda\times U)}$ and such that for each g in $B(\Lambda \times U)$, $g \circ H$ belongs to $B(\Lambda + U)$ and $\|g \circ H\|_{B(\Lambda+U)} \leqslant C\|g\|_{B(\Lambda\times U)}$.*

14.2. To prove Theorem X, we need to introduce some spaces of test-functions. The subspace $\mathscr{E}$ of $L^1(G)$ is defined by the condition that $f \in \mathscr{E}$ if the Fourier transform $\hat{f}$ of f is carried by a compact subset of Γ. Thc inversion theorem ([15], th. 1.5.1, p. 21) then shows that $\check{f}$ is the Fourier transform of $\hat{f}$ ($\check{f}(x) = f(-x)$); hence $f \in A(G)$. It is well known that $\mathscr{E}$ is dense in $L^1(G)$ and it follows from the Hahn–Banach theorem that $\mathscr{E}$ is dense in $\mathscr{C}_0(G)$. The norm of f in $\mathscr{E}$ will be $\sup_G |f| = \|f\|_\infty$.

If Ω is an open subset of Γ, $\mathscr{E}(\Omega)$ is the subspace of all $f \in \mathscr{E}$ whose spectra are compact subsets of Ω.

PROPOSITION 8: *Let Ω be an open subset of Γ and E the closure of Ω. Let f be a complex-valued continuous function on E. If a positive number b can be found such that, for each $\alpha \in \mathscr{E}(\Omega)$, $|\int_E f(\gamma)\, \hat{\alpha}(\gamma)\, d\gamma| \leqslant b\|\alpha\|_\infty$, then f belongs to $B(E)$ and $\|f\|_{B(E)} \leqslant b$.*

Let L be the linear form defined on $\mathscr{E}(\Omega)$ by $L(\alpha) = \int_E f(\gamma)\,\hat{\alpha}(\gamma)\,d\gamma$. The metric form of the Hahn–Banach theorem shows that L can be extended to a linear form on $\mathscr{C}_0(G)$ with the same norm. Hence there exists a complex-valued Radon measure μ on G such that $\|\mu\| \leqslant b$ and $L(\alpha) = \int_G \alpha(-x)\,d\mu\,(x)$ for all $\alpha \in \mathscr{E}(\Omega)$. The equalities $\int_E f(\gamma)\,\hat{\alpha}(\gamma)\,d\gamma = \int_\Gamma \hat{\mu}(\gamma)\,\hat{\alpha}(\gamma)\,d\gamma$ for each $\alpha \in \mathscr{E}(\Omega)$ and the continuity of f imply $\hat{\mu} =$ on Ω; hence $\hat{\mu} = f$ on E and $\|f\|_{B(E)} \leqslant b$.

14.3. We return to the proof of Theorem X,

(a) $\Rightarrow$ (c): We retain the previous notation and introduce some new notation:

(K, c) is suitable for Λ;

$0 < \varepsilon$ and $\varepsilon c < 1$;

V is an open neighbourhood of 0 in Γ such that $\sup\limits_{V\times K} |\chi_\gamma(x) - 1| \leqslant \varepsilon/2$;

U is the closure of V;

$\mathscr{E}(\Lambda \times V)$ is the normed space of all complex-valued bounded continuous functions α on $G \times G$, which can be written as finite sums $\alpha\,(x, y) = \sum_{\lambda\in\Lambda} \alpha_\lambda(x)\,\chi_\lambda(y)$, where $\alpha_\lambda \in \mathscr{E}(V)$ for each $\lambda \in \Lambda$. If $g \in B\,(\Lambda \times U)$ and $\alpha \in \mathscr{E}\,(\Lambda \times V)$, we define $\langle g, \alpha\rangle$ to be $\sum_{\lambda\in\Lambda} \int_\Gamma \hat{\alpha}_\gamma(\gamma)\,g\,(\lambda, \gamma)\,d\gamma$;

$\mathscr{E}\,(\Lambda + V)$ is the space $\mathscr{E}(\Omega)$ for $\Omega = \Lambda + V$; each element $\alpha \in \mathscr{E}\,(\Lambda + V)$ can be written as a finite sum $\sum_{\lambda\in\Lambda} \alpha_\lambda(x)\,\chi_\lambda(x)$, where $\alpha_\lambda \in \mathscr{E}(V)$ for each λ in Λ; *the corresponding element of $\mathscr{E}\,(\Lambda \times V)$ defined by $\sum_{\lambda\in\Lambda} \alpha_\lambda(x)\,\chi_\lambda(y)$ will be called* $\tilde{\alpha}$. If $g \in B\,(\Lambda + U)$ and $\alpha \in \mathscr{E}\,(\Lambda + V)$, $\langle g, \alpha\rangle = \int_\Gamma g(\gamma)\,\hat{\alpha}(\gamma)\,d\gamma$: this is a definition of the scalar product.

To get (c), we begin with a $g \in B\,(\Lambda \times U)$; Proposition 8 will be applied to prove that $g \circ H \in B\,(\Lambda + U)$. For each $\alpha \in \mathscr{E}\,(\Lambda + V)$, $\tilde{\alpha} \in \mathscr{E}\,(\Lambda \times V)$ and Theorem IX gives $\|\tilde{\alpha}\|_\infty \leqslant C\|\alpha\|_\infty$. We clearly have $\langle g \circ H, \alpha\rangle = \langle g, \tilde{\alpha}\rangle$. Hence $|\langle g \circ H, \alpha\rangle| \leqslant C\|g\|_{B(\Lambda\times U)}\|\alpha\|_\infty$. Proposition 8 tells us that $g \circ H \in B\,(\Lambda + U)$ and $\|g \circ H\|_{B(\Lambda+U)} \leqslant C\|g\|_{B(\Lambda\times U)}$.

If $g \in B\,(\Lambda \times U)$ has compact support, g belongs to $A\,(\Lambda \times U)$ and $\|g\|_{A(\Lambda\times U)} = \|g\|_{B(\Lambda\times U)}$ (see Proposition 6). In this case, $g \circ H$ also has compact support. Hence $\|g \circ H\|_{A(\Lambda+U)} \leqslant C\|g\|_{A(\Lambda\times B)}$; if g is any element of $A\,(\Lambda \times U)$, g is the limit of a sequence (g_k) of elements of $A\,(\Lambda \times U)$ with compact supports and the corresponding $g_k \circ H$ converge to $g \circ H$ in $A\,(\Lambda + U)$.

(c) $\Rightarrow$ (b): Trivial.

(b) $\Rightarrow$ (a): Any restriction algebra is semi-simple. The general theory of Banach algebras ([11], th. 4.1, p. 213; or [15], th. D5, p. 263) shows that the homomorphism $f \to f \circ H$ from $A\,(\Lambda \times U)$ to $A\,(\Lambda + U)$ is continuous. Hence $\|f \circ H\|_{A(\Lambda \times U)} \leqslant C\|f\|_{A(\Lambda + U)}$. Again using the duality between $\mathscr{E}(\Lambda + U)$ and $B(\Lambda + U)$ and between $\mathscr{E}(\Lambda \times U)$ and $B(\Lambda \times U)$, we get the inequality $\|\tilde{\alpha}\|_\infty \leqslant C\|\alpha\|_\infty$. Theorem IX now shows that Λ is a coherent set of frequencies.

15. Improvement on the preceding results when Λ is harmonious

15.1. In the applications, Theorems IX and X are useless: they do not give any information about the size of the compact neighbourhood U of 0 in Γ. It would be convenient to take U to be the largest possible. We shall see that this is almost the case when Λ is an harmonious set.

Before giving any precise result, Theorem VIII will be restated in the following way:

PROPOSITION 9: *Let G be a metrizable and separable l.c.a. group, Γ the dual group and $\Lambda \subset \Gamma$ a coherent set of frequencies. There exist a compact neighbourhood U of 0 in Γ and a linear operator L, whose domain and range will be specified in a moment, such that*

(a) *If f is a complex-valued bounded continuous function on G whose spectrum lies in $\Lambda + U$, $g = L(f)$ is an almost-periodic function whose spectrum lies in Λ;*

(b) *L is continuous: there is a constant C such that* $\sup_G |g| \leqslant C \sup_G |f|$

(c) $|g(x) - f(x)| \leqslant C|x|\,\|f\|_\infty$;

(d) *For every λ in Λ, the spectrum of g lies in $\{\lambda\}$ whenever the spectrum of f lies in $\lambda + U$;*

(e) *There is a real continuous function ω on G, vanishing at infinity, such that for each f, $\|g\|_\infty \leqslant \|\omega f\|_\infty$.*

Before proving Proposition 9, a remark should be made. Each coherent set of frequencies is uniformly discrete: there is a compact neighbourhood U of 0 in Γ such that for distinct λ in Λ, the corresponding $\lambda + U$ are disjoint. Each complex valued bounded continuous function f on G whose spectrum is a compact subset of $\Lambda + U$ can be written $\sum_{\lambda \in \Lambda} f_\lambda(x)\, \chi_\lambda(x)$, where the spectrum of each f_λ, $\lambda \in \Lambda$, lies in U.

To start with, we show that (a), (b), (c), (d) and (e) hold for any complex valued bounded continuous function f on G whose spectra lie in $\Lambda + U$ if they hold for slowly-perturbed finite trigonometric sums $\sum_{\lambda \in \Lambda} f_\lambda(x)\, \chi_\lambda(x)$ for which the spectrum of each f_λ lies in U.

Finally, (e) will be proved in this last case.

15.2. Reduction to slowly-perturbed finite sums

Let $(k_j)_{j\geqslant 1}$ be a sequence of elements of $L^1(G)$ such that $\|k_j\|_1 = 1$, $\hat{k}_j$ is compactly supported, and $\hat{k}_j \to 1$ uniformly on each compact subset of Γ ([15], th. 2.6.8, p. 53). Put $f^j = f * k_j$. Then

(a) $f^j = \sum_{\lambda\in\Lambda} f^j_\lambda \chi_\lambda$; the sum is finite and the spectrum of f^j_λ lies in U;

(b) $\|f^j\|_\infty \leqslant \|f\|_\infty$;

(c) $f^j \to f$, as $j \to +\infty$, uniformly on each compact subset of G.

Therefore (ωf^j), $j \geqslant 1$, is a Cauchy sequence for the sup-norm over G and, if (e) holds for perturbed trigonometric sums, $(g^j)_{j\geqslant 1}$ is also a Cauchy sequence for the sup-norm over G. Hence g^j tends uniformly on G to an almost-periodic function g. We put $g = L(f)$; (a) is satisfied and the verifications of (b) and (c) are trivial.

15.3. The case of slowly-perturbed trigonometric sums and the conclusion of the proof of Proposition 9

Let V be a compact neighbourhood of 0 in Γ contained in the interior of U, where U is defined by Theorem IX, and let W be a compact neighbourhood of 0 in Γ sufficiently small to satisfy $V + W \subset U$.

Let a be any function in the test-class $\mathscr{E}$ introduced in Section 14 such hat $a(0) = 1$ and whose spectrum is contained in W.

LEMMA 12: *For each element f of $L^\infty(G)$, the Fourier transform of the product af is the function $\hat{a} * \hat{f}$ in $A(\Gamma)$.*

Proof: af belongs to $L^1(G)$ since a does, and the Fourier transform of af belongs to $A(\Gamma)$. We put $\alpha = \hat{a}$ and to compute the convolution of the distribution $\hat{f}$ with α, we test on a function β in $A(\Gamma)$. Then $\int_\Gamma (\hat{f} * \alpha)\, \beta\, d\gamma = \int_\Gamma (\check{\alpha} * \beta)\, \hat{f}\, d\gamma : \check{\alpha}(\gamma) = \alpha(-\gamma)$ and the equality is the definition of convolution. By the definition of $\hat{f}$, $\int_\Gamma (\check{\alpha} * \beta)\, \hat{f}\, d\gamma = \int_G a\hat{\beta} f\, dx$. Hence the Fourier transform of af is $\hat{a} * \hat{f}$.

Returning to Proposition 9, let $\sigma(f)$ be the spectrum of f. Then the spectrum of af is contained in $\sigma(f) + W$. If $f = \sum_{\lambda\in\Lambda} f_\lambda \chi_\lambda$ is a perturbed trigonometric sum and if the spectrum of each f_λ is contained in V, $af = \sum_{\lambda\in\Lambda} af_\lambda \chi_\lambda$ is a perturbed trigonometric sum and the spectrum of each af_λ is contained in U. Theorem IX shows that the pure trigonometric sum $g = \sum_{\lambda\in\Lambda} af_\lambda(0)\chi_\lambda = \sum_{\lambda\in\Lambda} f_\lambda(0)\, \chi_\lambda(x)$ satisfies $\sup_G |g| \leqslant C \sup_G |af|$ and $|g(x) - (af)(x)| \leqslant C|x|\, \|af\|_\infty$. To get $|g(x) - f(x)| \leqslant D|x|\, \|f\|_\infty$, it suffices to show that $|a(x) - 1| \leqslant C_1|x|$. Since $\hat{a}$ has compact support, this last inequality is a corollary of Proposition 7.

15.4. A counterexample

Even if Λ is a set of integers and $U = [0, l]$, it is hopelessly difficult to compute the supremum l_0 of positive l for which Theorem IX is true. This supremum can be different from the supremum l_1 of positive l for which the intervals $[\lambda, \lambda + l]$ are pairwise disjoint.

If Λ is the set of all integers of the form $\lambda_j = 10j$, $j \in \mathbf{Z}$ and $j \neq 2^n$ for all $n \geqslant 0$, together with all integers of the form $\lambda_j = 10 \cdot 2^n - 1$, $j = 2^n$, then $l_1 = 9$ while the conclusion of Theorem IX is false with $l = 1$.

To see why the conclusion of Theorem IX is false in this case, consider any continuous $\pi/5$ periodic function f. The Fourier series of f can be written $f(x) \sim \sum_{-\infty}^{+\infty} c(j) \exp i10jx$ and the spectrum of f is contained in $\Lambda + [0, 1]$. If part (b) of Theorem IX was valid, there would be a continuous 2π-periodic function $g(x)$ with the following two properties: $g(x) \sim \sum_{-\infty}^{+\infty} c(j) \exp i\lambda_j x$ and $\|g\|_\infty \leqslant C\|f\|_\infty$. But $f - g$ is the lacunary series $\sum_{n\geqslant 0} c(2^n)$ $(\exp i\, 10 \cdot 2^n x - \exp i\,(10 \cdot 2^n - 1)\, x)$. By a classical result on Sidon sequences (Chapter VI, Theorem II), there is an absolute constant δ such that $\|f - g\|_\infty \geqslant \delta\,(\sum_{n\geqslant 0} |c(2^n)|)$. This would imply $\delta\,(\sum_{n\geqslant 0} |c(2^n)|) \leqslant (c + 1)\,\|f\|_\infty$ for any $\pi/5$-periodic function. But any sequence $(x_n)_{n\geqslant 0}$ in l^2 is the sequence of Fourier coefficients $c(2^n)$ of a $\pi/5$-periodic function ([24], th. 7.1, ch. XII).

15.5. A fundamental theorem

With a very slight modification of the range of the operator L, the preceding difficulty disappears completely.

Roughly speaking, Theorem XI states that each slowly-perturbed trigonometric sum $\sum_{\lambda\in\Lambda} f_\lambda(x)\, \chi_\lambda(x)$ can be approximated by a pure trigonometric sum $\sum_{\lambda\in\Lambda} P_\lambda(x)\, \chi_\lambda(x)$, where the frequencies of each trigonometric sum P_λ belong to a fixed finite subset F of the compact set E containing the spectrum of each f_λ.

(We have introduced new frequencies to get a better approximation; Λ is harmonious and no restrictions have to be made on E.)

THEOREM XI: *Let G be a separable and metrizable l.c.a. group, Γ the dual group and Λ a harmonious subset of Γ.*

For each compact subset E of Γ, there exist a finite subset F of E, a positive constant C and a linear operator L, whose domain and range will be specified in a moment such that

(a) *if f is a complex-valued bounded continuous function whose spectrum lies in $\Lambda + E$, $g = L(f)$ is an almost-periodic function whose spectrum lies in $\Lambda + F$;*

(b) $\sup_G |g| \leqslant C \sup_G |f|$; *moreover there is a real continuous function ω on G, vanishing at infinity, such that* $\sup_G |g| \leqslant \sup_G |\omega f|$;

(c) *for each x in G,* $|g(x) - f(x)| \leqslant C|x| \sup_G |f|$; *moreover* $|g(x) - f(x)| \leqslant |x| \sup_G |\omega f|$;

(d) *for each λ in Λ, if the spectrum of f lies in $\lambda + E$, the spectrum of g lies in $\lambda + F$.*

Before proving Theorem XI, a remark should be made: since F is finite, the almost-periodic functions whose spectra lie in $\Lambda + F$ are in a class of almost-periodic function because $\Lambda + F$ is harmonious. Roughly speaking, the aim of theorem XI is to replace a bad function f by a regular one g without serious modification of the spectrum.

15.6. *Summary of the proof*

The idea of the proof is to consider Γ as a dense subgroup of the Bohr compactification $\tilde{\Gamma}$ of Γ and, using a partition of unity on $\tilde{\Gamma}$ by means of the regular functions, to divide Λ into a finite number of pieces Λ_j^* for which Proposition 9 applied to the pair (E, Λ_j^*) will be true.

We recall that the Bohr compactification $\tilde{\Gamma}$ of Γ is the dual group of G_d, where G_d is G with the discrete topology; $A(\tilde{\Gamma})$ is the Banach algebra of complex valued continuous functions ψ on $\tilde{\Gamma}$ whose Fourier series are absolutely convergent. A continuous homomorphism $h : \Gamma \to \tilde{\Gamma}$ with a dense range is defined by the following rule: for each $\gamma \in \Gamma$, $h(\gamma)$ is the character χ_γ on G. Then, if $\psi \in A(\tilde{\Gamma})$, $\psi \circ h$ is an almost periodic function on Γ whose Fourier series is absolutely convergent.

15.7. *A set F_1*

If Λ is harmonious in Γ, let $M(\varepsilon)$, for each positive ε, be the relatively dense subset of G defined by $\sup_\Lambda |\chi_\lambda(x) - 1| \leqslant \varepsilon$. Let Λ_2 be the set of all γ in Γ such that $|\chi_\gamma(x) - 1| \leqslant 2\varepsilon$ for each positive ε and each x in $M(\varepsilon)$. Then Λ_2 is harmonious by Theorem I of Chapter II and $(E - E) \cap \Lambda_2 = F_1$ is finite. It is clear that $F_1 = -F_1$ and $0 \in F_1$. Finally let $\bar{\Lambda}$ be the set of all γ in Γ such that $|\chi_\gamma(x) - 1| \leqslant \varepsilon$ for all positive ε and each x in

$M(\varepsilon)$; $\bar{\Lambda}$ is harmonious by Theorem I of Chapter II. The above definition of F_1 seems artificial but F_1 plays a fundamental role when Λ is mapped into the Bohr compactification $\tilde{\Gamma}$ of Γ.

Bohr compactification, U, Λ^*, U^*. Let $h: \Gamma \to \tilde{\Gamma}$ be the homomorphic injection of Γ in $\tilde{\Gamma}$ and let U be the closure of $h(\Lambda)$ in $\tilde{\Gamma}$. The inverse image under h of the compact subset U of $\tilde{\Gamma}$ is again an harmonious subset of Γ contained in $\bar{\Lambda}$ as the following observation shows:

For each positive ε, each $t \in M(\varepsilon)$ and each $\chi \in U$, we have $|\chi(t) - 1| \leqslant \varepsilon$.

The inequalities $|\chi(t) - 1| \leqslant \varepsilon$, where t runs over $M(\varepsilon)$ define a closed subset of $\tilde{\Gamma}$ containing $h(\Lambda)$ and hence containing U. For each γ in $h^{-1}(U)$ we have $|\chi_\gamma(t) - 1| \leqslant \varepsilon$ $(t \in M(\varepsilon))$ and $h^{-1}(U) \subset \bar{\Lambda}$.

Let V be the compact subset $h(E)$ of $\tilde{\Gamma}$. The following lemma shows that the number of decompositions of an element χ of $\tilde{\Gamma}$ as a sum $\xi + \eta$, $\xi \in U$, $\eta \in V$ is finite; the finite set concerned here is precisely F_1.

LEMMA 13: *If an element χ of $\tilde{\Gamma}$ can be written $\chi = \xi + \eta = \xi' + \eta'$, where ξ and ξ' belong to U and η and η' belong to V, then $\xi - \xi' = \eta' - \eta$ belongs to $h(F_1)$.*

Proof: By the definition of V, we have $\eta' - \eta = h(\gamma)$, where $\gamma \in E - E$. On the other hand, for each $t \in M(\varepsilon)$, $|\xi(t) - 1| \leqslant \varepsilon$ and $|\xi'(t) - 1| \leqslant \varepsilon$, which imply $|(\xi - \xi')(t) - 1| \leqslant 2\varepsilon$. Then $|\chi_\gamma(t) - 1| \leqslant 2\varepsilon$ $(t \in M(\varepsilon))$ and $\gamma \in E - E$ show that $\gamma \in F_1$, since this is the definition of F_1.

In what follows, the compact subset $U^* = U + h(F_1)$ plays an important role; $h^{-1}(U^*) = h^{-1}(U) + F_1$ is contained in the harmonious set $\Lambda^* = \bar{\Lambda} + F_1$ (Chapter II, Section 3, Theorem II).

15.8. A partition of Λ

Let E_1 be a compact subset of Γ whose interior contains $E + F_1$. We shall divide the harmonious set Λ into a finite number of pieces, denoted by Λ_j^*, $1 \leqslant j \leqslant n$, in such a way that Proposition 9 is true for the pair (Λ_j^*, E_1), where Λ_j^* and E_1 take the places of Λ and U respectively in Proposition 9.

Let K be a compact neighbourhood of 0 in G such that, for each x in K and each γ in E_1, $|\chi_\gamma(x) - 1| \leqslant \frac{1}{11}$. Let M^* be the relatively dense subset of G defined by the condition that, for each x in M^* and each λ in Λ^*, $|\chi_\lambda(x) - 1| \leqslant \frac{1}{11}$, and let T be a finite subset of G such that $T + K + M^* = G$.

For each χ in $\tilde{\Gamma}$, let ω_χ (resp. ω_χ^*) be the set of all η in $\tilde{\Gamma}$ such that,

$|\chi(t) - \eta(t)| < \frac{1}{33}$ (resp. $|\chi(t) - \eta(t)| \leqslant \frac{1}{11}$) for each t in T. Finally let ω be the set of all η in $\tilde{\Gamma}$ such that $|\eta(t) - 1| \leqslant \frac{1}{33}$ for all $t \in T$.

As χ runs over U, the family consisting of the corresponding ω_χ is an open covering of U. There exist an integer n and a finite covering of U by $\omega_1, \ldots, \omega_n$, where $\omega_j = \omega_\chi$ for $\chi = \chi_j \in U$, $1 \leqslant j \leqslant n$. We define $U^* = U + h(F_1)$, $U_j^* = U^* \cap \omega_j^*$, where $\omega_j^* = \omega_\chi^*$ for $\chi = \chi_j$ and ω_j^* is a compact subset of $\tilde{\Gamma}$. Finally let Λ_j^* be the set of all λ in Γ such that $h(\lambda) \in U_j^*$. *The set* Λ_j^* *is contained in* Λ^* *and the union* $\bigcup_{1 \leqslant j \leqslant n} \Lambda_j^*$ *contains* Λ.

LEMMA 14: *For each j such that* $1 \leqslant j \leqslant n$, *Proposition* 9, *applied to the pair* (Λ_j^*, E_1) *is true.*

Since Λ_j^* is contained in Λ^*, we have $|\chi_\lambda(x) - 1| \leqslant \frac{1}{11}$ for each λ in Λ_j^* and each x in M^*. On the other hand, let a_j be an element of Λ_j^*, fixed once and for all. For every χ and χ' in ω_j^* and every t in T, we have $|\chi(t) - \chi_j(t)| \leqslant \frac{1}{11}$ and $|\chi'(t) - \chi_j(t)| \leqslant \frac{1}{11}$ and hence $|\chi(t) - \chi'(t)| \leqslant \frac{2}{11}$. This implies $|\chi_t(\lambda - a_j) - 1| \leqslant \frac{2}{11}$ for all λ in Λ_j^* and all t in T (we recall that for γ in Γ and t in G, $\chi_t(\gamma) = \chi_\gamma(t)$).

Finally we have $|\chi_x(\lambda - a_j) - 1| \leqslant \frac{4}{11}$ for every x in $T + M^*$ and every λ in Λ_j^*. Proposition 7 shows that each x in $T + M^*$ is an $\frac{8}{11}$-almost period of each almost-periodic function whose spectrum lies in $\Lambda_j^* - a_j$; hence the pair $(K, \frac{11}{3})$ is suitable for $\Lambda_j^* - a_j$. The argument of Theorem IX then shows that $(E_1, \Lambda_j^* - a_j)$ is a harmonic discretization of $\Lambda_j^* - a_j$; hence (E_1, Λ_j^*) is a harmonic discretization of Λ_j^* $(1 \leqslant j \leqslant n)$.

15.9. *Functions* φ_j, $1 \leqslant j \leqslant n$

Let V be the compact subset $h(E)$ of $\tilde{\Gamma}$, U_j the closure of $U \cap \omega_j$ in $\tilde{\Gamma}$; then U_j is contained in U and each χ in U_j satisfies the inequalities $|\chi(t) - \chi_j(t)| \leqslant \frac{1}{33}$, $t \in T$. Let Ω_j, $1 \leqslant j \leqslant n$, be an open subset of $\tilde{\Gamma}$ containing the compact subset $U_j + V$ of $\tilde{\Gamma}$ and small enough to satisfy the following condition: for any k in $\{1, \ldots, n\}$, Ω_j does not intersect Ω_k if $U_j + V$ does not intersect $U_k + V$. The family Ω_j, $1 \leqslant j \leqslant n$, is an open covering of the compact subset $U + V$ of $\tilde{\Gamma}$.

Let $\psi_1, \ldots, \psi_n$, be n continuous functions on $\tilde{\boldsymbol{R}}$, whose Fourier series are absolutely convergent and which satisfy the two conditions:

(1) the support of ψ_j is contained in Ω_j;

(2) $\psi_1 + \cdots + \psi_n = 1$ on a neighbourhood of $U + V$.

The regularity of the self-adjoint algebra $A(\tilde{\Gamma})$ enables us to obtain such partitions of unity.

We define $\varphi_j = \psi_j \circ h$ for each j in $\{1, \ldots, n\}$. These φ_j are almost-periodic functions on $\tilde{\Gamma}$ with absolutely convergent Fourier series; hence each φ_j is the Fourier transform of a discrete bounded complex-valued measure σ_j on G. We have $\varphi_1 + \cdots + \varphi_n = 1$ on a neighbourhood of $\Lambda + E$ and Wiener's theory (Chapter IV, Section 12) shows that each complex-valued bounded, continuous function f on G whose spectrum lies in $\Lambda + E$, is a sum $f = f * \sigma_1 + \cdots + f * \sigma_n$.

LEMMA 15: *With the preceding notation, the spectrum of each* $f * \sigma_j$, $1 \leqslant j \leqslant n$, *is contained in* $\Lambda_j^* + E_1$.

The spectrum of $f * \sigma_j$ is the support of the Fourier transform $\hat{f}\hat{\sigma}_j = \hat{f}\varphi_j$. To prove Lemma 15, it is sufficient to prove that if γ belongs to $\Lambda + E$, which contains the support of $\hat{f}$, and to the support of φ_j, then γ belongs to $\Lambda_j^* + E_1$.

The union $\bigcup_{1 \leqslant j \leqslant n} U_j$ is equal to U and $h(\Lambda + E)$ is contained in $U + V$. If we put $\chi = h(\gamma)$, we can find an integer k in $\{1, \ldots, n\}$ such that $\chi \in U_k + V$. We can then write $\chi = u + v$, where $u \in U_k$ and $v \in V$.

Since γ belongs to the support of φ_j, χ belongs to the support of ψ_j; hence $\chi \in \Omega_j$. If $U_j + V$ does not intersect $U_k + V$, Ω_j does not intersect $U_k + V$; this is impossible since χ belongs to these two subsets. Hence $U_j + V$ intersects $U_k + V$.

Lemma 13 shows that we can find elements ξ in U_j, ξ' in U_k such that $\xi - \xi' \in h(F_1)$. Put $u^* = u + \xi - \xi'$. We shall prove that u^* belongs to U_j^*.

Since u^* belongs to $U^* = U + h(F_1)$, it is sufficient to show that u^* belongs to the closed set ω_j^* ($U_j^* = U^* \cap \omega_j^*$ and ω_j^* is defined like ω_j but is 'three times larger'). But u and ξ' belong to U_k, and are therefore in the closure $\bar{\omega}_k$ of ω_k; hence $u - \xi'$ is in $\bar{\omega}_k - \bar{\omega}_k = \bar{\omega} - \bar{\omega}$. Since ξ belongs to ω_j, u^* belongs to ω_j^*.

Finally, by hypothesis, χ belongs to $h(\Gamma)$; so does each v in $V = h(E)$ and u as $\xi - \xi'$ are in $h(\Gamma)$. Hence the element u^* of U_j^* can be written $u^* = h(\lambda^*)$, where λ^* is in Λ_j^*. Let us put $\xi' - \xi = h(f)$, $f \in F_1$, and $v = h(s)$, $s \in E$. The decomposition $\chi = u^* + \xi' - \xi + v$ gives $\gamma = \lambda^* + f + s$; hence γ belongs to $\Lambda_j^* + F_1 + E$, which is contained in $\Lambda_j^* + E_1$.

15.10. A finite set F

The introduction of the finite set F and of a suitable partition of E_1 allows to describe what happens if for some λ^* in Λ^* and λ in Λ, the translated set $\lambda^* + E_1$ intersects $\lambda + E$.

Firstly, $\lambda - \lambda^*$ belongs to $\Lambda^* - \Lambda^*$, which is harmonious, and to $E_1 - E$; hence $\lambda - \lambda^*$ belongs to $(E_1 - E) \cap (\Lambda^* - \Lambda^*)$ which is a finite subset A of Γ. We consider the translated sets E_α of E by all α in A. For each s in E_1, there exists an open neighbourhood J_s of s small enough to satisfy the following conditions: for each α in A, s belongs to E_α if J_s intersects E_α and $(J_s - E) \cap (\Lambda^* - \Lambda^*) \subset A$.

We can find a finite covering of E_1 by a finite number $J_1, \ldots, J_q$ of these J_s corresponding to the elements $s_1, \ldots, s_q$ of E_1.

Definition 11: *With the preceding notation, F is the finite set of all differences* $s_k - \alpha$, $1 \leqslant k \leqslant q$, $\alpha \in A$, *which belong to E.*

We can state

Lemma 16: *For each* $\lambda^* \in \Lambda^*$, *each k in* $\{1, \ldots, q\}$ *and each* λ *in* Λ, *if* $\lambda^* + J_k$ *intersects* $\lambda + E$, *then* $\lambda^* + s_k$ *belongs to* $\lambda + F$.

We define elements $\zeta_1, \ldots, \zeta_q$ of $A(\Gamma)$ by the following two conditions: $\zeta_1 + \cdots + \zeta_q = 1$ on a neighbourhood of E_1 and the support of each ζ_k is contained in J_k.

15.11. Functions $Z_{j,k}$

We define on the subset $\Lambda_j^* + E_1$ of Γ, a complex-valued continuous function $Z_{j,k}$ by $Z_{j,k}(t) = \zeta_k(s)$ if $t = \lambda^* + s$, where $\lambda^* \in \Lambda_j^*$ and $s \in E_1$. Theorem X shows that

Lemma 17: *There is a complex-valued bounded Radon measure* $\mu_{j,}{}^{k}$ *whose Fourier transform restricted to* $\Lambda_j^* + E_1$ *is* $Z_{j,k}$.

Operator L: Proposition 9 is true for the pair (Λ_j^*, E_1). We can find a linear operator $L_{j,k}$, $1 \leqslant j \leqslant n$, $1 \leqslant k \leqslant q$, mapping complex-valued bounded continuous functions f on G whose spectrum lies in $\Lambda_j^* + J_k'$ ($J_k' = E_1 \cap J_k$) into almost-period functions $L_{j,k}(f)$ whose spectra lie in $\Lambda_j^* + s_k$; $L_{j,k}$ has the following properties:

(3) $\sup_G |L_{j,k}(f)| \leqslant C \sup_G |f|$;

(4) $|(L_{j,k}(f))(x) - f(x)| \leqslant C|x| \, \|f\|_\infty$;

(5) if the spectrum of f is contained in $\lambda^* + E_1$, $\lambda^* \in \Lambda_j^*$, the spectrum of $L_{j,k}(f)$ is $\lambda^* + s_k$;

(6) there is a real continuous function $\omega_{j,k}$ on G vanishing at infinity such that $\|L_{j,k}(f)\|_\infty \leqslant \|\omega_{j,k} f\|_\infty$.

We now consider an f whose spectrum is contained in $\Lambda + E$. We define f_j, $1 \leqslant j \leqslant n$, by $f_j = f * \sigma_j$; the spectrum of f_j is a subset of

$\Lambda_j^* + F_1 + E$, which is contained in the interior of $\Lambda_j^* + E_1$. For this reason we can define $f_{j,k}$ to be $f_j * \mu_{j,k}$. We have $f_j = \sum_{1 \leqslant k \leqslant q} f_{j,k}$ and hence $f = \sum f_{j,k}$ where the sum is over all pairs (j, k) such that $1 \leqslant j \leqslant n$ and $1 \leqslant k \leqslant q$.

The linear operator $L_{j,k}$ maps $f_{j,k}$ into $g_{j,k}$. Finally g is, by definition the sum of all the $g_{j,k}$, $1 \leqslant j \leqslant n$, $1 \leqslant k \leqslant q$ and we put $g = L(f)$.

It is plain that g is an almost periodic function and that L is continuous for the L^∞ norms over G.

What is the spectrum of g? If the spectrum of f is contained in $\Lambda + E$, Lemma 15 shows that the spectrum of $f_j = f * \sigma_j$ is contained in $\Lambda_j^* + E_1$. The definition of $\mu_{j,k}$ shows that the spectrum of $f_{j,k} = f_j * \mu_{j,k}$ is contained in $\Lambda_j^* + J_k$. But $f_{j,k} = f * \sigma_j * \mu_{j,k}$. Hence the spectrum of $f_{j,k}$ is also contained in $\Lambda + E$.

Finally the spectrum of $f_{j,k}$ is contained in the union of all the $\lambda^* + J_k$, $\lambda^* \in \Lambda_j^*$, which intersect $\Lambda + E$; in this case, Lemma 16 shows that $\lambda^* + s_k$ belongs to $\Lambda + F$. The spectrum of each $g_{j,k}$ is contained in $\Lambda + F$ and hence so is the spectrum of g.

In the same way, we show that if the spectrum of f is contained in $\lambda + E$, that of g is contained in $\lambda + F$. The inequality (c) of Theorem XI is obtained by addition of the corresponding inequalities (4) applied to $f_{j,k}$ and $L_{j,k}(f_{j,k}) = g_{j,k}$.

Theorem XI is thus proved.

16. A new characterization of Pisot numbers

16.1. Let θ be a real number greater than one and Λ_θ the set of all finite sums $\sum_{k \geqslant 0} \varepsilon_k \theta^k$, where $\varepsilon_k = 0$ or 1. It will be shown that Theorem XI applied to Λ_θ cannot hold if θ is not a Pisot number.

THEOREM XII: *The following two properties of θ are equivalent:*

(a) *θ is a Pisot number;*

(b) *for each compact set E of real numbers there exist a constant C and a linear operator L, whose domain and range will be given in a moment, such that*

(1) *if f is a complex-valued bounded continuous function whose spectrum lies in $\Lambda_\theta + E$, $g = L(f)$ is an almost-periodic function whose spectrum lies in $\Lambda_\theta + E$; moreover $\|g\|_\infty \leqslant C\|f\|_\infty$ and $g(0) = f(0)$;*

(2) *if the spectrum of f lies in $\lambda + E$ ($\lambda \in \Lambda$), the spectrum of g also lies in $\lambda + E$.*

(a) ⇒ (b): If θ is a Pisot number, Λ_θ is harmonious and Theorem XII is a weak version of theorem XI.

(b) ⇒ (a): The proof is based on the existence of a compact set E of real numbers with the following properties:

(1) the vector space X generated by E over $\mathbf{Q}$ has Lebesgue measure 0;

(2) 0 belongs to E;

(3) there exists a complex valued continuous function φ on $\mathbf{R}$ vanishing at infinity, whose spectrum is contained in E and such that $\varphi(0) = 1$.

The existence of such a set will be proved in Chapter VIII, Section 10.

16.2. Summary of the proof

Let M_θ be the set of all finite sums $\sum_{k \geqslant 0} p_k \theta^k$, where $p_k \in \mathbf{Z}$. Multiplying E by a suitable real number if necessary, we can assume that 0 is the only element of M_θ lying in the vector space generated by E over $\mathbf{Q}$. In this case the vector space of almost-periodic functions whose spectra lie in $\Lambda + E$ has remarkable properties and (b) implies the conclusion of Proposition 9. We then use the existence of a continuous function on $\mathbf{R}$, vanishing at infinity whose spectrum lies in E, to show that the validity of (b) implies that Λ_θ is a coherent set of frequencies.

Hence θ is a Pisot number.

16.3. Details of the proof

Let X be the vector space generated by E over $\mathbf{Q}$.

Reduction to the case $M_\theta \cap X = \{0\}$. Let Y be the set of all real numbers y such that $yM_\theta \cap X \neq \{0\}$. Then Y is contained in $\bigcup_{\lambda \in M_\theta} \lambda^{-1}X$; since M_θ is countable, Y has Lebesgue measure zero and so there is a real number $y \neq 0$ such that $yM_\theta \cap X = \{0\}$. If were place E by $y^{-1}E$, then $M_\theta \cap X = \{0\}$ and the other properties of E are preserved.

16.4. Almost-periodic functions whose spectra lie in $\Lambda_\theta + E$

The compact sets $\lambda + E$, $\lambda \in \Lambda_\theta$ are pairwise disjoint since $M_\theta \cap X = \{0\}$. A trigonometric sum P whose frequencies belong to $\Lambda + E$ can be written $\sum_{\lambda \in \Lambda} P_\lambda \chi_\lambda$, where the spectrum of each P_λ is contained in E. For each real number x, the mapping H_x from $\Lambda + E$ to $\mathbf{T}$ defined by $H_x(\lambda + t) = \exp 2\pi i x \lambda$ is a weak character on $\Lambda + E$. Hence $P(H_x)$ is defined by

$\sum_{\lambda\in\Lambda} P_\lambda(0)\,\chi_\lambda(x)$ and $|P(H_x)| \leqslant \|P\|_\infty$. The trigonometric sum $Q(x) = P(H_x)$ satisfies $\|Q\|_\infty \leqslant \|P\|_\infty$ and the linear mapping T defined by $T(P) = Q$ can be extended to the space of all almost-periodic functions whose spectra lie in $\Lambda + E$.

16.5. Let φ be the complex-valued continuous function on **R** vanishing at infinity, whose spectrum is contained in E, and such that $\varphi(0) = 1$. We apply (b) to the perturbed trigonometric sum $f = P\varphi$, where P is any trigonometric sum whose frequencies belong to Λ; we get an almost periodic function g with the following properties:

(i) if $P = \sum_{\lambda\in\Lambda} c_\lambda\chi_\lambda$, $g = \sum_{\lambda\in\Lambda} g_\lambda\chi_\lambda$, where each g_λ is an almost-periodic function whose spectrum lies in E;

(ii) $g_\lambda(0) = c_\lambda$;

(iii) $\|g\|_\infty \leqslant C\|P\|_\infty$.

We now define Q to be $T(g)$; then $Q(x) = \sum_{\lambda\in\Lambda} d_\lambda\chi_\lambda$, $d_\lambda = g_\lambda(0)$ and $\|Q\|_\infty \leqslant \|T(g)\|_\infty$. Then $Q = P$ and we get $\|P\|_\infty \leqslant C\|P\varphi\|_\infty$. Since φ vanishes at infinity on **R**, Λ is a coherent set of frequencies.

17. Slowly-perturbed trigonometric sums in the real case

17.1. Let $w : \mathbf{R} \to [1, +\infty]$ be an even log-concave function such that

$$\int_{-\infty}^{+\infty} \frac{\log w(x)}{1 + x^2}\, dx$$

is finite. For instance, $w(x) = \exp |x|^\alpha$ is such a function if $\alpha < 1$.

THEOREM XIII: *Let w be as above. Then, for each coherent set of frequencies Λ there exist a positive ε and a constant $C > 0$ such that*

(1) *if $f : \mathbf{R} \to \mathbf{C}$ is a continuous function which is $0\,(w(t))$ at infinity and whose Fourier series is $\sum_{\lambda\in\Lambda} a_\lambda(t) \exp 2\pi i\lambda t$ and if the spectrum of each a_λ is contained in $[-\varepsilon, \varepsilon]$, then, for each real t_0, $f(t_0, t) = \sum_{\lambda\in\Lambda} a_\lambda(t_0) \exp 2\pi i\lambda t$ is almost periodic;*

(2) $\sup_{\mathbf{R}} |f(t_0, t)| \leqslant C \sup_{\mathbf{R}} |f(x)/w\,(x - t_0)|$;

(3) $|f(t_0, t) - f(t)| \leqslant C\,|t - t_0| \sup_{\mathbf{R}} |f(x)/w\,(x - t_0)|$.

Proof: The Denjoy–Carleman's theorem ([11], p. 114, th. 2.4) shows that for each positive ε, there is a real continuous function a supported by $[-\varepsilon, \varepsilon]$, whose integral is 1 and such that $\hat{a} = 0\,(1/w)$ at infinity.

We now reproduce the proof of Proposition 9, to get Theorem XIII.

17.2. We end this chapter by describing the connection between the local approximation property given by statement (3) of Theorem XIII and the arithmetical structure of Λ.

Let $\theta > 1$ be a real number, Λ the set of all finite sums $\sum_{k \geqslant 0} \varepsilon_k \theta^k$, $\varepsilon_k = 0$ or 1, and Λ_k the set of the 2^k sums $\sum_0^{k-1} \varepsilon_j \theta^j$, $\varepsilon_j = 0$ or 1.

PROPOSITION 10: *If θ is not a Pisot number and ε is a positive real number, statement* (3) *of Theorem* XIII *does not hold for Λ.*

Proof: In fact we shall prove the stronger result that (3) fails in any neighbourhood of t_0. Let $a : \mathbf{R} \to \mathbf{R}$ be a continuous function, vanishing at infinity, not identically 0, such that $a(0) = 0$, and whose spectrum is contained in $[-\varepsilon, \varepsilon]$. Let $P_k(t) = 2^{-k} \sum_{\lambda \in \Lambda_k} \exp 2\pi i \lambda t$. Then $\sup_{\mathbf{R}} |a(t) P_k(t)| \to 0$ as $k \to +\infty$ as was proved in Section 5.2. Let

$$f(t) = a(t+h)\, P_k(t+h)$$

for some $h \neq 0$. The corresponding $f(0, t)$ is $a(h)\, P_k(t+h)$. If (3) were true for $t_0 = 0$ on the interval $[-h, h]$, we should get

$$\sup_{[-h,h]} |a(h)\, P_k(t+h)| = a(h) \to 0 \text{ as } k \to +\infty;$$

if $a(h) \neq 0$ this cannot happen and, since a is an entire function, we can find such an h as close to 0 as we like.

18. Notes

The main result of this chapter was Theorem XI. In the real case, we considered the real line as a dense subgroup of a suitable compact group G in such a way that our discrete harmonious set Λ was contained in a thin compact subset of G.

This idea has been used before by J.F. Méla in [65] and [66] to study the so-called 'interpolation sets for almost periodic functions'.

INTERLUDE

One of the most fascinating problems in harmonic analysis is to find connections between the *additive properties* of a set Λ of frequencies and the *functional properties* of the corresponding space of almost periodic functions whose spectra are contained in Λ.

For Λ a finite set of real numbers independent over $\mathbf{Q}$, we obtained the following connection in Chapter I: we can rapidly predict the size of trigonometric sums whose frequencies belong to Λ if and only if Λ is strongly independent over $\mathbf{Z}$ (Chapter I, Theorem IX).

In the general context of l.c.a. groups, we can also get striking results. We first recall some definitions:

(1) A subset Λ of a l.c.a. group G is discrete if each point λ of Λ is isolated in Λ for the relative topology;

(2) A subset Λ of G is relatively dense if there exists a compact subset K of G such that each translate of K intersects Λ;

(3) A set E of bounded continuous functions $f: G \to \mathbf{C}$ has a complete set of a priori ε-almost periods if there exists a relatively dense subset $M(\varepsilon)$ of G such that, for each τ in $M(\varepsilon)$ and each f in E, $\sup_G |f(x + \tau) - f(x)| \leqslant \varepsilon \sup_G |f|$;

(4) A set E of bounded continuous functions $f: G \to \mathbf{C}$ is locally trivial at 0 if there exists a compact neighbourhood V of 0 such that each continuous function $g: V \to \mathbf{C}$ is the restriction to V of a function f in E. In this case we say that E is trivial on V.

With these definitions we can state the following:

THEOREM I: *Let G be a metrizable and separable l.c.a. group, Γ the dual group, Λ a subset of Γ, and $E(\Lambda)$ the space of all almost periodic functions $f: G \to \mathbf{C}$ whose spectra lie in Λ.*

The following two properties are equivalent:

(a) *Λ is discrete, Λ is relatively dense in Γ and there is a finite subset F of Γ such that $\Lambda - \Lambda$ is contained in $\Lambda + F$;*

(b) *for each positive ε, there is a complete set of a priori ε-almost periods for all f in $E(\Lambda)$ and $E(\Lambda)$ is locally trivial.*

Before giving the proof, some examples should be noted.

If $G = \mathbf{R}$, $\Gamma = \mathbf{R}$ and $\Lambda = \mathbf{Z}$, $E(\Lambda)$ is the space of all continuous 1-periodic functions $f: \mathbf{R} \to \mathbf{C}$; $E(\Lambda)$ is trivial on each compact subset V of $]0, 1[$. More interesting examples will be studied in Chapter V.

If $G = \mathbf{Q}_p$, $\Gamma = \mathbf{Q}_p$ and Λ is the subset of all mp^n in $\mathbf{Q}_p$ such that $0 \leqslant mp^n \leqslant 1$, where m and n are otherwise arbitrary rational integers, then $E(\Lambda)$ is trivial on each compact subset V of $\mathbf{Q}_p$ whose Haar measure is less than the measure of $\mathbf{Z}_p$ (Theorem IX of Chapter V).

Proof of (a) $\Rightarrow$ (b); Λ is a relatively dense harmonious subset of Γ: for some compact subset K of Γ, $\Lambda + K = \Gamma$. Theorem II of Chapter IV shows that for each positive ε there is a relatively dense set of ε-almost periods for all f in $E(\Lambda)$.

The proof of the last statement of (b) is an easy adaptation of the proof of part (b) in Theorem IX of Chapter V. For this reason, the reader is referred to Chapter V, Section 8.7.

Proof of (b) $\Rightarrow$ (a): The existence, for each positive ε, of an a priori complete set $M(\varepsilon)$ of ε-almost periods for all f in $E(\Lambda)$ implies that Λ is harmonious.

To prove that Λ is relatively dense, from which (a) will follow, we use the following lemma:

LEMMA 1: *For each $n \geqslant 1$, let Δ_n be the set of all γ in Γ whose distance from 0 exceeds n. Then for any $h: \Gamma \to \mathbf{C}$ in $A(\Gamma)$, the norm of h in the restriction algebra $A(\Delta_n)$ tends to 0.*

Proof: Obvious, since the compactly supported h in $A(\Gamma)$ are dense in $A(\Gamma)$. We return to the implication (b) $\Rightarrow$ (a). If $E(\Lambda)$ is trivial on a compact neighbourhood V of 0 in G, a theorem of Banach applied to the surjective restriction operator $T: E(\Lambda) \to C(V)$ yields a constant C such that, for each complex valued continuous φ in $C(V)$, there exists an f in $E(\Lambda)$ such that $\sup_G |f| \leqslant C \sup_V |\varphi|$ and $f|_V = \varphi$.

It follows from duality that for each complex valued Radon measure μ carried by V and each bounded complex valued Radon measure ν, the equalities $\hat{\mu}(\lambda) = \hat{\nu}(\lambda)$ for all $\lambda \in \Lambda$ imply $\|\mu\| \leqslant C\|\nu\|$.

Let h be the Fourier-transform of an integrable function I, not identically zero, which is carried by V; then $h \in A(\Gamma)$.

If Λ is not relatively dense in G, there is for each $n \geqslant 1$, a γ_n in Γ whose distance from Λ exceeds n.

Let $d\mu_n = \gamma_n(x)\, I(x)\, dx$. Lemma 1 shows that there is a sequence ν_n of complex Radon measures such that $\|\nu_n\| \to 0$ and $\hat{\mu}_n(\lambda) = \hat{\nu}_n(\lambda)$ for all λ in Λ. But $\|\mu_n\| = \|I\|_1$ and this contradicts $\|\mu_n\| \leqslant C\|\nu_n\|$.

CHAPTER V

SPECIAL TRIGONOMETRIC SERIES (COMPLEX METHODS)

In the real and the p-adic case, precise and interesting results about coherent sets of frequencies need complex methods. In Section 1 we define the Laplace or the complex Fourier-transform of a distribution with compact support. The Paley–Wiener theorem gives precise information on the set of zeros of such a transform (Sections 2 and 3). In Section 9, we present the theory of mean-periodic functions on the line. Studying coherent sets of frequencies is equivalent to investigating the behaviour at infinity of all mean periodic functions whose spectrum is a given set of real numbers.

Trigonometric series whose frequencies belong to the set of all entire parts of $k\theta$, where $k \geqslant 1$ and θ is an irrational number greater than one have quite fascinating properties which are studied in Section 6.

But perhaps the most beautiful results concern the trigonometric series defined on the p-adic field $\mathbf{Q}_p$ by $f(x) = \sum_{\lambda \in \Lambda} a_\lambda \exp 2\pi i\lambda x$, where $\lambda = mp^n$, for rational integers m and n such that $0 < mp^n < 1$. These series have almost all the properties of 2π-periodic functions on the line and other better ones (Section 8). The harmonious set Λ and its homothetical copies $p^k\Lambda$, $k \geqslant 1$, are of outstanding importance in the atomization of distributions in $\mathbf{Q}_p$, just as arithmetic progressions are used to discretize problems on the real line.

1. The Laplace transform

1.1. The space of tempered distributions

Following the notation of [21], $\mathscr{S}$ is the Frechet space of all complex valued functions α on $\mathbf{R}$ all of whose derivatives, together with α itself, decay rapidly at infinity: for each $j \geqslant 0$ and each $k \geqslant 0$, $(1 + |x|)^j \varphi^{(k)}(x) \to 0$ as $|x| \to +\infty$. The dual space $\mathscr{S}'$ is the space of tempered distributions. For each $T \in \mathscr{S}'$ there is an integer n and a constant $C > 0$ such that, for every $\alpha \in \mathcal{S}$ satisfying

$$|(1 + |x|)^n \varphi^{(k)}(x)| \leqslant C, \quad 0 \leqslant k \leqslant n, \tag{5.1}$$

we have $|T(\alpha)| \leqslant 1$.

The Fourier transformation $\alpha \to \hat{\alpha}$ is an isomorphism of $\mathcal{S}$ and the Fourier-transform of each T in $\mathcal{S}'$ is defined by the rule $\langle T, \hat{\alpha} \rangle = \langle \hat{T}, \alpha \rangle$.

An element T of $\mathcal{S}'$ is supported by $[0, +\infty[$ if for each $\alpha \in \mathcal{S}$ which is 0 in a neighbourhood of $[0, +\infty[$, we have $\langle T, \alpha \rangle = 0$. From now on we shall confine our attention to those $T \in \mathcal{S}'$ supported by $[0, +\infty]$.

1.2. The domain of definition of the Laplace transform

Let $p = u + iv$ be a complex number. The function $t \to e^{-pt}$ does not belong to $\mathcal{S}$, but *if α is an indefinitely differentiable function which is* 1 *on* $[-1, +\infty[$ *and* 0 *on* $]-\infty, -2]$, then $\alpha(t)\, e^{-pt}$ belongs to $\mathcal{S}$ for each $p \in \mathbf{C}$ such that $u = \mathcal{R}e p > 0$. We can *define $\langle T, e^{-pt} \rangle$ to be $\langle T, \alpha(t)\, e^{-pt} \rangle$ whenever T is supported by* $[0, +\infty[$.

From now on, we write $\int_0^\infty e^{-pt}\, T(t)\, dt$ instead of $\langle T, e^{-pt} \rangle$.

The Laplace transform can be defined on a larger space of distributions. For instance consider the function $\exp \sqrt{t}$, $t \geqslant 0$. For each $p \in \mathbf{C}$ whose real part is positive, the integral $\int_0^\infty \exp(-pt + \sqrt{t})\, dt$ converges and defines the Laplace transform of this function.

Let $\mathcal{D}$ be the space of all complex valued indefinitely differentiable functions on the line with compact support. A distribution is a linear form T over $\mathcal{D}$ such that for each $m \geqslant 1$ there is an integer $n \geqslant 0$ and a constant $C > 0$ with the property that for all φ in $\mathcal{D}$ supported by $[-m, m]$ the inequalities $\sup\limits_{[-m,m]} |\varphi^{(j)}(t)| \leqslant C, 0 \leqslant j \leqslant n$, imply $|T(\varphi)| \leqslant 1$. If this is the case we write $T \in \mathcal{D}'$.

The domain of definition of the Laplace transform is the subspace $\mathcal{L}$ of $\mathcal{D}'$ consisting of all distributions T carried by $[0, +\infty]$ *such that, for each* $\varepsilon > 0$, *$e^{-\varepsilon t}\, T$ belongs to $\mathcal{S}'$.*

1.3. Properties of the Laplace transform

PROPOSITION 1: *For all T in $\mathcal{L}$, the Laplace transform $\Phi(p) = \int_0^\infty e^{-pt} T(t)\, dt$ is holomorphic in the open half-plane $\mathcal{R}e p > 0$.*

Proof: Let $0 < \varepsilon < \mathcal{R}e p$. Then $e^{-pt} = e^{-\varepsilon t} e^{-(p-\varepsilon)t}$, $e^{-\varepsilon t}\, T(t) \in \mathcal{S}'$ and $\mathcal{R}e\,(p - \varepsilon) > 0$. The integral is thus well defined.

To prove that $\Phi(p)$ is holomorphic, we apply the same trick and we are thereby reduced to examining only the case where T belongs to $\mathscr{S}'$. If $\mathscr{R}e p_0 > 0$, then with α defined as above, $\alpha(t)\,(\exp(-pt) - \exp(-p_0 t))/p - p_0$ tends to $-t\alpha(t)\exp(-p_0 t)$ in the topology of $\mathscr{S}$. Hence $\Phi(p)$ is differentiable over the field $\mathbf{C}$ at the point p_0 and $\Phi'(p)$ is the Laplace transform of $-tT(t)$.

THEOREM I: *Let T be a distribution carried by* $[0, +\infty[$ *such that, for each positive* ε, $e^{-\varepsilon t}T$ *is a tempered distribution. Then the Laplace transform* $\Phi(p)$ *of T is a holomorphic function in the open half plane with the following property: there is an integer* $n \geqslant 0$ *such that for each positive* ε, *we can find a constant* C_ε *such that* (5.2) $|\Phi(p)| \leqslant C_\varepsilon |p|^n$ *for* $\mathscr{R}e p \geqslant \varepsilon$. *Conversely each holomorphic function in the half-plane* $\mathscr{R}e p > 0$ *satisfying* (5.2) *is the Laplace transform of a unique distribution T carried by* $[0, +\infty[$ *such that* $e^{-\varepsilon t}$ *T belongs to* $\mathscr{S}'$ *for each positive* ε.

The proof of the first half of Theorem I begins with a lemma; here T is any distribution carried by $[0, +\infty[$.

LEMMA 1: *There exist an integer* $n \geqslant 0$, *a continuous complex valued function f carried by a compact subset of* $[0, +\infty[$ *and a distribution S carried by* $[1, +\infty[$ *such that* $T = D^n(f) + S$ *where* D^n *is the nth (distributional) derivative.*

Proof of the lemma: Let β be any indefinitely differentiable function with compact support, which is equal to 1 on a neighbourhood of $[0, 1]$. We write $T = \beta T + (1 - \beta)\,T$.

The distribution βT is carried by a compact set. Hence there exist an integer $n \geqslant 0$ and a continuous function g such that $g(t) = 0$ if $t \leqslant 0$, $g(t) = c$ if $t \geqslant t_0$ and $\beta T = D^n(g)$. Let γ be any indefinitely differentiable function with compact support, which is equal to 1 on a neighbourhood of $[0, t_0]$. We define $f = \gamma g$; then $D^n(f) = D^n(g)$ on a neighbourhood of $[0, t_0]$ and $T = D^n(f) + S$ is the required decomposition of T.

Proof of the first part of Theorem I: Let Φ_1 be the Laplace transform of f and Φ_2 the Laplace transform of S. Then Φ_1 is an entire function of p, bounded on $\mathscr{R}e p \geqslant 0$. The Laplace transform of $D^n(f)$ is $p^n \Phi_1(p)$ and satisfies (5.2). On the other hand, let α be an indefinitely differentiable function which is 0 if $t \leqslant \frac{1}{4}$ and 1 if $t \geqslant \frac{3}{4}$. Then for each $\varepsilon > 0$ and each $n \geqslant 0$, there is a constant $c = c(n, \varepsilon)$ such that each function $\varphi(t) = \alpha(t)\exp -pt$ ($\mathscr{R}e p \geqslant \varepsilon$) satisfies (5.1). This implies $|\langle S, \varphi \rangle| \leqslant C_\varepsilon$ if

$S \in \mathscr{S}'$; if $S \in \mathscr{L}$, we apply the preceding argument to $S(t) \exp(-\varepsilon t/2)$ with $\varepsilon/2$ instead of ε.

The first part of Theorem I is thus proved.

In the second half of Theorem I, the unknown distribution T will be a limit of convolutions $T * f$ as f runs over a sequence f_k of indefinitely differentiable functions of the form $f_k(t) = kg(kt)$, where g is supported by $[0, 1]$ and $\int g(t)\,dt = 1$. From now on, we write f instead of f_k.

Let $F(p)$ be the Laplace transform of f. The Laplace transform of $D^k f$ is $p^k F(p)$. Now the Laplace transform of a complex valued continuous function carried by a compact subset of $[0, +\infty[$ is bounded. Hence $\sup_{\mathscr{Re}\, p \geqslant 0} |p^k F(p)| \leqslant \mathbf{C}_k$.

The product $\Phi_1(p) = F(p)\,\Phi(p)$ has the following behaviour at infinity: for each positive $\varepsilon > 0$ and each $n \geqslant 0$, $|\Phi_1(p)| \leqslant C(n, \varepsilon)\,|p|^{-n}$ when $\mathscr{Re}\, p \geqslant \varepsilon$.

LEMMA 2: *For each real t, the integral* $(1/2\pi) \int_{-\infty}^{+\infty} \Phi_1(u + iv)\, e^{t(u+iv)}\, dv = T_f(t)$ *exists and is independent of* $u > 0$. *This integral vanishes if* $t \leqslant 0$.

Proof of the lemma: Let $u_1 < u_2$ be two such u. We apply Cauchy's theorem on the rectangle whose vertices are $(u_1, -L)$, $(u_2, -L)$, (u_2, L) and (u_1, L). We get

$$i \int_{-L}^{L} \Phi_1(u_2 + iv)\, e^{t(u_2+iv)}\, dv - i \int_{-L}^{L} \Phi_1(u_1 + iv)\, e^{t(u_1+iv)}\, dv$$

$$+ \int_{u_1}^{u_2} \Phi_1(u - iL)\, e^{t(u-iL)}\, du - \int_{u_1}^{u_2} \Phi_1(u + iL)\, e^{t(u+iL)}\, du = 0.$$

The behaviour of Φ_1 at infinity ensures that the last two integrals tend to 0 as L tends to infinity. If $t \leqslant 0$ and $u \geqslant 1$, $|\int_{-\infty}^{+\infty} \Phi_1(u + iv)\, e^{t(u+iv)}\, dv| \leqslant C(2, 1) \int_{-\infty}^{+\infty} |u + iv|^{-2}\, dv$, which tends obviously to 0 as u tends to $+\infty$.

LEMMA 3: *For each fixed* $u > 0$, $e^{-tu} T_f(t) \in \mathscr{S}$. *The Laplace transform of* T_f *is* Φ_1.

Proof: The function $e^{-tu} T_f(t)$ is the inverse Fourier transform of $\Phi_1(u + iv)$. To get estimates of the derivatives of Φ_1, we apply Cauchy's formulas on the rectangle with vertices $(u/2, v/2)$, $(3u/2, v/2)$, $(3u/2, 3v/2)$ and $(u/2, 3v/2)$: the inequalities $|\Phi_1(u/2 + iv)| \leqslant C'(u, n)\,|v|^{-n}$ and $|\Phi_1(3u/2 + iv)| \leqslant C''(u, n)\,|v|^{-n}$ imply that, for each fixed u, each derivative of $\Phi_1(u + iv)$ with respect to v decays rapidly at infinity. Hence $v \to \Phi_1(u + iv)$ belongs to $\mathscr{S}$ and taking the inverse Fourier transform, $e^{-tu}\, T_f(t) \in \mathscr{S}$.

The Fourier-transform of $e^{-tu} T_f(t)$ is $\Phi_1(u + iv)$; the Laplace transform of T_f is therefore Φ_1.

We now replace f by $f_k(t) = kg(kt)$ and we write T_k instead of T_{f_k}.

LEMMA 4: *As k tends to infinity, T_k tends to a distribution T carried by $[0, +\infty[$ in the sense that for each indefinitely differentiable function h with compact support, $\langle T_k, h\rangle \to \langle T, h\rangle$ as $k \to +\infty$.*

Proof: Let ε be any positive real number. The Fourier transform of $e^{-\varepsilon t}T_k$ is $\Phi_1(\varepsilon + iv) = \Phi(\varepsilon + iv)\,\Psi_k(\varepsilon + iv)$, where Ψ_k is the Laplace transform of $kg(kt)$. We remark that $|\Psi_k| \leqslant A$ throughout the half-plane $\mathcal{R}e\, u \geqslant 0$ for some A independent of k and that $\Psi_k \to 1$ uniformly on compact subsets of this half-plane.

The function $e^{\varepsilon t}h(t)$ is the Fourier transform of a function $r(v)$ in $\mathcal{S}$. Hence $\langle T_k, h\rangle = \langle e^{-\varepsilon t}T_k, e^{\varepsilon t}h\rangle = \int_{-\infty}^{+\infty} \Phi(\varepsilon + iv)\,\Psi_k(\varepsilon + iv)\,r(v)\,dv \to \int_{-\infty}^{+\infty} \Phi(\varepsilon + iv)\,r(v)\,dv$ as $k \to +\infty$ (we have $\int_{-\infty}^{+\infty} |\Phi(\varepsilon + iv)|\,|r(v)|\,dv < +\infty$ since Φ grows slowly at infinity and r decays rapidly at infinity and Lebesgue's dominated convergence theorem shows that the limit exists).

If a distribution T is defined on $\mathbf{R}$ by the statement that the inverse Fourier transform of $\Phi(\varepsilon + iv)$ is $e^{-\varepsilon t}\,T(t)$, T would appear to depend on ε. But the preceding argument gives, $\lim_{k\to+\infty} \langle T_k, h\rangle = \langle T, h\rangle$. Hence T does not depend on ε, T is carried by $[0, +\infty[$ and the Laplace transform of T is Φ.

2. The Paley-Wiener theorem

2.1. We list a number of corollaries of theorem I which are interesting in themselves. But first a definition is needed:

DEFINITION 1: *A complex valued function f defined in the complex plane is of exponential type l if, for each positive ε, there exists a constant $C(\varepsilon)$ such that* $|f(z)| \leqslant C(\varepsilon)\exp[(l + \varepsilon)\,|z|]$ for all z.

A similar definition can be given for functions of a real variable.

We say that f is of exponential type if f is of exponential type l for some l in $]0, +\infty[$.

THEOREM II: *Let S be a distribution carried by $[-l, l]$. Then the Laplace transform $F(z)$ of S is an entire function in the complex plane with the following two properties:*

(a) *F is of exponential type l;*

(b) *there exist an integer* $n \geqslant 0$ *and a constant* C *such that* $|F(iy)| \leqslant C(1 + |y|)^n$. *Conversely each entire function* $F(z)$ *satisfying* (a) *and* (b) *is the Laplace transform of a distribution carried by* $[-l, l]$.

2.2. Proof of the first part of Theorem II

If S is carried by $[-l, l]$, we can find, for each positive ε, two continuous functions f and g carried by $[-l, l + \varepsilon]$ and an integer $n \geqslant 0$ such that $S = D^n(f) + g$. The proof is the same as that of Lemma 1. Let $\Phi(z)$ be the Laplace transform of f, G that of g. Then $F(z) = z^n\Phi(z) + G(z)$.

$\Phi(z) = \int_{-l}^{l+\varepsilon} \exp(-zt) f(t)\, dt$ implies $|\Phi(z)| \leqslant \int_{-l}^{l+\varepsilon} \exp(-xt)\, |f(t)|\, dt \leqslant C(\varepsilon) \exp(l + \varepsilon)\, |x|$. The same inequality holds for $G(z)$. Hence

$$|F(z)| \leqslant C(\varepsilon)\,(1 + |z|^n) \exp(l + \varepsilon)\, |x|, \tag{5.3}$$

which implies (a) and (b).

2.3. To prove the converse, let $F(z)$ be an entire function satisfying the inequality (5.3) for each positive ε. Then $\exp[-(l + \varepsilon) z]\, F(z)$ is holomorphic in $\mathcal{R}e z \geqslant 0$ and is bounded by $C(\varepsilon)(1 + |z|^n)$. Theorem I shows that $\exp[-(l + \varepsilon) z] F(z)$ is the Laplace transform of a distribution T carried by $[0, +\infty[$. Hence $F(z)$ is the Laplace transform of a distribution S carried by $[-l - \varepsilon, +\infty[$. In the same way, it can be proved that S is carried by $]-\infty, l + \varepsilon]$. The uniqueness of S shows that it is carried by $[-l, l]$.

It remains to show that (a) and (b) imply (5.3). We use the following lemma of Phragmen–Lindelöf:

LEMMA 5: *Let* Y *be a closed sector in the plane whose opening is less than* π *radians, and let* $f(z)$ *be an holomorphic function on a neighbourhood of* Y. *If* f *is of exponential type on* Y *and if* $|f(z)| \leqslant 1$ *on the frontier of* Y, *then* $|f| \leqslant 1$ *on the whole of* Y.

Proof: Since this statement is invariant under rotation, we can assume that Y is defined by $-\alpha \leqslant \operatorname{Arg} z \leqslant \alpha < \pi/2$. If f tends to 0 when $z \in Y$ and $|z| \to +\infty$, the maximum modulus principle gives $|f(z)| \leqslant 1$ for all z in Y.

If $f(z)$ does not tend to zero at infinity, we apply the preceding argument to $f_\varepsilon(z) = \exp(-\varepsilon z^a) f(z)$, where $a > 1$ and $\alpha a \leqslant \pi/2$. We have $|\exp(-\varepsilon z^a)| = \exp(-\varepsilon\, \mathcal{R}e\,(z^a)) \leqslant \exp(-\varepsilon\, |z|^a \cos \alpha a)$. Hence $|f_\varepsilon(z)| \leqslant 1$ for each frontier point of Y and f_ε tends to zero when $z \in Y$ and $|z| \to +\infty$.

The preceding argument gives $|f_\varepsilon(z)| \leqslant 1$ for all positive ε and all z in Y. Letting ε tend to 0, we get $|f(z)| \leqslant 1$.

COROLLARY 1: *If F is holomorphic in a neighbourhood of $x \geqslant 0$, $y \geqslant 0$, if F is of exponential type and if $|F(x)| \leqslant C \exp lx$, $|F(iy)| \leqslant C(1 + |y|)^n$, then $|F(x + iy)| \leqslant C(1 + |z|)^n \exp lx$.*

Proof: We apply Lemma 5 to $f(z) = C^{-1}(z + i)^{-n} F(z) \exp -lz$. The proof of Paley–Wiener's theorem is complete.

We have proved the following improvement of Paley–Wiener's theorem. If F is an entire function of exponential type (not specified), if there exist an integer $n \geqslant 0$ and a constant A such that $|F(iy)| \leqslant A(1 + |y|)^n$, and if for some positive l there is for each positive ε a constant C_ε such that $|F(x)| \leqslant C_\varepsilon \exp(l + \varepsilon)|x|$, then F is the Laplace transform of a distribution carried by $[-l, l]$.

3. Repartition of roots of entire function of exponential type

In all of what follows, F will be an entire function of exponential type l, so that for each positive ε, there is a constant $A(\varepsilon)$ such that $|F(z)| \leqslant A(\varepsilon) \exp(l + \varepsilon)|z|$ for any complex number z. Let $a_1, \ldots, a_n, \ldots$ be the sequence of all roots of F, ordered in such a way that $|a_n| \leqslant |a_{n+1}|$, and with each root occurring in the sequence $(a_n)_{n \geqslant 1}$ a number of times equal to its multiplicity.

DEFINITION 2: *For each positive real number r, $n(r)$ is the number of roots of F in the closed disc $|z| \leqslant r$; alternatively, $n(r)$ is the* sup *of $n \geqslant 1$ such that $|a_n| \leqslant r$. The function $n(r)$ is defined on $[0, +\infty[$, is integer-valued and is a step function.*

THEOREM III: *If E is an entire function of exponential type l, then* $\limsup R^{-1} \int_0^R n(r)/r \, dr \leqslant l$.

We start with the well-known Jensen's formula: if $F(z)$ is holomorphic on a neighbourhood of $|z| \leqslant R$ and if $0 < |a_1| \leqslant |a_2| \leqslant \cdots \leqslant |a_n|$ is the sequence of roots of F, then

$$\log|F(0)| + \int_0^R \frac{n(r)}{r} \, dr = \frac{1}{2\pi} \int_0^{2\pi} \log|F(Re^{i\theta})| \, d\theta .$$

The proof is given in [16], th. 4.1, p. 181.

We return to Theorem III. If $F(0) = 0$, $F(z) = z^k F_1(z)$, where $F_1(0) \neq 0$ and if $n_1(r)$ is the number of roots of F_1 in the closed disc $|z| \leqslant r$, we have

$n(r) = k + n_1(r)$. Once Theorem III is proved for F_1, this relation at once gives the result for F.

We apply Jensen's formula to F_1 to get that for each positive ε and each $R \geqslant 1$

$$\log |F_1(0)| + \int_0^R \frac{n(r)}{r} \, dr \leqslant \log A(\varepsilon) + (l + \varepsilon) R.$$

Hence

$$\varlimsup_{R \to +\infty} \frac{1}{R} \int_0^R \frac{n(r)}{r} \, dr \leqslant l + \varepsilon.$$

Since this inequality holds for each positive ε, Theorem III is proved.

Remark: If $n(r)/r$ tends to a limit D as r tends to infinity, so does $R^{-1} \int_0^R n(r)/r \, dr$ and Theorem III gives $D \leqslant l$.

In general, let $D = \varliminf_{r \to +\infty} n(r)/r$. Then $\varlimsup_{R \to +\infty} R^{-1} \int_0^R n(r)/r \, dr \geqslant D$ and Theorem III gives $D \leqslant l$.

COROLLARY 2: *Let $(\lambda_k)_{-\infty < k < +\infty}$ be an increasing sequence of real numbers such that $\lim_{|k| \to +\infty} \lambda_k/k = 1/D$ exists. If $l < \pi D$, every distribution S carried by $[-l, l]$ whose Fourier transform vanishes at all the λ_k, $k \in \mathbf{Z}$ is the zero distribution.*

Proof: Apply the Paley–Wiener theorem and Theorem III.

4. Bernstein's inequality

THEOREM IV: *Let φ be a complex valued bounded continuous function on the line whose spectrum*[1] *lies in $[-l, l]$, or equivalently, a complex valued bounded continuous function on the line which is the restriction to* $\mathbf{R}$ *of an entire function of exponential type l.*

Then $\sup_{\mathbf{R}} |\varphi'| \leqslant l \sup_{\mathbf{R}} |\varphi|$.

In particular for each pair of real numbers x' and x'', $|\varphi(x') - \varphi(x'')| \leqslant l \, |x'' - x'| \sup_{\mathbf{R}} |\varphi|$.

Proof: The equivalence of the two definitions of φ is a consequence of Paley–Wiener's theorem.

Let $T/4$ be any real number greater than l and $p(t)$ the T-periodic function defined by $p(t) = 4T^{-1}t$ if $|t| \leqslant T/4$ and $p(t) = -4T^{-1}t + 2$ if

---

[1] In Sections 4 and 5, Fourier-transforms are defined by $\hat{f}(x) = \int_{-\infty}^{+\infty} e^{-ixt} f(t) \, dt$ to be consistent with the definition of Laplace transforms.

$T/4 \leqslant t \leqslant 3T/4$; p is the well-known sawtooth function of physics. Let $q(t) = p(t - T/4)$; the function q is positive-definite and has an absolutely convergent Fourier series: $q(t) = \sum_0^\infty c_k \cos(2k\pi t/T)$, $c_k \geqslant 0$, $\sum_0^\infty c_k = 1$. Hence $p(t) = \sum_{-\infty}^{+\infty} d_k \exp(2k\pi it/T)$, where $\sum_{-\infty}^{+\infty} |d_k| = 1$. If σ is the discrete measure $\sum_{-\infty}^{+\infty} d_k \delta(x + 2k\pi T^{-1})$ where δ is the Dirac measure, then $\hat{\sigma} = p$. Let φ be any complex valued bounded continuous function on the line whose Fourier-transform $S = \hat{\varphi}$ is carried by $[-l, l]$. The Fourier-transform of φ' is $itS(t) = i(T/4)p(t)S(t)$. Hence $\varphi' = i(T/4)\sigma * \varphi$ and $\|\varphi'\|_\infty \leqslant (T/4)\|\sigma\|\|\varphi\|_\infty = (T/4)\|\varphi\|_\infty$. Since this inequality is true for each $T/4 > l$, Bernstein's inequality is proved.

5. Other inequalities of Bernstein type

THEOREM V: *Let $(\lambda_k)_{-\infty<k<+\infty}$ be an increasing sequence of real numbers such that $\lim_{k\to+\infty} (\lambda_{j+k} - \lambda_j)/k = 1/D$ uniformly over $j \in \mathbf{Z}$. For each $l < \pi D$, there is a positive constant C such that for any complex valued bounded continuous function φ whose spectrum lies in $[-l, l]$, we have $\sup_{\mathbf{R}} |\varphi| \leqslant C \sup_{-\infty<k<+\infty} |\varphi(\lambda_k)|$.*

5.1. *Proof of Theorem V*

Theorem V will be deduced from Corollary 1 of Theorem III and standard compactness arguments that we now describe. A set Λ of real numbers is called regular if there exists a $\delta > 0$ such that each interval of length δ contains at most one point of Λ and if there exists an $L > 0$ such that each interval of length L contains at least one point of Λ. A realization of Λ is a presentation of Λ as an increasing sequence λ_k, $-\infty < k < +\infty$, of real numbers. We have $\delta \leqslant \lambda_{k+1} - \lambda_k \leqslant L$ and all other realizations of Λ are of the form $\mu_k = \lambda_{k-k_0}$ $(-\infty < k < +\infty)$. A sequence Λ_m, $m \geqslant 1$, of regular sets of real numbers tends to a regular set Λ if there is a realization $\lambda_{m,k}$, $-\infty < k < +\infty$ of each Λ_m and a realization λ_k, $-\infty < k < +\infty$, of Λ such that $\lambda_{m,k} \to \lambda_k$ for each k. We do not require uniformity of convergence with respect to k.

LEMMA 6: *Let Λ be a regular set of real numbers and x_m, $m \geqslant 1$, a sequence of real numbers. There exist a regular set Λ' of real numbers and a subsequence x'_m of x_m such that $\Lambda - x'_m$ tends to Λ' when m tends to infinity.*

Proof: Let $\lambda_{m,k}$ be a realization of $\Lambda - x_m$ such that, $\lambda_{m,k} \geqslant 0$ for $k \geqslant 0$ and $\lambda_{m,k} < 0$ for $k < 0$. We have $\lambda_{m,0} \leqslant L$, $\lambda_{m,-1} \geqslant -L$ and hence

$0 \leqslant \lambda_{m,k} \leqslant (k+1)L$ if $k \geqslant 0$ and $kL \leqslant \lambda_{m,k} < 0$ if $k < 0$. Cantor's diagonal process yields an infinite increasing sequence $m(j)$ of integers such that $\lambda_{m(j),k} \to \lambda'_k$ for each $k \geqslant 0$. For all $k \in \mathbf{Z}$, $0 < \delta \leqslant \lambda_{m,k+1} - \lambda_{m,k} \leqslant L$; hence $0 < \delta \leqslant \lambda'_{k+1} - \lambda'_k \leqslant L$ and the set Λ' of all λ'_k, $-\infty < k < +\infty$ is a regular set of real numbers.

The preceding proof shows that if Λ has a realization λ_k, $k \in \mathbf{Z}$, such that $\lim\limits_{k \to +\infty} (\lambda_{j+k} - \lambda_j)/k = 1/D$ uniformly over $j \in \mathbf{Z}$, so does Λ': for each $\varepsilon > 0$, there is an integer $k(\varepsilon)$ such that $k \geqslant k(\varepsilon)$ implies $(D^{-1} - \varepsilon)\,k \leqslant \lambda_{j+k} - \lambda_j \leqslant (D^{-1} + \varepsilon)\,k$. But we can find an integer l depending only on m such that for each k in $\mathbf{Z}$, $\lambda_{m,k} = \lambda_{k+l} - x_m$. Hence for each j in $\mathbf{Z}$, each $m \geqslant 1$ and each $k \geqslant k(\varepsilon)$, we have $(D^{-1} - \varepsilon)\,k \leqslant \lambda_{m,j+k} - \lambda_{m,j} \leqslant (D^{-1} + \varepsilon)\,k$. Passing to the limit, we get $(D^{-1} - \varepsilon)\,k \leqslant \lambda'_{j+k} - \lambda'_j \leqslant (D^{-1} + \varepsilon)\,k$ for $j \in \mathbf{Z}$ and $k \geqslant k(\varepsilon)$, which is the required inequality.

5.2. To prove Theorem V, let E be the Banach space of complex valued continuously differentiable functions on $[-2l, 2l]$ and $\mathscr{C}$ the convex set of all distributions S carried by $[-l, l]$ whose Fourier-transforms $\hat{S}$ satisfy the inequality $\sup\limits_{\mathbf{R}} |\hat{S}| \leqslant 1$. Then $\mathscr{C}$ is a compact subset of the dual space E^* of E with respect to the weak-star topology $\sigma(E^*, E)$. For each real t the mapping from $\mathscr{C}$ to the complex field defined by $S \to \hat{S}(t)$ is continuous.

If a sequence $(S_m)_{m \geqslant 1}$ of distributions carried by $[-l, l]$ exists such that $\sup\limits_{\mathbf{R}} |\hat{S}_m| = 1$ while $\sup\limits_{-\infty < k < +\infty} |\hat{S}_m(\lambda_k)| \leqslant 1/m$, let x_m be a real number such that $|\hat{S}_m(x_m)| \geqslant \frac{1}{2}$. Let T_m be the distribution $e^{-ix_m x}\,T$; T_m belongs to $\mathscr{C}$, $|\langle T_m, 1\rangle| \geqslant \frac{1}{2}$ and $|\hat{T}_m(\lambda_k - x_m)| \leqslant 1/k$. We can find an increasing sequence m_j of integers such that $\Lambda - x_{m(j)} \to \Lambda'$ and $T_{m(j)} \to T'$ as j tends to infinity.

Bernstein's theorem shows that for any two real numbers x' and x'' and any S in $\mathscr{C}$, $|\hat{S}(x') - \hat{S}(x'')| \leqslant |x' - x''|$. Applying this inequality to $x' = \lambda'_k \in \Lambda'$ and $x'' = \lambda_{m,k} \in \Lambda - x_m$, whose limit is λ'_k, we get, writing m instead of m_j, $\hat{T}_m(\lambda'_k) \to 0$ as $m \to +\infty$; hence $\hat{T}'(\lambda'_k) = 0$ for all k in $\mathbf{Z}$. The set Λ' of all λ'_k has a uniform density D and $D < \pi l$. Corollary 2 implies $T' = 0$. This is impossible since $|\langle T, 1\rangle| = \lim\limits_{m \to +\infty} |\langle T_m, 1\rangle| \geqslant \frac{1}{2}$. Hence Theorem V is proved. For a better result, the reader is referred to [54].

6. Irrational numbers and special trigonometric series

6.1. Let θ be an irrational number >1 and, for each positive integer k, let $[\theta k]$ be the integer m defined by $0 < \theta k - m < 1$; m is the integral part of θk. We define $\Lambda(\theta)$ to be the set of all $[\theta k]$, $k \in \mathbf{Z}$.

DEFINITION 3: *For each irrational number $\theta > 1$, $\mathscr{C}(\theta)$ is the Banach space of all complex valued continuous 1-periodic functions whose spectra lie in the set $\Lambda(\theta)$ of all integral parts $[k\theta]$, $k \in \mathbf{Z}$. The Fourier series of f is $\sum_{-\infty}^{+\infty} a_k \exp 2\pi i\, [k\theta]\, t$.*

THEOREM VI *(properties of $\mathscr{C}(\theta)$):*

(1) *$\mathscr{C}(\theta)$ is a closed, translation invariant space of 1-periodic functions on the line;*

(2) *if Ω is any open subset of* $[0, 1]$ *of Lebesgue measure greater than* $1/\theta$, *each function in $\mathscr{C}(\theta)$ is completely defined by its values on Ω, as the following properties show:*

(a) $\sup\limits_{[0,1]} |f| \leqslant C(\Omega) \sup\limits_{\Omega} |f|$;

(b) *if f is p-times continuously differentiable on Ω, f is p times continuously differentiable throughout the real line;*

(c) *if f is analytic on Ω, f is analytic on the real line;*

(3) *if K is any compact subset of* $[0, 1]$ *of Lebesgue measure less than* $1/\theta$, *any complex valued continuous function on K is the restriction to K of some function in $\mathscr{C}(\theta)$.*

A p-adic version of Theorem VI will be given in Theorem IX. Theorem VI fails if θ is rational.

6.2. Sketch of the proof of (a)

The proof is based on the following remark: under the injective homomorphism h from $\mathbf{Z}$ to $\mathbf{T}$ defined by $h(j) = \exp 2\pi i j \theta^{-1}$, the image of $\mathbf{Z}$ is a dense subgroup of $\mathbf{T}$, and $\Lambda(\theta)$ is mapped onto a dense subset of the arc I of $\mathbf{T}$ defined by the parametric representation $z = \exp -2\pi i t$, $0 \leqslant t \leqslant 1/\theta$. If f in a function defined on $\mathbf{T}$ whose spectrum lies in $\Lambda(\theta)$ its composition with this homomorphism is a function g defined on $\mathbf{Z}$ whose spectrum lies in I and we are reduced to Theorem V.

6.3. A computation of densities

Before going into the details, some remarks should be made. An open subset Ω of $[0, 1]$ whose Lebesgue measure is greater than $1/\theta$ contains a finite union of intervals U whose Lebesgue measure is greater than $1/\theta$. Once (a) is proved for U, it is proved for Ω.

Let M be the set of all integers m in $\mathbf{Z}$ such that $h(m) \in U$.

For any sequence $(c_j)_{j \in \mathbf{Z}}$ of complex numbers, we put

$$\mathfrak{M}\,\{c_j;\, m \leqslant j \leqslant n\} = \frac{c_m + \cdots + c_n}{n - m + 1}.$$

With this terminology we can state two lemmas:

Lemma 7: *For any complex valued continuous function f defined on* $\mathbf{T}$, $\lim_{k \to +\infty} \mathfrak{M}\,\{f(h(j));\, m \leqslant j \leqslant m + k\} = (1/2\pi) \int_0^{2\pi} f(e^{i\theta})\, d\theta$, *uniformly over* $m \in \mathbf{Z}$.

Since $|1/2\pi \int_0^{2\pi} f(e^{i\theta})\, d\theta| \leqslant \|f\|_\infty$ and $|\mathfrak{M}\,\{f(h(j));\, m \leqslant j \leqslant m + k\}| \leqslant \|f\|_\infty$, it suffices to consider the dense subspace of finite trigonometric sums; by linearity it is sufficient to examine the case where $f(e^{i\theta}) = e^{ip\theta}$, $p \in \mathbf{Z}$. In this case an easy computation gives the result.

Lemma 8: *For any complex valued Riemann integrable function g on* $\mathbf{T}$ *the same result holds.*

Proof: By linearity we can confine our attention to real g. For each positive ε, there are two continuous functions f_1 and f_2 on $\mathbf{T}$ such that $f_1 \leqslant g \leqslant f_2$ and $(1/2\pi) \int_0^{2\pi} (f_2 - f_1)(e^{i\theta})\, d\theta \leqslant \varepsilon$. Hence

$$\lim_{k \to +\infty} \mathfrak{M}\,\{f_1(h(j));\quad m \leqslant j \leqslant m + k\} \leqslant$$

$$\liminf_{k \to +\infty} \mathfrak{M}\,\{g(h(j));\quad m \leqslant j \leqslant m + k\} \leqslant$$

$$\limsup_{k \to +\infty} \mathfrak{M}\,\{g(h(j));\quad m \leqslant j \leqslant m + k\} \leqslant$$

$$\lim_{k \to +\infty} \mathfrak{M}\,\{f_2(h(j));\quad m \leqslant j \leqslant m + k\}.$$

But the two extreme terms of this inequality differ from $(1/2\pi) \int_0^{2\pi} f(e^{i\theta})\, d\theta$ by an error term not exceeding ε (Lemma 7). Hence $\lim_{k \to +\infty} \mathfrak{M}\,\{g(h(j));\, m \leqslant j \leqslant m + k\} = (1/2\pi) \int_0^{2\pi} g(e^{i\theta})\, d\theta$ uniformly on $m \in \mathbf{Z}$.

Lemma 9: *The uniform density of the set M of integers is equal to the Lebesgue measure of U.* (Note that the Lebesgue measure of the whole of $\mathbf{T}$ is 1.)

Proof: We apply Lemma 8 to the characteristic function of U.

6.4. *Proof of* (a)

To prove (a) for all f in $\mathscr{C}(\theta)$, it suffices to prove (a) for finite trigonometric sums $P(t) = \sum c_k \exp 2\pi i\, [k\theta]\, t$. Moreover, each open subset Ω of $[0, 1]$ whose measure is greater than $1/\theta$ contains a finite union U of closed intervals, the sum of whose lengths is greater than $1/\theta$. If we prove (a) for U instead of Ω, the inequality with Ω will hold a fortiori.

Let $h : \mathbf{Z} \to \mathbf{T}$ be the homomorphism defined by $h(j) = \exp 2\pi i j\theta^{-1}$. The dual homomorphism $h^* : \mathbf{Z} \to \mathbf{T}$ is also defined by $h^*(j) = \exp 2\pi i j\theta^{-1}$. Since $h^*(\mathbf{Z})$ is a dense subgroup H of $\mathbf{T}$, we have $\sup_{[0,1]} |P| = \sup_{H} |P|$ and $\sup_{U} |P| = \sup_{U \cap H} |P|$.

Let M be the set of all integers $j \in \mathbf{Z}$ such that $h^*(j) \in U$ and let μ be the measure carried by $\mathbf{T}$ and assigning the mass c_k to the point $\exp -2\pi i\theta^{-1}[k\theta]$. An easy computation gives $\hat{\mu}(j) = P\,(h(^*j))$. To get (a), we have to prove that $\sup_{\mathbf{Z}} |\hat{\mu}(j)| \leqslant C \sup_{M} |\hat{\mu}(j)|$ for all finite linear combinations of Dirac measures concentrated in points of the form $\exp -2\pi i\theta^{-1}\, [k\theta]$.

We now make the following crucial observation: let I be the arc of $\mathbf{T}$ defined by the parametric representation $z = \exp - 2\pi i t$, $0 \leqslant t \leqslant 1/\theta$. Then $h(\Lambda)$ is contained in I; the image of the set $\Lambda(\theta)$ under the homomorphism h lies in an interval. Writing $[k\theta] = k\theta - r$ where $0 < r < 1$, we get $\theta^{-1}\, [k\theta] \equiv -\theta^{-1} r \pmod 1$, which proves our remark.

Using the local inverse of the momomorphism $t \to 2\pi i t$, we now replace I by the interval $J = [-(1/\theta), 0]$; μ becomes a measure ν carried by $-J$ and all that remains is to show that $\sup_{2\pi\mathbf{Z}} |\hat{\nu}(t)| \leqslant C \sup_{2\pi M} |\hat{\nu}(t)|$.

Theorem V shows that to get this inequality, it suffices to show that $2\pi M$ is a regular set whose uniform density is greater than $2\pi/\theta$. Lemma 9 shows that this is the case and (a) is proved.

Let Ω be an open subset of $[0, 1]$ of Lebesgue measure greater than $1/\theta$. If a function in $\mathscr{C}(\theta)$ is zero on Ω, this function is zero everywhere. If two functions of $\mathscr{C}(\theta)$ agree on Ω, they are equal everywhere.

6.5. *Proof of* (b) *and* (c)

Let f be a function of $\mathscr{C}(\theta)$ continuously differentiable on an open subset Ω of $[0, 1]$ whose Lebesgue measure is greater than $1/\theta$. Let U be a finite union of closed intervals contained in Ω the sum of whose lengths is greater than $1/\theta$. On U, the family of functions $g\,(x) = t^{-1}\, [f(x + t)$

$- f(x)]$ of $\mathscr{C}(\theta)$ tends uniformly to $f'(x)$ as t tends to 0. Theorem VI, (a), implies uniform convergence over the whole of $\mathbf{R}$; hence f is continuously differentiable throughout $\mathbf{R}$.

Similarly, if f is k times continuously differentiable on Ω and if f belongs to $\mathscr{C}(\theta)$, then f is k times continuously differentiable on $\mathbf{R}$ and $\sup_{\mathbf{R}} |f^{(k)}| \leqslant A \sup_{\Omega} |f^{(k)}|$, where A is independent of $k \geqslant 0$.

Assume that f is analytic on Ω and let U be as above. The inequalities $|f^{(k)}(x)| \leqslant C^k k!$ on U imply $|f^{(k)}(x)| \leqslant AC^k k!$ on the whole of $\mathbf{R}$ and f is analytic throughout $\mathbf{R}$.

6.6. *Two remarks*

Remark 1: *For each open subset Ω of* [0, 1] *whose Lebesgue measure is greater than* $1/\theta$, *there is a constant C such that for any function f in $\mathscr{C}(\theta)$, $\int_0^1 |f(x)|^2\, dx \leqslant C \int_\Omega |f(x)|^2\, dx$.*

Proof of this remark: Let U be a finite union of closed intervals of [0, 1] such that, for a positive ε, $U + [-\varepsilon, \varepsilon]$ is contained in Ω, and such that the Lebesgue measure of U is greater than $1/\theta$. Let g be any complex valued continuous function whose support lies in $[-\varepsilon, \varepsilon]$ and such that $\|g\|_2 = (\int_{-\varepsilon}^{\varepsilon} |g(t)|^2\, dt)^{1/2} \leqslant 1$. For each f in $\mathscr{C}(\theta)$, the convolution product $f * g$ lies in $\mathscr{C}(\theta)$ and, by part (a) of Theorem VI, for each real x,

$$|(f * g)(x)| \leqslant A \sup_U |(f * g)(t)| \leqslant A \left(\int_\Omega |f|^2\, dt \right)^{1/2}.$$

Taking the sup with respect to g, we obtain, for each real x,

$$\int_{x-\varepsilon}^{x+\varepsilon} |f|^2\, dt \leqslant A^2 \int_\Omega |f|^2\, dt$$

and

$$\int_0^1 |f|^2\, dt = \lim_{T \to +\infty} T^{-1} \int_0^T |f|^2\, dt \leqslant \frac{A^2}{2\varepsilon} \int_\Omega |f|^2\, dt.$$

Remark 2: *For each open subset Ω of* [0, 1] *whose Lebesgue measure is greater than* $1/\theta$, *there is a constant C such that for any complex valued Radon measure μ whose spectrum lies in $\Lambda(\theta)$, $\|\mu\| \leqslant C \int_\Omega d|\mu|(t)$.*

The proof of this second remark is similar to that of the preceding one: the set U and the positive real number ε are defined as above, the test function g is any continuous complex valued function supported by

$[-\varepsilon, \varepsilon]$ and such that $\sup_{[-\varepsilon,\varepsilon]} |g(t)| \leqslant 1$. Then

$$|(\mu * g)(x)| \leqslant A \sup_U |(\mu * g)(t)| \leqslant A \int_\Omega d|\mu|(t).$$

Taking the sup with respect to g, we have, for each real x, $\int_{x-\varepsilon}^{x+\varepsilon} d|\mu|(t) \leqslant A \int_\Omega d|\mu|(t)$ and the proof ends as above.

6.7. *Proof of the last part of Theorem VI*

Let K be a compact subset of $[0, 1]$ whose Lebesgue measure is less than $1/\theta$. We want to show that any continuous function on K is the restriction to K of a function in $\mathscr{C}(\theta)$.

Let $\mathscr{C}[0, 1]$ be the Banach space of all complex valued continuous functions f on $[0, 1]$ such that $f(0) = f(1)$, let $\mathscr{C}(\theta)$ be the closed subspace of $\mathscr{C}[0, 1]$ defined above and let F be the quotient space $\mathscr{C}[0, 1]/\mathscr{C}(\theta)$

Let E be the closed subspace of $\mathscr{C}[0, 1]$ consisting of all f in $\mathscr{C}[0, 1]$, which vanish on K.

We can define a continuous linear mapping T from E to F: T is the injection from E into $\mathscr{C}[0, 1]$ composed with the projection from $\mathscr{C}[0, 1]$ onto $\mathscr{C}[0, 1]/\mathscr{C}(\theta)$.

If T is surjective, Theorem VI *is proved.* In fact this assumption means that, for any function f in $\mathscr{C}[0, 1]$ there is a function g in $\mathscr{C}(\theta)$ and a function h vanishing on K such that $f = g + h$; g is the required extension of f restricted to K.

At this stage we use a theorem due to Banach:

LEMMA 10: *Let E and F be two Banach spaces and T a continuous linear map from E to F. Let E' and F' be the (strong) dual spaces of E and F and T' be the adjoint operator from F' to E'. Assume that there exists a constant $c > 0$ such that for each y' in F', $\|T'y'\|_{E'} \geqslant c\|y'\|_{F'}$. Then T is surjective.*

Proof of Lemma 10: Let B_1 be the unit ball of E and B_2 the unit ball of F. It suffices to show that $T(B_1)$ is dense in cB_2 since if this is true, each y in $T(B_1)$ can be written $y = T(x)$, where $\|x\|_E \leqslant c^{-1}\|y\|_F$ and writing any y in cB_2 as a series $\sum_1^\infty y_k$, where $y_k \in T(B_1)$ and $\sum_1^\infty \|y_k\|_F < +\infty$, the corresponding series $\sum_1^\infty x_k$ will converge in the Banach space B_1 to an x such that $T(x) = y$. To prove that $T(B_1)$ is dense in cB_2, let $\mathscr{C}$ be the closure of $T(B_1)$ in F; $T(B_1)$ is a convex subset of F invariant under multiplication by complex numbers of modulus $\leqslant 1$ and so is $\mathscr{C}$. Hence $\mathscr{C}$ is defined by a family of linear inequalities $|\langle y, y'\rangle| \leqslant 1$, where y' belongs to a subset A of F'. Since each y of the form $T(x)$, $\|x\| \leqslant 1$, belongs to $\mathscr{C}$, we have $\|T'(y')\|_{E'} \leqslant 1$; hence $\|y'\|_{F'} \leqslant c^{-1}$ for each y' in A and $\mathscr{C}$ contains cB_2.

To apply Lemma 10, E', F' and T' must be identified. Let M [0, 1] be the Banach space of all complex valued Radon measures on [0, 1] and $M(K)$ the subspace of all μ in M [0, 1] which are supported by K. Then the dual space of E is the quotient Banach space M [0, 1]/$M(K)$, whose elements are complex valued Radon measures μ on [0, 1] such that $\int_K d|\mu|(t) = 0$ and the norm of such a μ in this quotient space is the total variation of μ.

The dual space F' of F is the closed subspace of M [0, 1] defined by the condition that $\hat{\mu}(\lambda) = 0$ for all λ in $\Lambda(\theta)$: F' is the annihilator of $\mathscr{C}(\theta)$ in M [0, 1].

Let Λ' be the complement of Λ in $\mathbf{Z}$ and $\mathscr{T}$ the arc of $\mathbf{T}$ defined by the parametric relation $z = \exp -2\pi it$, $1/\theta < t < 1$. Then Λ' is the set of all integers j in $\mathbf{Z}$ such that $h(j) \in \mathscr{T}$ and everything that was done with Λ can be done with Λ'. As a special case, Remark 2 after Theorem VI gives here: for each open subset Ω of [0, 1] of Lebesgue measure greater than $1 - 1/\theta$, there is a constant C such that, for each μ in F', $\|\mu\| \leqslant C \int_\Omega d\,|\mu(t)|$. Applying this inequality to $\Omega = \complement_{\mathbf{T}} K$, we get $\|\mu\|_{F'} \leqslant C\|\mu\|_{E'}$ for each μ in F'; T is surjective and Theorem VI is now completely proved.

7. Special trigonometric series: the a-periodic case

Let a, b, α and β be four real numbers. Assume that a and b are different from 0, that b/a is irrational and that $\alpha b - a\beta \neq 0$. Let l be the homomorphism from $\mathbf{R}^2$ into $\mathbf{R}$ defined by $l(x, y) = \alpha x + \beta y$, let $\Gamma \subset \mathbf{Z}^2$ be the set of all pairs (m, n) of rational integers such that $|am + bn| \leqslant 1$ and put $\Lambda = l(\Gamma)$; Λ is a harmonious set of real numbers and it may be expected that almost-periodic functions whose spectra lie in Λ enjoy nice properties.

We shall study the Banach space $\mathscr{C}_\Lambda$ of all complex valued almost periodic functions f whose spectra lie in Λ.

Firstly, what are the compact subsets K of $\mathbf{R}$ where the functions of $\mathscr{C}_\Lambda$ are determined?

7.1. THEOREM VII: *Let $K \subset \mathbf{R}$ be a Riemann-integrable compact set of real numbers. If the measure of K is greater than $1/|a\beta - \alpha b|$, there exists a constant A such that for each f in $\mathscr{C}_\Lambda$,*

(a) $\sup_{x \in \mathbf{R}} |f(x)| \leqslant A \sup_{x \in K} |f(x)|$;

(b) *if* $f \in \mathscr{C}_\Lambda$ *is p-times continuously differentiable on K, then f is p-times continuously differentiable throughout the real line;*

(c) *if f is analytic on a neighbourhood of K, f is analytic throughout the real line;*

(d) *if K is a compact set of real numbers whose measure is less than* $1/|a\beta - \alpha b|$, (a) *and* (b) *are false.*

Before proving this theorem, some comments should be made. If ε is any positive real number and K is the union of the two intervals $I_0 = [0, |a\beta - \alpha b|^{-1} - \varepsilon]$ and $\mathscr{T}_m = [m - \varepsilon, m + \varepsilon]$, then, even if m is very large and the two intervals are very far from each other, the values of f on K determine f completely. On the other hand the knowledge of f on I_0 is not sufficient to reveal any substantial information. The role played by the very small and distant interval $\mathscr{T}_m$ is indeed curious.

The set Λ of all finite sums $\sum_{k\geqslant 0} \varepsilon_k \theta^k$, $\varepsilon_k = 0$ or 1, $\theta = (1 + \sqrt{5})/2$ can also be defined as the set of all positive real numbers $m + n\theta = \lambda$, $m \in \mathbf{Z}$, $n \in \mathbf{Z}$ such that $-1 \leqslant m + n\bar{\theta} \leqslant \theta$ where $\bar{\theta} = (1 - \sqrt{5})/2$. In this case the critical measure $1/|a\beta - \alpha b|$ is $\frac{1}{2} + 3\sqrt{5}/10$.

7.2. *Computations of densities*

To obtain a better understanding of the number $1/|a\beta - \alpha b|$, the following lemma will be useful:

LEMMA 11: *Each interval of real numbers of length T contains* $T\,|a\beta - \alpha b| + o(T)$ *points of* Λ; $o(T)/T$ *tends to* 0 *uniformly with respect to the position of the interval as T tends to infinity.*

To prove this lemma, we in fact consider a more general situation which is of interest in itself; the corresponding result will be proved for a large class of harmonious and relatively dense sets of real numbers.

Definition of models:

Let n be an integer greater than 1;

U, a compact subset of $\mathbf{R}^{n-1}$;

$a_1, \ldots, a_n$, n real numbers linearly independent over $\mathbf{Q}$;

I, the linear form defined over $\mathbf{R}^n$ by $I(x_1, \ldots, x_n) = a_1x_1 + \cdots + a_nx_n$;

$L_1, \ldots, L_{n-1}$, $n - 1$ other linear forms such that $L_1, \ldots, L_{n-1}$, I constitute a basis of the dual of $\mathbf{R}^n$;

$(L_1^*, \ldots, L_n^*)$ is the isomorphism of $\mathbf{R}^n$ which is the adjoint of the inverse of the isomorphism L of $\mathbf{R}^n$ defined by $\mathbf{x} \to (L_1(\mathbf{x}), \ldots, L_{n-1}(\mathbf{x}), I(\mathbf{x}))$,

$\mathbf{x} \in \mathbf{R}^n$: for each $\mathbf{x}$ and $\mathbf{y}$ in $\mathbf{R}^n$, $L_1^*(\mathbf{x}) L_1(\mathbf{y}) + \cdots + L_{n-1}^*(\mathbf{x}) L_{n-1}(\mathbf{y}) + L_n^*(\mathbf{x}) I(\mathbf{y}) = \mathbf{x} \cdot \mathbf{y}$.

We assume that the coefficients of L_n^* are linearly independent over $\mathbf{Q}$.

Under these assumptions, we construct a set of real numbers which is called a model (Chapter II, Section 6); Λ is the set of all real $\lambda = a_1 k_1 + \cdots + a_n k_n$ for all $\mathbf{k} = (k_1, \ldots, k_n)$ in $\mathbf{Z}^n$ such that the element $(L_1(\mathbf{k}), \ldots, L_{n-1}(\mathbf{k}))$ of $\mathbf{R}^{n-1}$ belongs to U.

7.3. *A Poisson formula with remainder term for a model (computation of densities)*

In the following theorem U and Λ are defined just as above; ε is any positive real number and U_ε is the set of all $\mathbf{x}$ in $\mathbf{R}^{n-1}$ whose distance from U does not exceed ε. Finally Λ_ε is defined using $L_1, \ldots, L_{n-1}$, I and U_ε.

THEOREM VIII: *Assume that U is Riemann integrable, i.e. that the frontier of U has Lebesgue measure zero. For each positive ε there exists a complex measure μ_ε such that*

(1) $\sup\limits_{-\infty < x < +\infty} \int_x^{x+1} d|\mu_\varepsilon|(t) < +\infty$;

(2) *the Fourier transform $\mathfrak{F}\mu_\varepsilon$ of μ_ε is also a measure; $\mathfrak{F}\mu_\varepsilon$ is equal to the sum $\sum_{\lambda \in \Lambda} \delta(x - \lambda) + R_\varepsilon(x)$*, where $\delta(x)$ *is the unit mass concentrated in* $\{0\}$;

(3) *the remainder term $R_\varepsilon(x)$ has the form $\sum_{\lambda \in \Lambda_\varepsilon \setminus \Lambda} \alpha(\lambda)\, \delta(x - \lambda)$ where $0 \leqslant \alpha(\lambda) \leqslant 1$;*

(4) $\overline{\lim\limits_{T \to +\infty}}\ 1/2T \int_{x-T}^{x+T} d\,|R_\varepsilon(x)| \to 0$ as $\varepsilon \to 0$, *uniformly over all real x.*

This means that the Fourier transform of μ_ε is the sum of unit masses concentrated in each point of Λ together with an error term. This error term consists of positive masses, less than one, concentrated in points of $\Lambda_\varepsilon \setminus \Lambda$[1]. The mean value of this error term is arbitrarily small as ε tends to zero.

Now to the proof. We start by defining a function α which will be extended to the whole of Λ_ε by the rule $\alpha(\lambda) = 1$ if $\lambda \in \Lambda$. Let φ be a C^∞ function on $\mathbf{R}^{n-1}$, which is 1 on U and 0 outside U_ε. If $\lambda = a_1 k_1 + \cdots + a_n k_n$ we put $\mathbf{k} = (k_1, \ldots, k_n) \in \mathbf{Z}^n$, $\alpha(\lambda) = \varphi(L_1(\mathbf{k}), \ldots, L_{n-1}(\mathbf{k}))$ and β is a C^∞ function defined on $\mathbf{R}^n$ by $\beta(\mathbf{x}) = \varphi(L^1(\mathbf{x}), \ldots, L_{n-1}(\mathbf{x}))$.

Let $\nu = \sum_{\lambda \in \Lambda_\varepsilon} \alpha(\lambda)\, \delta(x - \lambda)$ where $\delta(x)$ is the unit mass concentrated in $\{0\}$ and let us compute $\mathfrak{F}\nu$. If f is any test function in $\mathscr{D}$, we have

$$\int_{\mathbf{R}} f(x)\, d\nu(x) = \sum_{\mathbf{Z}^n} \beta(k_1, \ldots, k_n) f(a_1 k_1 + \cdots + a_n k_n).$$

[1] $\Lambda_\varepsilon \setminus \Lambda$ is the complement of Λ in Λ_ε.

The function $\beta(x_1, \dots, x_n) f(a_1 x_1 + \cdots + a_n x_n)$ decays rapidly at infinity over $\mathbf{R}^n$; its Fourier transform $\varrho(x_1, \dots, x_n) = (\det L)^{-1} \hat{\varphi}(L_1^*, \dots, L_{n-1}^*) \hat{f}(L_n^*)$ also decays rapidly at infinity.

At this stage, the well-known Poisson formula will be applied.

If $\varrho : \mathbf{R}^n \to \mathbf{C}$ belongs to $\mathscr{S}$ and if $\hat{\varrho}(y_1, \dots, y_n) = \int_{\mathbf{R}^n} \exp -2\pi i (x_1 y_1 + \cdots + x_n y_n) \varrho(x_1, \dots, x_n)\, dx_1 \cdots dx_n$, then $\sum_{\mathbf{Z}^n} \varrho(\mathrm{k}) = \sum_{\mathbf{Z}^n} \hat{\varrho}(\mathrm{k})$.

In our case this equality gives

$$\int_{\mathbf{R}} f(x)\, d\nu(x) = \sum_{\mathbf{Z}^n} \varrho(\mathbf{k}) = \int_{\mathbf{R}} \hat{f}\, d\mu,$$

where μ is the atomic measure assigning the mass $(\det L)^{-1} \hat{\varphi}(L_1^*, \dots, L_{n-1}^*)(k_1, \dots, k_n)$ to the point $L_n^*(k_1, \dots, k_n)$. We have

$$\int_x^{x+1} d|\mu(t)| = (\det L)^{-1} \sum |\hat{\varphi}|(L_1^*, \dots, L_{n-1}^*),$$

where the latter sum is extended over all $\mathbf{k}$ in $\mathbf{Z}^n$ such that $L_n^*(\mathbf{k})$ belongs to $[x, x+1]$. Let θ be the characteristic function of $[0, 1]$. Then

$$\int_x^{x+1} d|\mu(t)| = (\det L)^{-1} \sum_{\mathbf{Z}^n} |\hat{\varphi}|(L_1^*, \dots, L_{n-1}^*)\, \theta(L_n^* - x).$$

Since the linear forms $L_1^*, \dots, L_n^*$ form a basis of the dual space of $\mathbf{R}^n$, there exists a $\xi \in \mathbf{R}^n$ such that $L_1^*(\xi) = \cdots = L_{n-1}^*(\xi) = 0$ and $L_n^*(\xi) = x$. Let ω be the function defined on $\mathbf{R}^n$ by $\omega = (\det L)^{-1} |\hat{\varphi}|(L_1^*, \dots, L_{n-1}^*)\, \theta(L_n^*)$. With this notation $\int_x^{x+1} d|\mu(t)| = \sum_{\mathbf{Z}^n} \omega(\mathbf{k} - \xi)$. Since ω decays rapidly at infinity, we have $\sup_{\xi \in \mathbf{R}^n} \sum_{\mathbf{Z}^n} \omega(\mathbf{k} - \xi) < +\infty$ and hence $\sup_{x \in \mathbf{R}} \int_x^{x+1} d|\mu(t)| < +\infty$.

We now have to prove (4). We need the following notion: for any compact subset V of $\mathbf{R}^{n-1}$, let Λ_V be defined by $(L_1, \dots, L_{n-1}, I)$ and V. We define $\overline{\text{dens}}\, \Lambda_V$ to be $\overline{\lim}_{T \to +\infty} (2T)^{-1} \sup_{x \in \mathbf{R}} (\text{Card}\, [x - T, x + T] \cap \Lambda_V)$.

LEMMA 12: $\overline{\text{dens}}\, \Lambda_V \leqslant (\det L)^{-1}$ meas V.

Proof: Let ψ be any indefinitely differentiable function defined on $\mathbf{R}$, whose support lies in $[0, 1]$ and such that $\int_0^1 \psi(t)\, dt = 1$. We construct, using V, a measure μ_ε just as we did with U. Then, since R_ε is a positive measure, (2) gives

$$\overline{\text{dens}}\, \Lambda_V \leqslant \lim_{T \to +\infty} \frac{1}{2T} \int_{x-T}^{x+T} (\mathfrak{F}\mu_\varepsilon) * \psi\, dx = \mu_\varepsilon\{0\}, \text{ uniformly over } x \in \mathbf{R}.$$

The last equality follows from the observation that $(\mathfrak{F}\mu_\varepsilon) * \psi$ is an almost periodic function whose Fourier coefficient for the frequency x is $\hat{\psi}(x)\mu_\varepsilon\{-x\}$.

Letting ε tend to zero we get Lemma 12.

Let V be the closure of $U_\varepsilon\backslash U$. We have $\overline{\text{dens}}\,(\Lambda_\varepsilon\backslash\Lambda) \leqslant (\det L)^{-1}$ meas $\overline{(U_\varepsilon/U)}$ and since the measure of the frontier of U is 0, $\lim\limits_{\varepsilon\to 0} \overline{\text{dens}}\,\Lambda_\varepsilon\backslash\Lambda = 0$. Since the masses of R_ε lie in $]0, 1[$ and are concentrated in points of $\Lambda_\varepsilon\backslash\Lambda$, (4) follows.

COROLLARY 3: *If U is Riemann integrable,* $\lim\limits_{T\to+\infty}(1/2T)$ Card $([x - T, x + T] \cap \Lambda_U) = (\det L)^{-1}$ meas U, *uniformly over all real* x.

7.4. *Proof of Theorem VII*

Let h be the homomorphism from $\mathbf{R}$ to $\mathbf{T}^2$ defined by $h(t) = (\exp 2\pi i\alpha t, \exp 2\pi i\beta t)$. To prove (a), it is sufficient to verify (a) when $f(t)$ is a finite sum $\sum_{\lambda\in\Lambda} a_\lambda \exp 2\pi i\lambda t$ since these sums are dense in $\mathscr{C}_\Lambda$ with respect to the topology of uniform convergence over the whole of $\mathbf{R}$. Such a sum can be written $P \circ h$, where P is defined on $\mathbf{T}^2 = \mathbf{R}^2/\mathbf{Z}^2$ by $P(x, y) = \sum_{(m,n)\in\Gamma} a_{m,n} \exp 2\pi i\,(mx + ny)$.

For all (x_0, y_0) in $\mathbf{R}^2$ and all real t, we have

$$P(x_0 + at, y_0 + bt) = \sum_{(m,n)\in\Gamma} a'_{m,n} \exp 2\pi i\,(am + bn)\,t = g(t),$$

where $a'_{m,n} = a_{m,n} \exp 2\pi i\,(mx_0 + ny_0)$.

This function $g(t)$ is the Fourier transform of a measure carried by $[-1, 1]$. Let $h' : \mathbf{R} \to \mathbf{T}^2$ be the affine map defined by $h'(t) = (\exp 2\pi i\,(x_0 + at), \exp 2\pi i\,(y_0 + bt))$. Since b/a is irrational, $h'(\mathbf{R})$ is dense in $\mathbf{T}^2$. Let Λ' be the set of all real λ such that $h'(\lambda) \in L = h(K)$; then Λ' is a model whose uniform density is $|a\beta - \alpha b|$ meas K and $h'(\Lambda')$ is dense in $h(K)$.

Theorem V gives $\sup\limits_{t\in\mathbf{R}} |g(t)| \leqslant A \sup\limits_{t\in\Lambda'} |g(t)| = A \sup\limits_{L} |P|$. Since $h'(\mathbf{R})$ is dense in $\mathbf{T}^2$, $\sup\limits_{t\in\mathbf{R}} |g(t)| = \sup\limits_{\mathbf{T}^2} |P|$. Since $h(\mathbf{R})$ is dense in $\mathbf{T}^2$, $\sup\limits_{t\in\mathbf{R}} |f(t)| = \sup\limits_{\mathbf{T}^2} |P|$; h restricted to K is a homeomorphism from K to L and this implies $\sup\limits_{K} |f| = \sup\limits_{L} |P|$. Hence $\sup\limits_{t\in\mathbf{R}} |f(t)| \leqslant A \sup\limits_{K} |f|$ and Theorem VII is proved.

Property (a) implies that, for each Riemann-integrable compact set K of real numbers whose measure exceeds $1/|a\beta - \alpha b|$, there exists a constant C such that, for each complex-valued almost-periodic function f

whose frequencies belong to Λ,

$$\lim_{T\to+\infty} \frac{1}{2T}\int_{-T}^{T} |f(t)|^2\,dt \leqslant C\int_K |f(t)|^2\,dt.$$

Landau proved in [54], th. 2 p. 39, that this last property implies meas $K \geqslant$ dens Λ. This is not the case here. Another method of proving this type of statements will be given in the p-adic case.

8. Special trigonometric series on p-adic fields

8.1. Let $\mathbf{Q}_p$ be the p-adic field, $\mathbf{Z}_p$ the subgroup of p-adic integers and A_p the ring of elements mp^n, $m \in \mathbf{Z}$, $n \in \mathbf{Z}$, of $\mathbf{Q}_p$. By an abuse of notation we shall identify A_p with a subring of $\mathbf{Q}$.

Let Λ be the subset of Q_p consisting of all mp^n in A_p such that $0 \leqslant mp^n < 1$. By Hensel's decomposition, each p-adic number x can be written uniquely as $x = \lambda + r$, where $\lambda \in \Lambda$, $r \in \mathbf{Z}_p$. Hence each coset $x + \mathbf{Z}_p$, $x \in \mathbf{Q}_p$, of the compact group $\mathbf{Z}_p$ meets Λ in a single point.

The Haar measure, denoted by dx, is normalized on $\mathbf{Q}_p$ by $|\mathbf{Z}_p| = 1$. For each x in $\mathbf{Q}_p$, we define $\exp 2\pi i x$ to be equal to $\exp 2\pi i \lambda$ if $\lambda \in A_p$ and $x - \lambda \in \mathbf{Z}_p$. With this notation, each continuous character on $\mathbf{Q}_p$ can be written in the form $x \to \exp 2\pi i x y$.

8.2. We denote by $\mathscr{C}_\Lambda$ the Banach space of all complex-valued almost-periodic functions on $\mathbf{Q}_p$, whose spectra lie in Λ. The set of all finite sums $P(x) = \sum_{\lambda\in\Lambda} a_\lambda \exp 2\pi i \lambda x$ is a dense subspace of $\mathscr{C}_\Lambda$.

THEOREM IX: *Let K be a Riemann integrable compact subset of $\mathbf{Q}_p$.*

(a) *If the Haar measure $|K|$ of K exceeds* 1, *there is a constant A such that or each f in $\mathscr{C}_\Lambda$, $\sup_{\mathbf{Q}_p} |f| \leqslant A \sup_K |f|$;*

(b) *If the Haar measure $|K|$ of K is less than* 1, *each complex-valued continuous function defined on K is the restriction to K of a function in $\mathscr{C}_\Lambda$. In particular, there is a non-zero function in $\mathscr{C}_\Lambda$, which vanishes identically on K;*

(c) *The subgroup $\mathbf{Z}_p$ of $\mathbf{Q}_p$, whose Haar measure is exactly* 1, *does not satisfy the conclusion of* (a).

For a more precise result, the reader is referred to [89].

8.3. Proof of (a)

Replacing K by a smaller compact subset if necessary, we may assume that K is a finite union of disjoint open balls B_j, where an open ball of $\mathbf{Q}_p$ is a coset of the form $x\mathbf{Z}_p + y$, $x \in \mathbf{Q}_p$, $y \in \mathbf{Q}_p$, and that the measure of K still exceeds 1.

Let Λ' be the set of real numbers defined by the rule: $\lambda \in \Lambda'$ if $\lambda \in A_p$ and furthermore if λ, regarded as an element of $\mathbf{Q}_p$, belongs to K.

LEMMA 13: *Λ' can be ordered in such a way as to become an increasing sequence $(\lambda'_k)_{-\infty<k<+\infty}$ of real numbers such that $\lim_{k\to+\infty} (\lambda'_{j+k} - \lambda'_j)/k = 1/|K|$ uniformly over $j \in \mathbf{Z}$.*

Proof of Lemma 13: An open ball of $\mathbf{Q}_p$ can always be written $B = x\mathbf{Z}_p + y$, where $x = p^j$, $j \in \mathbf{Z}$ and $y \in A_p$. Then $B \cap A_p$ considered as a set of real numbers, is a two-way infinite arithmetic progression of common difference $|x|_p^{-1} = |B|^{-1}$ and uniform density $|B|$. Lemma 13 follows from this observation.

We return to the proof of Theorem IX. Let P be any finite sum $\sum_{\lambda\in\Lambda} a_\lambda \exp 2\pi i x\lambda$ $(x \in \mathbf{Q}_p)$ and $\tilde{P}$ the corresponding trigonometric sum defined on $\mathbf{R}$ by $\tilde{P}(t) = \sum_{\lambda\in\Lambda} a_\lambda \exp 2\pi i t\lambda$. The frequencies of $\tilde{P}$ belong to $]0, 1[$. Since A_p is dense in $\mathbf{Q}_p$ and also in $\mathbf{R}$, we have $\sup_{\mathbf{Q}_p} |P| = \sup_{A_p} |P| = \sup_{A_p} |\tilde{P}| = \sup_{\mathbf{R}} |\tilde{P}|$. Similarly, as $\Lambda' \cap K$ is dense in K, $\sup_K |P| = \sup_{\Lambda'} |P| = \sup_{\Lambda'} |\tilde{P}|$. Since Λ' is regular and has uniform density $|K|$, Theorem V gives $A \sup_{\Lambda'} |\tilde{P}| \geqslant \sup_{\mathbf{R}} |\tilde{P}|$ and the proof is complete.

8.4. Necessary and sufficient conditions for $\sup_{\mathbf{Q}_p} |f| \leqslant A \sup_K |f|$

To prove (c), we first give the following characterization of compact subsets K of $\mathbf{Q}_p$ such that $\sup_K |f|$ and $\sup_{\mathbf{R}} |f|$ are equivalent norms for functions f in $\mathscr{C}_\Lambda$.

PROPOSITION 2: *Let $\Lambda = A_p \cap]0, 1[$, let K be a compact subset of $\mathbf{Q}_p$, let $\Lambda' = A_p \cap K$ considered as a set of real numbers and let A be a positive constant. The following two statements are equivalent:*

(a) *for each almost periodic function $f: \mathbf{Q}_p \to \mathbf{C}$ whose frequencies lies in Λ, $\sup_{\mathbf{Q}_p} |f| \leqslant A \sup_K |f|$;*

(b) *for each bounded continuous function $\varphi: \mathbf{R} \to \mathbf{C}$ whose spectrum lies in* $[0, 1]$, $\sup_{\mathbf{R}} |\varphi| \leqslant A \sup_{\Lambda'} |\varphi|$.

Assuming, for the moment, that Proposition 2 is proved, let us see why statement (c) of Theorem IX follows.

If $K = \mathbf{Z}_p, \Lambda' = \mathbf{Z}$ and $\varphi(x) = \exp 2\pi ix - 1$ is such that the spectrum of φ lies in [0, 1] and $\varphi = 0$ on $\mathbf{Z}$ while $\sup_{\mathbf{R}} |\varphi| = 2$. For this reason $\mathbf{Z}_p$ does not satisfy the conclusion of (a) in Theorem IX.

In the same way, it can be shown that if K is the union of a finite number of disjoint open balls of $\mathbf{Q}_p$ and if $|K| = 1$, then K does not satisfy the conclusion of (a) in Theorem IX.

8.5. *Proof of Proposition* 2

The interest of Proposition 2 is that it establishes an equivalence between properties of functions on the p-adic field and properties of functions on the real line.

The proof of Proposition 2 depends on a series of lemmas:

LEMMA 14: *Let* $a : \mathbf{R} \to \mathbf{R}$ *be the function defined by* $a(t) = \sup(0, 1 - |2t - 1|)$; *a is the Fourier transform of the function* $b : \mathbf{R} \to \mathbf{R}$ *defined by* $b(x) = (\exp i\pi x)(1 - \cos \pi x)/(\pi^2 x^2)$. *Let* σ *be the atomic measure carried by* A_p *and assigning to each point t in* A_p *the mass* $b(t)$. *Then the Fourier transform of* σ, *in the sense of distributions of* $\mathbf{Q}_p$, *is also an atomic measure* τ, *carried by* Λ *and assigning to each* λ *in* Λ *the mass* $a(\lambda)$.

Remark: The measure σ is not bounded, but an elementary computation gives $|\sigma| (x + \mathbf{Z}_p) = 1$ for each x in $\mathbf{Q}_p$.

Proof: Let $D \subset \mathbf{Q}_p \times \mathbf{R}$ be the discrete and relatively dense subgroup of all pairs (t, t), $t \in A_p$. The annihilator of D is the discrete subgroup D' of $\mathbf{Q}_p \times \mathbf{R}$ of all pairs $(-t, t)$, $t \in A_p$.

A class $\mathscr{S}$ of test functions can be defined on $\mathbf{Q}_p$ in the following way; $f: \mathbf{Q}_p \to \mathbf{C}$ belongs to $\mathscr{S}$ if there exist n disjoint open balls $B_1, \ldots, B_n$ and n complex numbers $a_1, \ldots, a_n$ such that f restricted to B_j equals a_j and f equals 0 on $\mathbf{Q}_p \backslash \bigcup_1^n B_j$. The class $\mathscr{S}$ is invariant under Fourier transforms.

To compute the Fourier transform $\hat{\sigma}$ of σ, we consider the integral $\int_{\mathbf{Q}_p} \hat{f}(x)\, d\sigma(x) = \sum_D \hat{f}(x)\, b(y)$, where the sum is over all pairs $(x, y) \in D$, $x \in \mathbf{Q}_p$, $y \in \mathbf{R}$.

The Poisson formula ([7], Vol. 2, Ch. VIII, Section 31) gives

$$\sum_D \hat{f}(x)\, b(y) = \sum_{D'} f(-x)\, \hat{b}(y) = \sum_D f(x)\, \hat{b}(y) = \int_{\mathbf{Q}_p} f(t)\, d\tau(t),$$

where τ is the measure supported by A_p and assigning the mass $a(t)$ to each t in A_p.

LEMMA 15: *With the notation of Proposition 2, the following two statements are equivalent:*

(a) $\sup_{\mathbf{Q}_p} |f| \leqslant A \sup_K |f|$;

(b) *for each t in A_p, there exists a complex Radon measure μ_t carried by K, whose norm does not exceed A and such that $\hat{\mu}_t(\lambda) = \exp 2\pi i\lambda t$ for each λ in Λ.*

In the proof it suffices to confine attention to the dense subspace of finite trigonometric sums whose frequencies belong to Λ.

(a) ⇒ (b): Let $\mathscr{E}$ be the vector space of all restrictions to K of finite trigonometric sums $P(x) = \sum_{\lambda\in\Lambda} c_\lambda \exp 2\pi i\lambda x$; the norm of P in $\mathscr{E}$ is $\sup_K |f|$. If (a) holds, we can define a linear form L_t over $\mathscr{E}$ for each t in A_p by $L_t(f) = f(t)$; the norm of L_t does not exceed A. By the metric form of the Hahn–Banach theorem, there is a complex Radon measure μ_t carried by K such that $L_t(f) = \int_K f \, d\mu_t$ for each f in $\mathscr{E}$; testing on $f(x) = \exp 2\pi ixt$ we get (b).

(b) ⇒ (a): Let P be any complex valued trigonometric sum whose frequencies lie in Λ. Then (b) gives $P(t) = \int_K P(-x)\, d\mu_t(x)$, which implies $|P(t)| \leqslant A \sup_K |P|$. Since A_p is dense in $\mathbf{Q}_p$ we have $\sup_{\mathbf{Q}_p} |P| = \sup_{A_p} |P| \leqslant A \sup_K |P|$.

LEMMA 16: *With the notation of Lemma 15, for each t in A_p, let ν_t be the restriction of μ_t to A_p and let ϱ_t be the remainder term,* i.e. $\varrho_t = \mu_t - \nu_t$; *then for each λ in Λ, $\hat{\mu}_t(\lambda) = \hat{\nu}_t(\lambda)$.*

Proof: Let σ be the atomic measure carried by A_p and defined as in Lemma 14. The convolution product $\mu_t * \sigma$ is well defined since one of the two terms has compact support; testing on functions in $\mathscr{S}$, we see that the Fourier transform of $\mu_t * \sigma$ is $\hat{\mu}_t\hat{\sigma}$, the product of a regular function with an atomic measure. If $\lambda \in \Lambda$, the mass of $\hat{\mu}_t\hat{\sigma}$ concentrated in λ is $a(\lambda) \exp 2\pi i\lambda t$, where $a(\lambda)$ denotes the value of a on λ considered as a real number, and if $\lambda \notin \Lambda$, the mass of $\hat{\mu}_t\hat{\sigma}$ concentrated in λ is 0. Hence $\hat{\mu}_t\hat{\sigma} = \sum_{\lambda\in\Lambda} a(\lambda) \exp 2\pi i\lambda t\, \delta(x-\lambda)$, where $\delta(x)$ is the unit mass concentrated in 0.

Let ε_t be the unit mass concentrated in t; then $\mu_t * \sigma = \varepsilon_t * \sigma$ since these two measures have the same Fourier-transform. But $\mu_t * \sigma = \nu_t * \sigma + \varrho_t * \sigma$ and $\nu_t * \sigma$ is carried by the group A_p, while $\varrho_t * \sigma$ is carried by the complement of A_p. Since $\varepsilon_t * \sigma$ is also carried by A_p, the only possibility is that $\varepsilon_t * \sigma = \nu_t * \sigma$ and $\varrho_t * \sigma = 0$. Lemma 16 is thus proved.

Note that $\|\nu_t\| \leqslant \|\mu_t\| \leqslant A$.

Proof of Proposition 2: The implication (b) $\Rightarrow$ (a) has already been proved in Section 8.3. To get (a) $\Rightarrow$ (b), we apply Lemma 16: for each $t \in A_p$, there exists an atomic measure ν_t carried by A_p, whose norm does not exceed A and such that $\hat{\nu}_t(\lambda) = \exp 2\pi i\lambda t$ for each λ in Λ. Regarding A_p as a dense subgroup of $\mathbf{R}$, ν_t becomes an atomic measure λ_t carried by Λ' and such that $\hat{\lambda}_t(x) = \exp 2\pi ixt$ for each x in $]0, 1[\cap A_p$. Since $[0, 1[\cap A_p$ is dense in $[0, 1]$, we have $\hat{\lambda}_t(x) = \exp 2\pi ixt$ for each x in $[0, 1]$.

Let $\varphi : \mathbf{R} \to \mathbf{C}$ be any bounded function whose spectrum lies in $[0, 1]$. If we can write $\overline{\varphi}(t) = \int_{\mathbf{R}} \overline{\varphi}(x)\, d\lambda_t(x) = \sum_{x \in \Lambda'} \overline{\varphi}(x)\, \lambda_t\{x\}$, there is no more to prove. Since $\|\lambda_t\| = \sum_{x \in \Lambda'} |\lambda_t\{x\}| \leqslant A$, we have $\sup_{\mathbf{R}} |\overline{\varphi}| = \sup_{A_p} |\overline{\varphi}| \leqslant A \sup_{\Lambda'} |\varphi|$. The following lemma concludes the proof:

LEMMA 17: *Let I be a compact interval of real numbers, μ a complex bounded Radon measure on* $\mathbf{R}$ *whose Fourier transform vanishes on I, and φ a bounded complex continuous function on* $\mathbf{R}$ *whose spectrum lies in I. Then* $\int_{\mathbf{R}} \overline{\varphi}(x)\, d\mu(x) = 0$.

In our context μ is the difference between λ_t and the unit mass concentrated in t. A dilation and a translation reduce the problem to the case where $I = [-1, 1]$. We define φ_n, $n \geqslant 1$, by $\varphi_n(x) = \varphi((1 - 1/n)\,x)$; the spectrum of φ_n lies in $[-1 + 1/n, 1 - 1/n]$ and Proposition 4, Section 12, Chapter IV, gives $\int_{\mathbf{R}} \overline{\varphi}_n(x)\, d\mu(x) = 0$. But $|\varphi_n| \leqslant \|\varphi\|_\infty$ and $\varphi_n(x) \to \varphi(x)$ as n tends to infinity. Lebesgue's dominated convergence theorem now gives $\int_{\mathbf{R}} \overline{\varphi}(x)\, d\mu(x) = 0$.

8.6. *Proof of part* (b) *of Theorem IX*

To start with, we show why the second part of (b) is a consequence of the first part. If $|K| < 1$ the union K' of K and of a point x in $\mathbf{Q}_p$, $x \in K$, has the same property as K; hence the function which is 1 on x and 0 on K is the restriction to K' of a function in $\mathscr{C}_\Lambda$ which vanishes on K without being identically 0.

First of all, we shall prove (b) if $K = \mathbf{Z}_p$ when Λ is replaced by the set $\Lambda(\varepsilon)$ of all $x = mp^n$ in $\mathbf{Q}_p$ such that $0 \leqslant x < 1 + \varepsilon$. The extension f of the given complex-valued continuous $g : \mathbf{Z}_p \to \mathbf{C}$ will be given by a convolution

$$f = g * \mu \tag{5.4}$$

with a discrete Radon measure μ carried by A_p and satisfying

$$\sup_{x \in \mathbf{Q}_p} \int_{x+\mathbf{Z}_p} d\,|\mu|(t) < +\infty; \tag{5.5}$$

the Fourier transform ν of μ, in the sense of distributions,

$$\nu = \mathfrak{F}\mu \quad \text{is a discrete measure carried by } \Lambda(\varepsilon) \tag{5.6}$$

and, if I is the characteristic function of $\mathbf{Z}_p$ and $\delta(t)$ is the unit mass concentrated in 0

$$I(t)\, d\mu\,(t) = \delta(t). \tag{5.7}$$

Next, by a suitable partition of unity on K, we reduce the problem to the first case.

8.7. Details of the proof: construction of μ

Let a be a C^∞ real function on $\mathbf{R}$ carried by $[0, 1 + \varepsilon]$ and such that $\sum_{j \in \mathbf{Z}} a\,(x - j) = 1$ identically. The Fourier transform a of $\hat{a}$ has the following properties: $\hat{a}$ belongs to the class $\mathscr{S}$ of Schwartz, $\hat{a}(0) = 1$ while $\hat{a}(j) = 0$ for all $j \in \mathbf{Z}, j \neq 0$. We put $b(x) = \hat{a}(-x)$.

Let μ be the atomic measure carried by A_p which assigns the mass $b(t)$ to each point t in A_p. As in Proposition 2, the Fourier-transform ν of μ, in the sense of distributions, is also an atomic measure, carried by $A_p \cap [0, 1 + \varepsilon]$ which assigns the mass $a(t)$ to each point t of A_p.

Since $\mathbf{Z}_p \cap A_p = \mathbf{Z}$, the properties of b restricted to $\mathbf{Z}$ at once give (5.7). The estimates (5.5) follow from the following two observations:

(a) $(x + \mathbf{Z}_p) \cap A_p$ is a translate of $\mathbf{Z}$;

(b) owing to the rapid decay of b at infinity, $\sup_{t \in \mathbf{R}} \sum_{j=-\infty}^{+\infty} |b\,(j + t)| < +\infty$.

From (5.7) it is obvious that $f = g$ on $\mathbf{Z}_p$.

8.8. Details of the proof: the partition of K

Since $|K| < 1$ there there exist two integers m and $n \geqslant 1$ such that K is contained in the union $\bigcup_1^m B_j$ of m open balls of measure p^{-n} and such that $mp^{-n} < 1$.

From now on we replace K by $\bigcup_{1}^{m} B_j$. For each j, there is a function $\varphi_j(x) = \exp(2\pi i p^{-n} x) - a_j$, which, restricted to B_j, is 0 and which restricted to each B_k, $k \neq j$, is a constant different from zero since each B_j is a translate of $p^n \mathbf{Z}_p$.

Let $0 < \varepsilon < 1 - mp^{-n}$. For each continuous $g : K \to \mathbf{C}$, let g_j be the restriction of g to B_j, and f_j the extension of g_j to an almost periodic function whose spectrum lies in $A_p \cap [0, p^{-n} + \varepsilon]$; f_j is given in terms of g_j by a simple modification of (5.4), where $\mathbf{Z}_p$ is replaced by $B_j = p^n \mathbf{Z}_p + x_j$.

We put $f = \sum_{j=1}^{m} d_j f_j \prod_{k \neq j} \varphi_k$, where d_j is a constant such that on B_j, $1 = d_j \prod_{k \neq j} \varphi_k$.

Then f equals f_j on each B_j, f is almost-periodic and the spectrum of f lies in $A_p \cap [0, p^{-n} + \varepsilon + (m-1) p^{-n}] \subset A_p \cap [0, 1[$.

Note that the extension $g \to f$ from $\mathscr{C}(K)$ into $\mathscr{C}_A$ is given by a continuous linear operator.

8.9. How to discretize distributions on $\mathbf{Q}_p$?

The extension operator defined in Section 8.6 can be used to atomize distributions taken in certain spaces that we now define.

The notation is that of Sections 8.6 and 8.7.

Let $\omega : [0, +\infty[\to [1, +\infty[$ be an increasing function and let $L_\omega^\infty(\mathbf{Q}_p)$ be the Banach space of all classes of measurable functions $f : \mathbf{Q}_p \to \mathbf{C}$ such that for a constant $C > 0$, $|f(x)| \leqslant C\omega(|x|_p)$ almost everywhere with respect to the Haar measure on $\mathbf{Q}_p$.

We define $\mathscr{S}(\mathbf{Q}_p)$ to be the vector space of all test functions $\varphi : \mathbf{Q}_p \to C$ which are compactly supported and locally constant. The Fourier-transform of such a φ also has these two properties.

For any $f \in L_\omega^\infty(\mathbf{Q}_p)$, the Fourier transform $S = \mathfrak{F} f$ of f, in the sense of distributions, is defined by

$$\langle S, \varphi \rangle = \int_{\mathbf{Q}_p} f(x)\, \hat{\varphi}(x)\, dx \quad (\varphi \in \mathscr{S}(\mathbf{Q}_p)).$$

Let E_ω be this Banach space of distributions; $\|S\|$ is the norm of $\hat{S}$ in $L_\omega^\infty(\mathbf{Q}_p)$.

A linear atomizing operator L can be constructed with the following properties:

(a) *for each distribution S in E_ω, $L(S)$ is a measure carried by $\Lambda_\varepsilon = A_p \cap [0, 1 + \varepsilon]$, where $\varepsilon > 0$ is fixed from now on;*

(b) *for each t in* Λ_ε, *the mass of* $L(S)$ *concentrated in t is* $a(t)\int_{t+\mathbf{Z}_p} S(x)\,dx$ *where a is defined as in Section* 8.7*;*

(c) *the Fourier transforms of S and of* $L(S)$ *coincide on* $\mathbf{Z}_p$*;*

(d) *for a constant* $C > 0$, $\|L(S)\| \leqslant C\|S\|$, *where the norms are computed in* E_ω.

In fact we put $S = \mathfrak{F}f$, $f \in L_\omega^\infty$, $\tilde{f} = (If) * \mu$ and $L(S) = \mathfrak{F}\tilde{f}$. The terminology is that of Section 8.6 and the easy verifications of (a), (b), (c) and (d) are left to the reader.

For each $k \geqslant 1$ and each Radon measure μ on $\mathbf{Q}_p$, let μ_k be the Radon measure $\mu(p^k x)$; if μ is atomic and if the mass of μ concentrated in $x \in \mathbf{Q}_p$ is $a(x)$, $\mu(p^k x)$ is also atomic and the mass of concentrated μ_k in $p^{-k}x$ is $a(x)$. Let I_k be the characteristic function of $p^{-k}\mathbf{Z}_p$.

A sequence L_k, $k \geqslant 1$, of linear atomizing operators can be found such that, for each S in E_ω, $L_k(S) = S_k$ is a nice sequence of approximations of S; each S_k is carried by the harmonious subset $p^k\Lambda_\varepsilon$ of $\mathbf{Q}_p$ and this explains the role played by harmonious subsets of $\mathbf{Q}_p$ in atomizing processes.

If $S \in E_\omega$, $S = \mathfrak{F}f$, where $f \in L_\omega^\infty(\mathbf{Q}_p)$. We define S_k to be $\mathfrak{F}f_k$, where

$$f_k = (I_k f) * \mu_k. \tag{5.8}$$

Then (5.5)–(5.7) imply that

(a) S_k is a measure carried by $p^k\Lambda_\varepsilon$;

(b) for each t in Λ_ε, the mass of $L_k(S)$ concentrated in $p^k t$ is $a(p^k t)\int_{p^k(t+\mathbf{Z}_p)} S(x)\,dx$;

(c) the Fourier-transform of S and of S_k coincide on $p^{-k}\mathbf{Z}_p$;

(d) for a constant C independent of S and of $k \geqslant 1$, $\|S_k\| \leqslant C\|S\|$ where the norms are computed in $E_\omega = \mathfrak{F}L_\omega^\infty$; notice that $p^{-k}\mathbf{Z}_p$ is an increasing sequence of open subgroups of $\mathbf{Q}_p$ whose union is $\mathbf{Q}_p$.

9. Mean periodic functions

The theory of mean periodic functions provides a remarkable interpretation of coherent sets of frequencies: Λ is a coherent set of frequencies if and only if each mean periodic function whose spectrum lies in Λ is bounded or, equivalently, if and only if each mean periodic function whose spectrum lies in Λ is almost periodic.

The exposition follows that of [47].

9.1. Engineers using Laplace transforms to solve linear differential equations of the type $T * f = 0$ where T is a derivation operator have noticed that this Laplace transform, originally defined in some right half plane, has a meromorphic extension F to the complex plane. Knowledge of the set $(z_k)_{k \geqslant 1}$ of poles of F and of the set $P_k\{1/(z - z_k)\}$ of principal parts of F at once gives the solution f as a finite sum $\sum_{k \geqslant 1} P_k(t) \exp z_k t$.

Using the same idea, we study convolution equations of the type $\mu * f = 0$, where μ is a given complex valued Radon measure carried by a compact interval $[a, b]$ of the line and f is an unknown complex valued continuous function on the line. A solution of this equation is called a mean periodic function.

The main theorem is the following: to get all solutions of $\mu * f = 0$ it is sufficient to consider all solutions f of the type $P(t) \exp \lambda t$, where P is a polynomial and λ a complex number, all finite sums of these particular solutions and all limits of such finite sums for the topology of uniform convergence on compact subsets of the line.

Similar results can be found in [60].

9.2. The space $\mathscr{C}(\mathbf{R})$

From now on, a function will be a complex valued continuous function on the line, without any growth condition at infinity; $\mathscr{C}(\mathbf{R})$ is the linear space of all such functions. The topology of $\mathscr{C}(\mathbf{R})$ is that of uniform convergence on compact subsets of $\mathbf{R}$.

A measure will be a complex valued Radon measure carried by a compact subset of the line. The linear space of such measures will be denoted by $\mathfrak{M}_0(\mathbf{R})$. Then $\mathfrak{M}_0(\mathbf{R})$ is the dual space of $\mathscr{C}(\mathbf{R})$.

A mean periodic function f is defined by one of the two equivalent properties:

(a) f belongs to $\mathscr{C}(\mathbf{R})$ and the vector space spanned by all translates of f is not dense in the whole of $\mathscr{C}(\mathbf{R})$;

(b) f is a solution of a convolution equation $\mu * f = 0$, where μ belongs to $\mathfrak{M}_0(\mathbf{R})$.

From our point of view it is more exciting to start with such a convolution equation and to regard f as an unknown function.

9.3. The Laplace transform of f

Following the engineers, we solve $\mu * f = 0$ by taking the Laplace transform of f. We shall prove that this Laplace transform has a meromorphic extension F to the complex plane. The determination of the poles of F and of the principal parts of F at these poles will give the approximation of f by sums $\sum_{k \geqslant 0} P_k(t) \exp z_k t$.

In defining the Laplace transform of f, the first idea is to consider the integral $\int_0^\infty f(t) \exp -zt\, dt$. This is not satisfactory since a mean periodic function can have a very rapid growth at infinity (Section 9.14). Furthermore this integral does not furnish the meromorphic extension.

Our definition of F is more unnatural: let $f^+(t) = f(t)$ if $t \geqslant 0, f^+ = 0$ if $t < 0, f^-(t) = f(t)$ if $t < 0$ and $f^-(t) = 0$ if $t \geqslant 0$. Consider the element g of $\mathscr{C}(\mathbf{R})$ defined by $g = f^+ * \mu = -f^- * \mu$. The first definition of g shows that g is supported by $[a, +\infty[$ and the second one that g is supported by $]-\infty, b]$ when f is supported by $[a, b]$. Hence we can define the Laplace transform of g, $G(z) = \int_a^b g(t) \exp(-zt)\, dt$; G is an entire function of exponential type. Let $M(z) = \int_a^b \exp(-zt)\, d\mu(t)$.

DEFINITION 4: *The Laplace transform of the mean periodic function f is the meromorphic function* $F(z) = G(z)/M(z)$.

To prove the consistency of this definition, we show that F is independent of the choice of μ such that $f * \mu = 0$. If $f * \mu' = 0$, $\mu' \in \mathfrak{M}_0(\mathbf{R})$, we define $g' = f^+ * \mu' = -f^- * \mu'$ and we have $g' * \mu = g * \mu'$; hence $G'(z)\, M(z) = G(z)\, M'(z)$ and $F(z) = F'(z)$.

If f has a Laplace transform in the ordinary sense in $\mathscr{R}ez > x_0$, this Laplace transform $\mathscr{L}(f)$ equals F on $\mathscr{R}ez > x_0$.

In fact

$$
\begin{aligned}
G(z) &= \int_a^b g(t) \exp -tz\, dt = \int_a^\infty g(t) \exp -tz\, dt \\
&= \int_a^\infty \exp -tz\, dt \int_a^b f^+(t-s)\, d\mu(s) \\
&= \int_a^b d\mu(s) \int_a^\infty f^+(t-s) \exp -tz\, dt \\
&= \int_a^b \exp -zs\, d\mu(s) \int_0^\infty f^+(u) \exp -zu\, du \\
&= M(z)\, \mathscr{L}(f)(z).
\end{aligned}
$$

9.4. Our program is to determine f by means of the set of poles of F. We need the following theorem:

THEOREM X: *If the Laplace transform F of the mean periodic function f is an entire function in the complex plane, then $f = F = 0$.*

Before proving this theorem, we note the following: an entire function can be the Laplace transform of a complex valued continuous function on the line, but not of a mean periodic function (mean periodic functions are very special continuous functions on the line).

9.5. The proof of Theorem X depends on a series of easy lemmas and on a deeper proposition.

LEMMA 18: *Let f be a mean periodic function and σ an element of $\mathfrak{M}_0(\mathbf{R})$. Then the Laplace transform of the mean periodic function $f * \sigma$ and the product of the Laplace transform of f with $\int_{\mathbf{R}} \exp -zt \, d\sigma(t)$ differ by an entire function.*

Proof: We have $f^+ * \sigma = (f * \sigma)^+ + r$, where r is a compactly supported bounded function. Assume that $f * \mu = 0$ and put $g = f^+ * \mu$, $h = (f * \sigma)^+ * \mu$. Then $g * \sigma = h + r * \mu$, which gives $G(z)\,S(z) = H(z) + R(z)\,M(z)$, where $R(z) = \int_{\mathbf{R}} \exp(-zt)\, r(t)\, dt$ is an entire function. The Laplace transform of f is $G(z)/M(z)$ and that of $f * \sigma$ is $G(z)\,S(z)/M(z) - R(z)$, which completes the proof.

LEMMA 19: *Let μ be a complex valued Radon measure carried by $[a, b]$ and λ a complex number such that $\int_a^b \exp -\lambda t \, d\mu(t) = 0$. Then there is a complex valued function m with bounded variation on $[a, b]$ which is 0 outside $[a, b]$ and such that, for each complex $z \neq \lambda$,*

$$\int_a^b \exp(-zt) m(t)\, dt = \frac{1}{z - \lambda} \int_a^b \exp -\lambda t \, d\mu(t).$$

Proof: Replacing $d\mu(t)$ by $\exp(-\lambda t)\, d\mu(t)$, $m(t)$ by $\exp(-\lambda t)\, m(t)$ and $z - \lambda$ by Z, we can assume that $\lambda = 0$. We simply take $m(s) = \int_{-\infty}^{s} d\mu(t)$.

From now on we shall write μ_λ instead of $m(t)\, dt$.

LEMMA 20: *With the hypothesis of Lemma 19, let f be a mean periodic function, F the Laplace transform of f and λ a complex number which is not a pole of F. Then $f * \mu = 0$ implies $f * \mu_\lambda = 0$.*

Proof: As in the proof of Lemma 19, we may restrict our attention to the case $\lambda = 0$. Then using the notation of Lemma 19, $f * m$ has a distributional derivative equal to $f * \mu = 0$. Hence the continuous function $f * m$ is equal to a constant a. If $a = 0$, the lemma is proved; if $a \neq 0$, the Laplace transform of $f * m$ is a/z. Lemma 18 shows that 0 is a pole of F, which by hypothesis is not the case.

PROPOSITION 3: *Let μ be an N-times differentiable function carried by $[a, b]$, $N \geqslant 4$. Let $(\lambda_k)_{k \geqslant 0}$ be the sequence of all zeros of the entire function $M(z) = \int_a^b \exp(-zt)\,\mu(t)\,dt$. There exist an integer $n \geqslant 0$ and a complex constant c such that*

$$-it\mu = c\mu + n\mu_0 + \sum_{k \geqslant 1}\left(\mu_{\lambda_k} + \frac{1}{\lambda_k}\mu\right), \qquad (5.9)$$

where μ_λ is defined as in Lemma 19 *and where the series is uniformly convergent on $[a, b]$.*

Proof: The idea is to take Fourier transforms of the two sides of (5.9) and to establish convergence in $L^1(-\infty, +\infty)$ for these Fourier-transforms. Returning to the original functions, we obtain uniform convergence.

First of all, let us write down the Laplace transform of each side of (5.9). We obtain

$$M'(z) = aM(z) + \frac{k}{z}M(z) + \sum_{k \geqslant 1}\left(\frac{M(z)}{z - \lambda_k} + \frac{M(z)}{\lambda_k}\right). \qquad (5.10)$$

But (5.10) is a consequence of Hadamard's factorization of entire functions of exponential type: $M(z) = Ce^{cz}z^n\prod_1^\infty(1 - z/\lambda_k)$; taking the logarithmic derivatives we get (5.10) with uniform convergence on compact subsets of **C**.

In proving convergence in $L^1(-\infty, +\infty)$ of the restrictions to the imaginary axis, the following lemma will be useful:

LEMMA 20: *Let*

$$X_k(z) = \frac{M(z)}{z - \lambda_k} + \frac{M(z)}{\lambda_k} = \frac{zM(z)}{\lambda_k(z - \lambda_k)}.$$

We have for a constant C and all real y, $|y| \geqslant 1$,

$$|X_k(iy)| \leqslant C/|y|^{N-2}|\lambda_k|^2. \qquad (5.11)$$

The sequence of zeros of an entire function of exponential type satisfies $\sum_1^\infty |\lambda_k|^{-2} < +\infty$ ([16], th. 8.2, p. 328) and we have $N \geqslant 4$ (the value of N is given by the regularity of μ). Hence Lemma 20 gives Proposition 3.

To prove Lemma 20 we have to consider three cases:

(a) $|z - \lambda_k| > |\lambda_k|/2$ implies $|X_k(z)| \leqslant 2|z|\,|M(z)|\,|\lambda_k|^{-2}$ by the second form of X_k;

(b) $|\lambda_k|/2 \geqslant |z - \lambda_k| > 1$ implies $|X_k(z)| \leqslant 2|z|^2\,|M(z)|\,|\lambda_k|^{-2}$, since $|z| \geqslant |\lambda_k|/2$;

(c) if $|z - \lambda_k| < 1$ and $|\lambda_k| \geqslant 2$, then $|M(z)/(z - \lambda_k)| \leqslant \sup_{|z'-z|=2} |M(z')/(z' - \lambda_k)|$, since $M(z)/(z - \lambda_k)$ is an entire function. Hence $|X_k(z)| \leqslant 2\,|z|^2/|\lambda_k|^2 \sup_{|z'-z|=2} |M(z')|$, where we have here $|z' - \lambda_k| \geqslant 1$ and $|z/\lambda_k| \leqslant 2\,|z/\lambda_k|^2$ by the assumptions on z and λ_k. But from a direct evaluation of $M(z) = \int_a^b \exp -zt\mu\,(t)\,dt$, we obtain $|M(z)| \leqslant C\,e^{c_1|x|}/|z|^N$, where $z = x + iy$. Substituting this estimate in the inequalities obtained in (a), (b), (c) we get (5.11) with perhaps another constant C. Note, finally, that if $|\lambda_k| < 2$ and $|z - \lambda_k| < |\lambda_k|/2$ we get $|z| \leqslant 3$ and there is nothing to prove.

9.6. *Proof of Theorem XI*

Assume that the Laplace transform of f is an entire function and that $f * \mu = 0$; convolving this equation with a regular compactly supported function, it may be assumed that μ is N-times continuously differentiable.

Let $(\lambda_k)_{k\geqslant 0}$ be the sequence of zeros of $M(z) = \int_{\mathbf{R}} \exp(-zt)\,\mu(t)\,dt$. Lemma 20 shows that $f * \mu_{\lambda_k} = 0$ for all $k \geqslant 0$.

By the uniform convergence of (5.9), we get $f * (t\mu) = 0$. Iterating this, we obtain $f * (P\mu) = 0$ for all polynomials P. Let t_0 be a real number such that $\mu(t_0) \neq 0$. Then the Dirac mass ε_{t_0} concentrated in t_0 is the limit of a sequence $P_k\mu$ in the weak-star topology $\sigma(\mathfrak{M}_0(\mathbf{R}), \mathscr{C}(\mathbf{R}))$. Passing to the limit, we get $f * \varepsilon_{t_0} = 0$ and $f = 0$.

9.7. THEOREM XI: *Let μ be a complex valued Radon measure carried by a compact set of real numbers. The set of all solutions of $f * \mu = 0$, where f is a complex valued continuous function on the line, is given by the following set of rules:*

(a) *all solutions h of $h * \mu = 0$ of the type $h(t) = P(t)\exp \lambda t$, where P is a polynomial and λ a complex number must be found;*

(b) *all finite sums $g = \sum_{k\geqslant 1} P_k(t) \exp \lambda_k t$ of solutions found in* (a) *must be written,*

(c) *all limits f of convergent sequences g_j, $j \geqslant 1$ of solutions found in* (b) *must be taken where the topology is that of uniform convergence on compact subsets of the line.*

Before giving the proof, a remark should be made. The degree m of P and the complex number λ concerned in (a) can be found as follows: λ is a zero of the entire function $M(z) = \int_{\mathbf{R}} \exp -zt \, d\mu(t)$ and m is the multiplicity of this zero.

9.8. Now to the proof. Let E be the space of all functions g defined in (b); E and the closure E of $\bar{E}$ are translation invariant; $\bar{E}$ is determined by the following rule: the function h belongs to $\bar{E}$ if, for each $\nu \in \mathfrak{M}_0(\mathbf{R})$ whose Laplace transform $N(z)$ at least vanishes whenever the Laplace transform $M(z)$ of μ vanishes, the convolution $h * \nu$ is identically 0.

We have to prove that f belongs to $\bar{E}$, and to this end we show that $k = f * \nu$ is identically 0. This is the case if the Laplace transform $K(z)$ of k is an entire function (Theorem X).

By Lemma 18, $K(z) - F(z)\,N(z)$ is an entire function. But, by definition (Section 9.3) of $F(z)$, $F(z)\,M(z)$ is an entire function. Since $N(z)$ vanishes at least whenever $M(z)$ vanishes, $N(z)/M(z)$ is also entire. Hence $K(z)$ is entire, $k = 0$ and $f \in \bar{E}$.

9.9. The spectrum and the Fourier series of a mean periodic function

Let f be a mean periodic function, F the Laplace transform of f, $(z_j)_{j \geqslant 1}$ the sequence of poles of F, $P_j\,(1/(z - z_j))$ the principal part of the expansion of F in a neighbourhood of z_j and m_j the multiplicity of the pole z_j or the degree of P_j.

DEFINITION 4: *With the above notation, putting $z_j = 2\pi i \lambda_j$, we define the spectrum of f to be set of all pairs (λ_j, m_j), $j \geqslant 1$.*

DEFINITION 5: *The Fourier series of f is the formal series $\sum_{k \geqslant 1} P_j(t) \exp 2\pi i \lambda_j t$.*

The argument used in Section 9.8 gives the following result:

THEOREM XII: *For the topology of uniform convergence on compact subsets of* $\mathbf{R}$, *each mean periodic function is a limit of finite sums $f_k(t) = \sum_{j \geqslant 1} P_{j,k}(t) \exp 2\pi i \lambda_j t$.*

The degree of $P_{j,k}$ equals the degree m_j of P_j, and the coefficients of $P_{j,k}$ tend to those of P_j as $k \to +\infty$.

To prove this last remark, the following lemma will be needed:

LEMMA 21: *Let Λ be the spectrum of a mean periodic function f and $(f_j)_{j \geqslant 1}$ a sequence of mean periodic functions whose spectra lie in Λ and* *which converges to f uniformly on each compact set of real numbers. Then the Laplace transforms F_j of f_j tend to the Laplace transform F of f in the following sense: on each compact subset K of complex numbers not containing a pole of F, F_j tends uniformly to F and if λ is a pole of F of order m, then λ is a pole of F_j of order $\leqslant m$ and the coefficients of the Taylor expansion of F_j in a neighbourhood of λ tend to those of F.*

Proof: Let μ be an element of $\mathfrak{M}_0(\mathbf{R})$ such that $f * \mu = 0$ and let $M(z) = \int \exp(-zt)\, d\mu(t)$. Let $g = f^+ * \mu$, $g_j = f_j^+ * \mu$, $G(z) = \int \exp(-zt)\, g(t)\, dt$, $G_j(z) = \int \exp(-zt)\, g_j(t)\, dt$. Then $f_j * \mu = 0$ since the spectrum of f_j is contained in that of f. The Laplace transform of f is $G(z)/M(z)$ and that of f_j is $G_j(z)/M(z)$. The supports of g_j are contained in a fixed interval $[a, b]$ and g_j tends to g uniformly on $[a, b]$. Hence $G_j(z) \to G(z)$ uniformly on compact subsets of $\mathbf{C}$, and lemma 21 is proved.

9.10. *How to construct mean periodic functions?*

Let Λ be a sequence of pairs $(\lambda_k, m_k)_{k \geqslant 1}$ where $\lambda_k \in \mathbf{C}$ and $m_k \in \mathbf{N}$. Is it possible to construct a mean periodic function whose spectrum is Λ? Beurling and Malliavin [30] give a complete answer to this question. We shall state the results. The problem can be divided into two parts.

THEOREM XIII: *There is a mean periodic function whose spectrum is Λ if and only if there is an entire non zero function $A(z)$ defined in the complex plane and satisfying $A(\lambda) = 0$ for each $\lambda \in \Lambda$ and*

$$|A(x)| \leqslant C(1 + |x|)^N \text{ for suitable } C > 0 \text{ and } N \geqslant 0 \quad (x \in \mathbf{R})$$

$$|A(z)| \leqslant C \exp l|z| \quad \text{for suitable } C > 0 \text{ and } l > 0 \quad (z \in \mathbf{C}).$$

Proof: The condition is necessary. If f is mean periodic there is a complex measure μ carried by some compact interval such that $\mu * f = 0$. By definition of the spectrum $A(z) = \int_{\mathbf{R}} \exp(-2\pi izt)\, d\mu(t)$ vanishes on Λ.

We define the complex Fourier transform of a compactly supported distribution S on the line to be $A(z) = \int_{\mathbf{R}} \exp(-2\pi izt)\, S(t)\, dt$.

The condition is sufficient. By Paley–Wiener's theorem, A is the complex Fourier-transform of a distribution S carried by $[-l, l]$. Convolving S with a suitable test function φ in $\mathscr{D}$, we may assume that S is an element of $\mathscr{D}$, not equal to 0, whose complex Fourier transform vanishes on Λ. It remains to construct f as a series $\sum_1^\infty a_k t^{m_k} \exp 2\pi i\lambda_k t$ $(\Lambda = (\lambda_k, m_k)_{k \geqslant 1})$, where $a_k > 0$ tends to zero rapidly enough to ensure uniform convergence on compact subsets of $\mathbf{R}$. Hence $f * S = 0$ and f is a mean periodic function.

9.11. It remains to study sets of zeros of entire functions of exponential type having a slow growth on the real axis. For a precise statement of Beurling–Malliavin's theorem, a new definition will be needed.

DEFINITION 6: *The mean period associated with a set Λ of pairs $(\lambda_k, m_k)_{k \geqslant 1}$ ($\lambda_k \in \mathbf{C}$, $m_k \geqslant 0$) is the infimum $p(\Lambda)$ of lengths of intervals $[a, b]$ such that there exists a non-zero complex Radon measure μ carried by $[a, b]$ whose complex Fourier-transform*[1] *vanishes on Λ.*

Remark 1: *If $[a, b]$ is an interval whose length is less than $p(\Lambda)$, each complex valued continuous function defined on $[a, b]$ is the limit of a sequence of finite sums $\sum_{k \geqslant 1} P_k(t) \exp 2\pi i \lambda_k t$, where $d^0 P_k \leqslant m_k$.*

Proof: Apply the Hahn–Banach theorem.

Remark 2: *If $[a, b]$ is an interval whose length is greater than $p(\Lambda)$, each mean periodic function f whose spectrum lies in Λ and which vanishes on $[a, b]$ vanishes on the whole real line.*

Proof: Let μ be a measure carried by $[a', b']$, $b' - a' < b - a$, such that the complex Fourier-transform of μ vanishes on Λ. The problem is clearly translation invariant and so we may assume that $a = 0$. Let f^+ be defined by $f^+(t) = f(t)$ if $t \geqslant 0$ and $f^+(t) = 0$ if $t < 0$. Since f vanishes on $[0, b]$ and since μ is carried by $[a', b']$ whose length is less than b, $f^+ * \mu = f * \mu = 0$. Using the definition of the Laplace transform F of f, we get $F = 0$. Hence $f = 0$ by Theorem I.

As a corollary we have that any mean periodic function whose spectrum lies in Λ is fully determined by its values on a compact subset depending only on Λ. It is interesting to compute the length of such an interval of definition. Beurling and Malliavin's theorem gives the exact value of this length in terms of the exterior density of Λ, which we shall now define.

9.12. A density due to Beurling and Malliavin

Let Ω be a set of open intervals on the line (the notation $\omega \in \Omega$ denotes intervals of this set), and let $T(\Omega)$ be the union of all squares $T(\omega)$ of $\mathbf{R}^2$ one of whose diagonals is on $\omega \in \Omega$: if x is the middle point of ω and $2l$ the length of ω, $T(\omega)$ is defined by $|Y| + |X - x| \leqslant l$.

A set Ω of intervals is said to be negligable if $\int_{T(\Omega)} (1 + x^2 + y^2)^{-1} \, dx \, dy < +\infty$. (Note that for any set of open intervals, $T(\Omega)$ is an open set of squares and is a measurable set.)

---

[1] The complex Fourier-transform of μ is $\hat{\mu}$: $\mathbf{C} \to \mathbf{C}$ defined by $\hat{\mu}(z) = \int_a^b \exp - 2\pi i z t \, d\mu(t)$.

For example, if there is an $l > 0$ such that the lengths of the $\omega \in \Omega$ are all less than l, Ω is negligable. Roughly speaking, Ω is negligable means that it is improbable that Ω contains arbitrarily large intervals.

A complex valued (not necessarily bounded) Radon measure μ on the real line is said to be regular and has the density a if, for each positive ε, the set Ω_ε of all open intervals ω of length $|\omega|$ such that $||\omega|^{-1} \int_\omega d\mu - a| \geqslant \varepsilon$ is a negligable set of intervals.

We then write $a = A\,(d\mu)$.

To each set Λ of pairs $(\lambda_k, m_k)_{k \geqslant 1}$ of real numbers together with multiplicities, we attach a positive measure dN_Λ defined by $\sum_{k \geqslant 1} m_k\, \delta\,(x - \lambda_k)$, where $\delta(x)$ is the unit mass concentrated in 0 and we assume that $k \to \lambda_k$ is an injective map.

Definition 7: *The exterior density of* Λ, $d_e(\Lambda)$ *is* $\inf\,(A\,(d\nu))$ *over regular positive measures* $d\nu$ *such that* $d\nu \geqslant dN_\Lambda$.

Theorem XIV: *With the above notation,*

(1) *The set* Λ *of pairs* $(\lambda_k, m_k)_{k \geqslant 1}$ *is the set of frequencies of a mean periodic function* $f(t)$ *on the line if and only if* $d_e(\Lambda)$ *is finite;*

(2) *The mean period associated with* Λ *is* $d_e(\Lambda)$.

The proof of this result is deep and difficult. Since this theorem will not be used in the sequel, the reader is referred to [30].

An easy corollary can be given. We recall that a set Λ of real numbers is called regular if there exists a $d > 0$ such that each interval of length d contains at most one point of Λ and a $D > 0$ such that each interval of length D contains at least one point of Λ.

Corollary 4: *For a regular* Λ, *the mean period* $p(\Lambda)$ *of* Λ *lies in the interval* $[1/D, 1/d]$.

Proof: Consider $\mathbf{R}$ as the union of all intervals $[kd, (k+1)\,d[$, $k \in \mathbf{Z}$ and construct a measure $d\nu$ as follows: if $[kd, (k+1)\,d[\cap \Lambda = \{\lambda\}$, unit mass is concentrated in λ and if $[kd, (k+1)\,d[$ does not intersect Λ, unit mass is concentrated in kd. Hence $d\nu \geqslant dN_\Lambda$ and $d\nu$ is regular and has the density $1/d$. We thus obtain $p(\Lambda) \leqslant 1/d$.

To get $p(\Lambda) \geqslant 1/D$, we merely apply Theorem III.

9.13. *Mean periodic functions whose spectra are contained in a given* Λ

From now on, Λ is assumed to be uniformly discrete, i.e. there exists a $d > 0$ such that each interval of length d contains at most one point of Λ. We assume also that Λ does not present multiplicity (all $m_k = 0$).

$\mathscr{C}_\Lambda$ is then the Frechet space of all mean periodic functions whose spectra lie in Λ; the topology $\mathscr{C}$ is that of uniform convergence on compact sets of real numbers. It is remarkable that the vector space of all mean periodic functions is not complete in the topology $\mathscr{C}$; the vector space of all mean periodic functions whose spectra are contained in a given Λ for which $p(\Lambda) < +\infty$ is complete for $\mathscr{C}$ since there exists a fixed measure μ with compact support such that $\mu * f = 0$ for all such f.

9.14. Behaviour at infinity

All f in $\mathscr{C}_\Lambda$ are fully determined by their values on any interval I whose length $|I|$ exceeds $p(\Lambda)$. The fundamental problem is now the following: is there a method of computing or estimating the values of f outside I in terms of the values taken by f on I?

The answer is given by the following theorem:

THEOREM XV: *Let $\omega : \mathbf{R} \to [1, +\infty[$ be a function and Λ a regular set of real numbers: each interval of length d contains at most one point of Λ and each interval of length D contains at least one point of Λ. The following assertions are equivalent:*

(a) *each mean periodic function f whose spectrum lies in Λ is $0\,(\omega(t))$ as $|t|$ tends to infinity;*

(b) *there exist a constant C and a compact set K of real numbers such that, for each mean periodic function f whose spectrum lies in Λ and for all real t, $|f(t)| \leqslant C\omega(t) \sup_K |f|$.*

Proof: Let $\mathscr{B}$ be the Banach space of all complex valued continuous functions which are $0\,(\omega(t))$ as $|t| \to +\infty$. The norm of such an f in $\mathscr{B}$ is $\sup_{\mathbf{R}} |\omega^{-1} f|$. Let $I: \mathscr{C}_\Lambda \to \mathscr{B}$ be the injection defined by (a). The topology of $\mathscr{C}_\Lambda$ is defined by the fundamental sequence of neighbourhoods of 0, $V_k = \{f \in \mathscr{C}_\Lambda : \sup_{[-k,k]} |f| \leqslant 2^{-k}\}$. The closed graph theorem shows that I is continuous. Hence the inverse image under I of the unit ball of $\mathscr{B}$ contains a V_k for sufficiently large k and this gives (b). The converse implication is trivial.

Example: Let $\theta > 2$ be a real number and Λ_θ the set of all finite sums $\sum_{k \geqslant 0} \varepsilon_k \theta^k$, $\varepsilon_k = 0$ or 1. If θ is a Pisot number, all mean periodic functions whose spectra lie in Λ_θ are almost periodic functions as we shall prove later. If θ is not a Pisot number, mean periodic functions whose spectra lie in Λ_θ are known as soon as they are known on an arbitrarily small interval $[0, l]$ but it is impossible to give estimates for these functions in terms of their values on a compact set of real numbers.

To substantiate these two remarks, two observations can be made:

(a) if $\theta > 2$ each interval $[x - T, x + T]$ of real numbers contains $o(T)$ elements of Λ and the little o is uniform with respect to x;

(b) there exists[1] a sequence $P_k(t)$ of finite trigonometric sums whose frequencies belong to Λ such that $P_k(0) = \sup_{\mathbf{R}} |P_k(t)| = 1$ while P_k tends uniformly to 0 on each compact set of real numbers not containing 0. Assertion (b) in Theorem XV cannot be true, since for $s \notin K$, $t \to P_k(t - s)$ converges uniformly to 0 on K and takes the value 1 at s.

THEOREM XVI: *Let Λ be a regular set of real numbers. The following three assertions are equivalent:*

(a) *each mean periodic function whose spectrum lies in Λ is bounded;*

(b) *each mean periodic function whose spectrum lies in Λ is almost periodic;*

(c) *Λ is a coherent set of frequencies.*

Proof: (a) $\Rightarrow$ (c) by Theorem XV. Since trigonometric sums are dense in $\mathscr{C}_\Lambda$ for the topology of uniform convergence on compact sets of real numbers (Theorem XII) and since by (c) this topology is equivalent to that of uniform convergence on the whole real line, each mean periodic function in $\mathscr{C}_\Lambda$ is almost periodic. Hence (c) $\Rightarrow$ (b). It is obvious that (b) $\Rightarrow$ (a).

9.15. The L^2 theory

We say that a set Λ of real numbers is uniformly discrete if there exists a $d > 0$ such that the distance between any two distinct points in Λ exceeds d.

THEOREM XVII: *There exist a $T > 0$ and two constants C_1 and $C_2 > 0$ depending only on $d > 0$ such that, for each mean periodic function f whose spectrum is simple and lies in Λ, we have*

$$C_1 \sum_{\lambda \in \Lambda} |a_\lambda|^2 \leqslant \int_{-T}^{T} |f(t)|^2\, dt \leqslant C_2 \sum_{\lambda \in \Lambda} |a_\lambda|^2,$$

where the Fourier series of f is $\sum_{\lambda \in \Lambda} a_\lambda \exp 2\pi i \lambda t$.

By Theorem XII, we can restrict our attention to finite trigonometric sums $P(t) = \sum_{\lambda \in \Lambda} a_\lambda \exp 2\pi i \lambda t$.

Let $\Delta : \mathbf{R} \to [0, +\infty[$ be defined by $\Delta(x) = \sup(0, 1 - 2|x|/d)$.

---

[1] See Lemma 6, Section 5 of Chapter IV.

Then Δ is the Fourier transform of a positive function b which is $0(t^{-2})$ as $|t| \to +\infty$.

Proof of:

$$\int_{-T}^{T} |P(t)|^2 \, dt \leqslant TC \sum_{\lambda \in \Lambda} |a_\lambda|^2 \quad \textit{for each} \quad T > 0. \tag{5.12}$$

Let $l > 0$ be such that $\inf_{[-l,l]} b(t) \geqslant \varepsilon > 0$ (if $\varepsilon \leqslant 1/4d$, such an l exists). Then Parseval's equality gives $\|bP\|_2^2 = \int_{-\infty}^{+\infty} |\sum_{\lambda \in \Lambda} a_\lambda \Delta(x - \lambda)|^2 \, dx = \|\Delta\|_2^2 \sum_{\lambda \in \Lambda} |a_\lambda|^2$. Hence $\varepsilon^2 \int_{-l}^{l} |P(t)|^2 \, dt \leqslant \|\Delta\|_2^2 \sum_{\lambda \in \Lambda} |a_\lambda|^2$. Since the space of trigonometric sums is translation-invariant, we get, for each $k \in \mathbf{Z}$, $\varepsilon^2 \int_{(2k-1)l}^{(2k+1)l} |P(t)|^2 \, dt \leqslant \|\Delta\|_2^2 \sum_{\lambda \in \Lambda} |a_\lambda|^2$ and adding a sufficiently large number of these relations we get (5.12).

Proof of:

$$C_1 \sum_{\lambda \in \Lambda} |a_\lambda|^2 \leqslant \int_{-T}^{T} |P(t)|^2 \, dt. \tag{5.13}$$

We start as before with $\|bP\|_2^2 = \sum_{k \in \mathbf{Z}} \int_{(2k-1)T}^{(2k+1)T} b^2(t) \, |P(t)|^2 \, dt = \|\Delta\|_2^2 \sum_{\lambda \in \Lambda} |a_\lambda|^2$. Since $b(t)$ is $0(t^{-2})$ as $|t| \to +\infty$, $\|bP\|_2^2 \leqslant \int_{-T}^{T} b^2(t) \, |P(t)|^2 \, dt + T^{-1} C_3 \sum_{\lambda \in \Lambda} |a_\lambda|^2$ (using 5.12). Since $b(t) \leqslant 1/2d$, we get $\int_{-T}^{T} |P(t)|^2 \, dt \geqslant 4d^2 (\|\Delta\|_2^2 - T^{-1} C_3) \sum_{\lambda \in \Lambda} |a_\lambda|^2$. If T is large enough, (5.13) is proved.

Notice that the L^2 theory is easy and has a metric form while the L^∞ theory (equivalence between $\sup_{[-T,T]} |f|$ and $\sup_{\mathbf{R}} |f|$) is deep and is connected with the arithmetical structure of Λ.

10. Notes

If Λ is a set of real numbers and if E is a Lebesgue-measurable subset of the line, we write $\Lambda \blacktriangleleft (E, 2)$ if there exists a constant C such that for each trigonometric sum $P(t) = \sum_{\lambda \in \Lambda} a_\lambda \exp 2\pi i \lambda t$, $\sum |a_\lambda|^2 \leqslant C \int_E |P|^2 \, dt \leqslant C^2 \sum |a_\lambda|^2$.

For the case where E is an interval, this problem has been widely studied and the infimum of the lengths of those E for which $\Lambda \blacktriangleleft (E, 2)$ can be computed [48].

If E is a finite union of intervals, the problem becomes quite difficult: $\Lambda \blacktriangleleft (E, 2)$ implies $|E| \geqslant \overline{\text{dens}}\,(\Lambda)$, where $|\ \ |$ denotes Lebesgue measure and the definition of the upper density of Λ and the proof are in [54]. The converse result is plainly false, even if $\Lambda = \mathbf{Z}$. The L^2 version of Theorem VII gave the first example for which $|E| > \text{dens}\,(\Lambda)$ implies $\Lambda \blacktriangleleft (E, 2)$.

It would be interesting to give other examples for which the latter implication is true and where the arithmetical structure of Λ is not so explicit.

For lacunary sequences Λ, some partial results have been obtained. For each Sidon set Λ and each Lebesgue measurable E of positive measure, $\Lambda \blacktriangleleft (E, 2)$ holds [31], although this hypothesis on Λ is not necessary for this conclusion [32].

An n-dimensional version of Theorem XI is given in [60].

CHAPTER VI

SPECIAL TRIGONOMETRIC SERIES (GROUP-THEORETIC METHODS)

This chapter is divided into two parts. The first one is devoted to topological Sidon sets Λ of real numbers: each bounded function $b:\Lambda \to \mathbf{C}$ is the restriction to Λ of the Fourier-transform of a complex bounded Radon measure μ on $\mathbf{R}$. For almost periodic functions f whose spectra lie in Λ, there is a diffusion of regularity: if f is continuously differentiable on an arbitrary small interval, the derivative of f exists on the whole line and is almost periodic. Sidon topological sets are stable coherent sets of frequencies; in general, they are not harmonious sets.

In the second part, we show that if $(\lambda_k)_{k \geqslant 1}$ is an increasing sequence of real numbers such that $\lambda_k \neq k$ for infinitely many $k \geqslant 1$ and $\lambda_k - k \to 0$ as $k \to +\infty$, the set Λ of λ_k, $k \geqslant 1$, never has this property. A continuously differentiable almost periodic function f whose spectrum lies in Λ and whose derivative is not bounded can exist; Λ is not a coherent set of frequencies.

The proof of the last result uses the determination of idempotent or semi-idempotent measures on a compact group.

A. Topological Sidon sets of real numbers

1. Definition and basic properties of topological Sidon sets

DEFINITION 1: *A set Λ of real numbers is a topological Sidon set if each complex-valued bounded function defined on Λ is the restriction to Λ of the Fourier transform of a complex valued bounded Radon measure μ on* $\mathbf{R}$.

THEOREM I: *Let Λ be a set of real numbers. The following two statements are equivalent:*

(a) *Λ is a topological Sidon set;*

(b) *there exist a compact set K of real numbers and a positive constant C such that, for all trigonometric sums $P(t) = \sum_{\lambda\in\Lambda} c_\lambda \exp 2\pi i\lambda t$, we have $\sum_{\lambda\in\Lambda} |c_\lambda| \leqslant C \sup\limits_K |P|$.*

Proof: (a) $\Rightarrow$ (b): The hypothesis on Λ yields a linear map $T: l^\infty(\Lambda) \to B(\Lambda)$ and the closed graph theorem shows that T is continuous. Hence there is a constant $C_1 > 0$ such that for each bounded function $b: \Lambda \to \mathbf{C}$, we can find a measure μ in $M(\mathbf{R})$ satisfying the two conditions $\|\mu\| \leqslant C_1 \|b\|_\infty$ and $\hat{\mu}(\lambda) = b(\lambda)$ for all λ in Λ.

Theorem VI of Chapter IV shows that Λ is a coherent set of frequencies. Hence there is a compact set K of real numbers and a positive constant C_2 such that, for each function b in $B(\Lambda)$, we can find a measure μ carried by K such that $\|\mu\| \leqslant C_2 \|b\|_{B(\Lambda)}$ and $\hat{\mu}(\lambda) = b(\lambda)$ for all $\lambda \in \Lambda$.

Combining these two facts, we find a compact set K of real numbers and a positive constant C such that, for each bounded function $b: \Lambda \to \mathbf{C}$ there is a measure μ carried by K whose norm does not exceed $C \sup_\Lambda |b|$ and whose Fourier transform restricted to Λ is b.

Now let $P(t)$ be a trigonometric sum $\sum_{\lambda \in \Lambda} c_\lambda \exp 2\pi i \lambda t$. We define $b: \Lambda \to \mathbf{C}$ by setting $b(\lambda) = 1$ if $c_\lambda = 0$ and $b(\lambda) = c_\lambda / |c_\lambda|$ otherwise, so that $|b(\lambda)| = 1$. Let μ be a measure carried by K whose norm is less than C and whose Fourier-transform restricted to Λ is b. Then $\int_K \bar{P}(t)\, d\mu(t) = \sum_{\lambda \in \Lambda} \bar{c}_\lambda \hat{\mu}(\lambda) = \sum_{\lambda \in \Lambda} |c_\lambda|$. Hence $\sum_{\lambda \in \Lambda} |c_\lambda| \leqslant C \sup_K |P|$.

(b) $\Rightarrow$ (a): Let $b: \Lambda \to \mathbf{C}$ be any bounded function. Let $\mathscr{E}$ be the normed space of the restrictions to K of all trigonometric sums $\sum_{\lambda \in \Lambda} c_\lambda \exp 2\pi i \lambda t$ with the norm $\sup_K |P|$. Then b defines a continuous linear form L on $\mathscr{E}$ by $L(P) = \sum_{\lambda \in \Lambda} \bar{c}_\lambda b(\lambda)$; from (b) we get $|L(P)| \leqslant \sum_{\lambda \in \Lambda} |a_\lambda|\ |b(\lambda)| \leqslant C \|b\|_\infty \sup_K |P|$. By the metric form of the Hahn–Banach theorem, L can be extended to a continuous linear form on the whole of $C(K)$. Hence we get a complex valued Radon measure μ such that $\|\mu\| \leqslant C \|b\|_\infty$ and $\hat{\mu}(\lambda) = b(\lambda)$ for all λ in Λ.

Remark: A Sidon set Λ in $\mathbf{R}$ can be defined by the following property: there is a constant C such that for each trigonometric sum $P(t) = \sum_{\lambda \in \Lambda} c_\lambda \exp 2\pi i \lambda t$ we have $\sum_{\lambda \in \Lambda} |c_\lambda| \leqslant C \sup_{\mathbf{R}} |P|$. A topological Sidon set Λ is then defined by the two conditions: Λ is a Sidon set and Λ is a coherent set of frequencies. We use the qualification 'topological' to distinguish 'topological Sidon sets' from Sidon sets, for example, the set of all 2^{-k}, $k \geqslant 0$, is a Sidon set but is not a topological Sidon set.

2. Examples of topological Sidon sets

PROPOSITION 1: *Let $(\lambda_k)_{k\geqslant 1}$ be an increasing sequence of positive real numbers such that $\lambda_{k+1} \geqslant 4\lambda_k$ for all $k \geqslant 1$. Then the set Λ of all λ_k, $k \geqslant 1$, is a topological Sidon set.*

The proof of this proposition depends on the following lemma:

LEMMA 1: *Let $(b_k)_{k\geqslant 1}$ be any sequence of complex numbers of absolute value 1 and $(\lambda_k)_{k\geqslant 1}$ an increasing sequence of positive real numbers such that $\lambda_{k+1} \geqslant 4\lambda_k$. Then there is a real number s such that $|s| \leqslant 1/\lambda_1$ and $|\exp 2\pi i\lambda_k s - b_k| \leqslant 1$ for each $k \geqslant 1$.*

Proof: The inequality $|\exp 2\pi i\lambda_k s - b_k| \leqslant 1$ is satisfied when s belongs to the union of closed intervals $I_{k,l}$ of lengths $1/3\lambda_k$ whose centers belong to a suitable coset of $\lambda_k^{-1}\mathbf{Z}$. Since $\lambda_{k+1} \geqslant 4\lambda_k$ it is very easy to see that each $I_{k,l}$ contains a complete interval $I_{k+1,l'}$. Hence each $I_{1,l}$ contains such a solution s.

Proof of Proposition 1: Let $P(t)$ be any trigonometric sum $\sum_{k\geqslant 1} c_k \exp 2\pi i\lambda_k t$. We define b_k by $b_k c_k = |c_k|$ if $c_k \neq 0$ and by $b_k = 1$ if $c_k = 0$. Then $P(s) = \sum_{k\geqslant 1}|c_k|\, z_k$, where $|z_k| = 1$ and $|z_k - 1| \leqslant 1$ which imply $\mathscr{Re}\, z_k \geqslant \frac{1}{2}$. Hence $\mathscr{Re}\, P(s) \geqslant \frac{1}{2}\sum_{k\geqslant 1}|c_k|$ and $\sum_{k\geqslant 1}|c_k| \leqslant 2 \sup_{|t|\leqslant \lambda_1^{-1}} |P(t)|$.

It will be proved later that any finite union Λ of topological Sidon sets is again a topological Sidon set whenever there is a positive d such that $\inf|\lambda - \lambda'| \geqslant d$, $\lambda \in \Lambda$, $\lambda' \in \Lambda$. Applying this result in our situation, we get the following theorem:

THEOREM II: *Let α be a real number greater than 1 and let $(\lambda_k)_{k\geqslant 1}$ be an increasing sequence of positive real numbers such that, for all $k \geqslant 1$, $\lambda_{k+1} \geqslant \alpha\lambda_k$. Then the set Λ of all λ_k, $k \geqslant 1$, is topological Sidon set.*

3. Construction of remarkable measures associated with topological Sidon sets

3.1. If Λ is a topological Sidon set, every bounded function $b: \Lambda \to \mathbf{C}$ is the restriction to Λ of an element $\tilde{b}$ of $B(\mathbf{R})$ but this statement does not give any information about the values of $\tilde{b}$ outside Λ. The following theorem shows that we can be far more precise about $\tilde{b}$.

THEOREM III: *Let Λ be a topological Sidon set of real numbers. There is a constant $C > 0$ such that for each positive ε, each positive η, and every bound-*

ed function $b : \Lambda \to \mathbf{C}$, we can find a measure μ with the following properties:

$$\|\mu\| \leqslant C\varepsilon^{-1} \sup_{\Lambda} |b| \tag{6.1}$$

$$\hat{\mu}(\lambda) = b(\lambda) \quad \textit{for all} \quad \lambda \in \Lambda \tag{6.2}$$

$$|\hat{\mu}(t)| \leqslant \varepsilon \sup_{\Lambda} |b| \quad \textit{if the distance from } t \textit{ to } \Lambda \textit{ exceeds } \eta. \tag{6.3}$$

Notice that the norm $\|\mu\|$ of μ does not depend on $\eta > 0$.

If Λ is a set of integers, then Λ is a topological Sidon set if and only if there is a constant C such that for every trigonometric sum $P(t) = \sum_{\lambda \in \Lambda} c_\lambda \exp 2\pi i \lambda t$, we have $\sum_{\lambda \in \Lambda} |c_\lambda| \leqslant C \sup_{[0,1]} |P|$. The proof of Theorem III then gives the following result. For some constant $C > 0$ and for each positive ε, there is a measure μ carried by $[0, 2\pi]$ such that $\|\mu\| \leqslant C\varepsilon^{-1}$, $\hat{\mu}(\lambda) = 1$ if $\lambda \in \Lambda$, and $|\hat{\mu}(j)| \leqslant \varepsilon$ if $j \in \mathbf{Z}, j \notin \Lambda$. If the norm of μ were independent of $\varepsilon > 0$, μ would tend to a weak limit ν as ε tended to 0. Such a measure ν is carried by $[0, 2\pi]$, $\hat{\nu}(\lambda) = 1$ when $\lambda \in \Lambda$ and $\hat{\nu}(j) = 0$ for all other integers j. The existence of ν is impossible by Theorem IX below.

3.2. In the proof, we can confide our attention to the case where $b(\lambda) = 1$ for all λ in Λ. Let μ_1 be the measure obtained in this case. Each bounded function $b : \Lambda \to \mathbf{C}$ is the restriction to Λ of the Fourier transform $\hat{\mu}_2$ of a measure μ_2 whose norm does not exceed $C_2\|b\|_\infty$. Theorem III is proved if we put $\mu = \mu_1 * \mu_2$.

Since Λ is countable, it can be presented as a sequence $(\lambda_k)_{k \geqslant 1}$. As a first step, Λ is replaced by $\Lambda_k = \{\lambda_1, \ldots, \lambda_k\}$ and a measure μ_k is constructed such that

(1) μ_k is carried by a fixed compact set K, independent of k;

(2) $\|\mu_k\| \leqslant C/\varepsilon$;

(3) $\hat{\mu}_k(\lambda) = 1$ for all $\lambda \in \Lambda_k$;

(4) $|\hat{\mu}_k(t)| \leqslant \varepsilon$ if the distance from t to Λ_k exceeds η.

Let $\mathscr{C}(K)$ be the Banach space of all complex valued continuous functions on K and let $\sigma(\mathfrak{M}(K), \mathscr{C}(K))$ be the topology of pointwise convergence of measures tested on such continuous functions. We can find a subsequence of the μ_k which converges to a measure μ satisfying conditions (1), (2) and (3) of Theorem III (with $b = 1$).

μ_k will now be constructed. Let (K_1, C_1) be suitable for Λ: this means that for each finite trigonometric sum $P(t) = \sum_{\lambda \in \Lambda} c_\lambda \exp 2\pi i \lambda t$ we have $\sum_{\lambda \in \Lambda} |c_\lambda| \leqslant C_1 \sup_{K_1} |P|$. Then (K_1, C_1) also has this property for Λ_k and, from now on, we write μ instead of μ_k and Λ instead of Λ_k.

3.3. A 'random' space Ω used in the proof of Theorem III

Let $\{-1, 1\}$ be the multiplicative group of two elements and $\Omega = \{-1,1\}^k$ the product of k copies of $\{-1, 1\}$. The Haar measure $d\omega$ on Ω is defined by the condition that each point of Ω is assigned the mass 2^{-k}. The dual group of Ω is the group G of all sequences $g = (\varepsilon_1, \ldots, \varepsilon_k)$ of 0 or 1, where addition is defined to be componentwise addition modulo 2. Each function $f: \Omega \to \mathbf{C}$ has a Fourier series

$$f(\omega) = \sum_{g \in G} c_g \chi_g(\omega) = \sum_{(\varepsilon_1, \ldots, \varepsilon_k)} c(\varepsilon_1, \ldots, \varepsilon_k)\, \omega_1^{\varepsilon_1} \cdots \omega_k^{\varepsilon_k}. \tag{6.4}$$

Parseval's relation gives

$$\sum_{\omega \in \Omega} |f(\omega)|^2 = 2^k \sum_{(\varepsilon_1, \ldots, \varepsilon_k)} |c(\varepsilon_1, \ldots, \varepsilon_k)|^2 \quad \text{or}$$

$$\int_\Omega |f(\omega)|^2\, d\omega = \sum_{(\varepsilon_1, \ldots, \varepsilon_k)} |c(\varepsilon_1, \ldots, \varepsilon_k)|^2. \tag{6.5}$$

3.4. 'Random' measures μ_ω used in the proof of Theorem III

Since (K_1, C_1) is suitable for Λ_k, we can find a measure ν_ω for each ω in Ω such that $\|\nu_\omega\| \leqslant C_1$, ν_ω is carried by K and $\hat{\nu}_\omega(\lambda_j) = \omega_j$ for $1 \leqslant j \leqslant k$. Let μ_ω be defined by $\mu_\omega = \int_\Omega \nu_{\omega/\omega'} * \nu_{\omega'}\, d\omega'$. This convolution is, in fact, a finite sum. Hence μ_ω is a measure; $\|\mu_\omega\| \leqslant C_1^2$, $\hat{\mu}_\omega(\lambda_j) = \int_\Omega \omega_j \omega_j'^{-1}\, \omega_j'\, d\omega' = \omega_j \int_\Omega d\omega' = \omega_j$ and, for all real t,

$$\hat{\mu}_\omega(t) = \sum_{(\varepsilon_1, \ldots, \varepsilon_k)} c(t; \varepsilon_1, \ldots, \varepsilon_k)\, \omega_1^{\varepsilon_1} \cdots \omega_k^{\varepsilon_k} \tag{6.6}$$

where $\varepsilon_j = 0$ or 1 and $\sum_{(\varepsilon_1, \ldots, \varepsilon_k)} |c(t; \varepsilon_1, \ldots, \varepsilon_k)| \leqslant C_1^2$.

This last inequality shows that the Fourier coefficients of the 'random' measure μ_ω are more regular than the Fourier coefficients of the 'random' measures ν_ω; the regularity is required with respect to $\omega \in \Omega$.

To prove this last inequality, note that

$$\hat{\nu}_\omega(t) = \sum_{(\varepsilon_1, \ldots, \varepsilon_k)} d(t; \varepsilon_1, \ldots, \varepsilon_k)\, \omega_1^{\varepsilon_1} \cdots \omega_k^{\varepsilon_k}.$$

By Parseval's formula $\sum_{(\varepsilon_1, \ldots, \varepsilon_k)} |d(t; \varepsilon_1, \ldots, \varepsilon_k)|^2 \leqslant C_1^2$. But $c(t; \varepsilon_1, \ldots, \varepsilon_k) = d^2(t; \varepsilon_1, \ldots, \varepsilon_k)$ which proves (6.6).

3.5. *Riesz products R_ω*

We now define 'random' Riesz products $R_\omega : \mathbf{R} \to \mathbf{C}$ by $R_\omega(t) = -1 - i + \prod_1^k (1 + \varepsilon\omega_j \cos \lambda_j t) + i \prod_1^k (1 + \varepsilon\omega_j \sin \lambda_j t)$.

The following statements are easily seen to be true.

(a) $R_\omega = -1 - i + A_\omega + iB_\omega$, where $A_\omega \geqslant 0$, $B_\omega \geqslant 0$; hence $|R_\omega| \leqslant A_\omega + B_\omega + 2$;

(b) $\int_\Omega |R_\omega(t)|\, d\omega \leqslant 4$;

(c) $R_\omega(t) = \varepsilon \sum_1^k \omega_j \exp i\lambda_j t + \varepsilon^2 \sum' r(\varepsilon_1, \ldots, \varepsilon_k, t)\, \omega_1^{\varepsilon_1} \cdots \omega_k^{\varepsilon_k}$, where $r(\varepsilon_1, \ldots, \varepsilon_k, t)$ is a product of two or more cosines added to i times a product of two or more sines and where the dash means that the last sum is extended over all sequences of 0's and 1's such that $\varepsilon_1 + \cdots + \varepsilon_k \geqslant 2$.

The Fourier-transform[1] of $R_\omega(t)$ (with respect to t) is the measure

$$\mathfrak{F}R_\omega = \varepsilon \sum_1^k \omega_j \delta(x - \lambda_j) + \varepsilon^2 \sum' \mu_{\varepsilon_1, \ldots, \varepsilon_k}\, \omega_1^{\varepsilon_1} \cdots \omega_k^{\varepsilon_k}, \tag{6.7}$$

where $\delta(x)$ is the unit mass concentrated in 0 and where $\mu_{\varepsilon_1, \ldots, \varepsilon_k}$ is an atomic measure whose norm does not exceed 2.

3.6. *A function f*

The function f is $1/2n$ on $[-n, n]$ and 0 outside. Then $\|f\|_1 = \hat{f}(0) = 1$, and if n is chosen large enough, $|\hat{f}(t)| \leqslant \varepsilon$ whenever $|t| \geqslant \eta$. With such a choice of n, we put $[-n, n] = K_2$.

3.7. *A formula for μ*

We define μ to be $\int_\Omega \mu_\omega * (R_\omega f)\, d\omega$. This convolution is taken in $\mathfrak{M}(\mathbf{R})$ and $\int_\Omega$ denotes a finite sum.

3.8. *Properties of μ*

μ is carried by $K_1 + K_1 + K_2 = K$ which is independent of k

$\|\mu\| \leqslant \int_\Omega \|\mu_\omega\|\, \|R_\omega f\|\, d\omega \leqslant C_1^2 \int_\Omega \|R_\omega f\|\, d\omega = C_1^2 \int_{\Omega\times\mathbf{R}} |R_\omega(t) f(t)|\, dt.$

Reversing the order of integration and using the inequality $\int_\Omega |R_\omega(t)|\, d\omega \leqslant 4$, we get $\|\mu\| \leqslant 4C_1^2$.

---

[1] The Fourier transform of f is $\hat{f}(x) = \int_{-\infty}^{+\infty} e^{-ixt} f(t)\, dt$.

Finally

$$\hat{\mu}(t) = \int_{\Omega} \hat{\mu}_{\omega}(t)\,(R_{\omega}f)\hat{}\,(t)\,d\omega. \tag{6.8}$$

The Fourier transform of $R_{\omega}f$ is given by

$$(R_{\omega}f)^{\wedge} = f^{\wedge} * \mathfrak{F}R_{\omega} = \varepsilon \sum_{1}^{k} \omega_j \hat{f}(t-\lambda_j) + \varepsilon^2 \sum{}' \omega_1^{\varepsilon_1} \cdots \omega_k^{\varepsilon_k} g_{\varepsilon_1,\ldots,\varepsilon_k}(t), \tag{6.9}$$

where $g_{\varepsilon_1,\ldots,\varepsilon_k} = \mu_{\varepsilon_1,\ldots,\varepsilon_k} * f$ and hence $\|g_{\varepsilon_1,\ldots,\varepsilon_k}\|_{\infty} \leqslant 2$; the last sum is extended over all sequences $(\varepsilon_1, \ldots, \varepsilon_k)$ of 0's and 1's such that $\varepsilon_1 + \cdots + \varepsilon_k \geqslant 2$.

Then (6.8) and (6.9) give

$$\hat{\mu}(t) = \varepsilon \sum_{1}^{k} \hat{f}(t-\lambda_j) \int_{\Omega} \hat{\mu}_{\omega}(t)\,\omega_j\,d\omega$$

$$+ \varepsilon^2 \sum{}' g_{\varepsilon_1,\ldots,\varepsilon_k}(t) \int_{\Omega} \hat{\mu}_{\omega}(t)\,\omega_1^{\varepsilon_1} \cdots \omega_k^{\varepsilon_k}\,d\omega \tag{6.10}$$

or

$$\hat{\mu}(t) = \varepsilon \sum_{1}^{k} \hat{f}(t-\lambda_j)\, c\,(t; 0, \ldots, 0, 1, 0, \ldots, 0)$$

$$+ \varepsilon^2 \sum{}' g_{\varepsilon_1,\ldots,\varepsilon_k}(t)\, c\,(t; \varepsilon_1, \ldots, \varepsilon_k). \tag{6.11}$$

If the distance from t to Λ_k exceeds η, we use (6.11). Since $|\hat{f}(t-\lambda_j)| \leqslant \varepsilon$ and $|g_{\varepsilon_1,\ldots,\varepsilon_k}(t)| \leqslant 2$, we have $|\hat{\mu}(t)| \leqslant 2\varepsilon^2 \sum_{\varepsilon_1,\ldots,\varepsilon_k} |c\,(t; \varepsilon_1, \ldots, \varepsilon_k)| \leqslant 2C_1^2\varepsilon^2$.

If $t = \lambda_j$, $1 \leqslant j \leqslant k$, then $\hat{\mu}_{\omega}(t) = \omega_j$. Using (6.10) we obtain $\hat{\mu}(t) = \varepsilon$.

Finally μ has the following properties: $\|\mu\| \leqslant 4C_1^2$, $\hat{\mu}(\lambda) = \varepsilon$ if $\lambda \in \Lambda_k$ and $|\hat{\mu}(t)| \leqslant 2C_1^2\varepsilon^2$ if $d\,(t, \Lambda_k) \geqslant \eta$; μ is carried by $K = K_1 + K_1 + K_2$ which is independent of k.

Replacing μ by $\varepsilon^{-1}\mu$ and then ε by $\varepsilon/C_1\sqrt{2}$, we obtain the required μ_k.

4. The union of two topological Sidon sets

From now on let Λ be a uniformly discrete set of real numbers. This means that there exists a $d > 0$ such that for all λ and λ' in Λ, $\lambda \neq \lambda'$ implies $|\lambda - \lambda'| \geqslant d$.

THEOREM IV: *If Λ is the union $\Lambda_1 \cup \Lambda_2$ of two topological Sidon sets, Λ is again a topological Sidon set.*

The proof depends on the following lemma:

LEMMA 2: *Let Λ be a set of real numbers, K a compact set of real numbers and C a positive real number, and assume that, for any function $b : \Lambda \to \mathbf{C}$, there is a complex valued Radon measure μ such that*

(a) *μ is carried by K;*

(b) $\|\mu\| \leqslant C$;

(c) $|\hat{\mu}(\lambda) - b(\lambda)| \leqslant \frac{1}{2}$ *for all λ in Λ.*

Then Λ is a topological Sidon set and for all trigonometric sums $P(t) = \sum_{\lambda \in \Lambda} c_\lambda \exp 2\pi i\lambda t$, we have $\sum_{\lambda \in \Lambda} |c_\lambda| \leqslant 2C \sup_K |P|$.

Proof: Define $b(\lambda)$ to be $c_\lambda/|c_\lambda|$ if $c_\lambda \neq 0$ and to be 1 if $c_\lambda = 0$. Then, for the corresponding μ given by the lemma, $\int_K \overline{P(t)}\, d\mu(t) = \sum_{\lambda \in \Lambda} \bar{c}_\lambda \hat{\mu}(\lambda) = \sum_{\lambda \in \Lambda} \bar{c}_\lambda (\hat{\mu}(\lambda) - b(\lambda)) + \sum_{\lambda \in \Lambda} |c_\lambda|$. Hence $\mathcal{Re} \int_K \bar{P}(t)\, d\mu(t) \geqslant \frac{1}{2} \sum_{\lambda \in \Lambda} |c_\lambda|$. But $\mathcal{Re} \int_K \bar{P}(t)\, d\mu(t) \leqslant C \sup_K |P|$, which completes the proof.

Proof of Theorem IV: Let $b : \Lambda \to \mathbf{C}$ be any function such that $|b(\lambda)| = 1$ for all $\lambda \in \Lambda$. Let b_1 (resp. b_2) be the restriction of b to Λ_1 (resp. Λ_2). Using Theorem III, we find two constants C_1 and C_2, two compact sets of real numbers K_1 and K_2 (not depending on b) and for each positive ε two measures μ_1 (depending on ε), ν_1 (depending on b_1), and two further measures μ_2 (depending on ε), ν_2 (depending on b_2) such that

(a) $\hat{\mu}_1 = 1$ on Λ_1, $|\hat{\mu}_1| \leqslant \varepsilon$ on Λ_2, $\|\mu_1\| \leqslant C_1\varepsilon^{-1}$;

(b) $\hat{\nu}_1 = b$ on Λ_1, $\|\nu_1\| \leqslant C_1$;

(c) $\hat{\mu}_2 = 1$ on Λ_2, $|\hat{\mu}_2| \leqslant \varepsilon$ on Λ_1 $\|\mu_2\| \leqslant C_2\varepsilon^{-1}$;

(d) $\hat{\nu}_2 = b$ on Λ_2, $\|\nu_2\| \leqslant C_2$.

We then put $\mu = \mu_1 * \nu_1 + \mu_2 * \nu_2$; μ is carried by $K = (K_1 + K_1) \cup (K_2 + K_2)$ and $\|\mu\| \leqslant \varepsilon^{-1}(C_1^2 + C_2^2)$.

On Λ_1, $|\hat{\mu}(\lambda) - b(\lambda)| \leqslant C_2\varepsilon$; on Λ_2, $|\hat{\mu}(\lambda) - b(\lambda)| \leqslant C_1\varepsilon$. If $C_1\varepsilon \leqslant \frac{1}{2}$ and $C_2\varepsilon \leqslant \frac{1}{2}$ the hypothesis of Lemma 2 is valid and Theorem IV is proved.

5. Topological Sidon sets and stability of coherent sets of frequencies

A set of real numbers Λ is a coherent set of frequencies if there is a compact set K of real numbers and a positive constant C such that for every trigonometric sum $P(t) = \sum_{\lambda \in \Lambda} c_\lambda \exp 2\pi i\lambda t$, $\sup_{\mathbf{R}} |P| \leqslant C \sup_K |P|$.

A set of real numbers Λ is uniformly discrete if $d_\Lambda = \inf\{|\lambda' - \lambda|; \lambda \in \Lambda, \lambda' \in \Lambda, \lambda' \neq \lambda\}$ is a positive real number.

Two uniformly discrete sets of real numbers Λ and Λ' are said to be adjacent if there is a 1-1 mapping $p:\Lambda \to \Lambda'$ such that $p(\lambda) - \lambda \to 0$ as $|\lambda| \to +\infty$.

A set of real numbers Λ is a stable coherent set of frequencies if each adjacent Λ', including Λ itself, is also a coherent set of frequencies.

THEOREM V: *Stable coherent sets of frequencies are Sidon topological sets and conversely.*

For any closed set E of real numbers, let $B(E)$ be the Banach algebra of all restrictions to E of the Fourier transforms of complex bounded Radon measures μ on $\mathbf{R}$; $\|\varphi\|_{B(E)} = \inf\{\|\mu\|; \hat{\mu}|_E = \varphi\}$. The proof is based on the following lemma:

LEMMA 3: *Let Λ be a topological Sidon set. Then there exist a positive l, $(2l<d_\Lambda)$, and a positive constant A such that, for each bounded function $b:\Lambda \to \mathbf{C}$, the function $\tilde{b}$ defined on $\Lambda + [-l, l]$ by $\tilde{b}(\lambda + t) = b(\lambda)$, $\lambda \in \Lambda$, $|t| \leqslant l$, belongs to $B(\Lambda + [-l, l])$ and satisfies $\|\tilde{b}\|_{B(\Lambda+[-l,l])} \leqslant A\|b\|_\infty$.*

This lemma is given by Theorem X of Chapter IV, since a topological Sidon set is a coherent set of frequencies.

The union of a topological Sidon set with a finite set is also a topological Sidon set. To prove the stability of topological Sidon sets, it may be assumed that for each $\lambda \in \Lambda$, $|\lambda - p(\lambda)| \leqslant l$.

Let $b':\Lambda' \to \mathbf{C}$ be any bounded function and let $b:\Lambda \to \mathbf{C}$ be $b' \circ p$. Then, in the notation of Lemma 3, b can be extended to a function $\tilde{b}:\Lambda + [-l, l] \to \mathbf{C}$ whose norm in $B(\Lambda + [-l, l])$ does not exceed $A\|b\|_\infty = A\|b'\|_\infty$. The restriction of $\tilde{b}$ to Λ' is b' and is the restriction of an element of $B(\mathbf{R})$ whose norm does not exceed $A\|b'\|_\infty$. Hence Λ' is a topological Sidon set.

Conversely, let us assume that Λ is a stable coherent set of frequencies. We can clearly find a set Λ' of real numbers with the following two properties: Λ' is linearly independent over the field $\mathbf{Q}$ of rationals and Λ' is adjacent to Λ. If we can prove that such a Λ' is a topological Sidon set, by the first part of the proof, Λ will also be a topological Sidon set.

LEMMA 4: *If Λ' is linearly independent over $\mathbf{Q}$ and is a coherent set of frequencies, then Λ' is a topological Sidon set.*

Proof: Each function $\chi:\Lambda' \to \mathbf{T}$ is a weak character on Λ'; hence, by Theorem VI of Chapter IV, $\chi \in B(\Lambda')$ and $\|\chi\|_{B(\Lambda')} \leqslant A$. But each function $b:\Lambda' \to \mathbf{C}$ which is bounded by 1 can be written $b = \frac{1}{2}(\chi_1 + \chi_2)$ for two such χ. Then $b \in B(\Lambda')$ and $\|b\|_{B(\Lambda')} \leqslant A$.

Lemma 3 can be improved.

THEOREM VI: *Let Λ be a topological Sidon set of real numbers and d_Λ the infimum of the distances between two distinct points of Λ. For each l in $]0, 2^{-1}d_\Lambda[$ there is a constant $C(l)$, which is an increasing function of l, such that any bounded function $b: \Lambda + [-l, l] \to C$ which is constant on each interval $[\lambda - l, \lambda + l]$ is the restriction to $\Lambda + [-l, l]$ of the Fourier transform of a measure whose norm does not exceed $C(l) \sup_\Lambda |b|$.*

Remark: The condition $2l < d_\Lambda$ ensures that the intervals $[\lambda - l, \lambda + l]$, $\lambda \in \Lambda$, are disjoint.

If $2l = d_\Lambda$, Theorem VI can fail. For instance the set Λ of all real numbers 2^k and $2^k + 1 + k^{-1}$ $(k \geqslant 2)$ is a topological Sidon set. For this set $d_\Lambda = 1$ and Theorem VI cannot be true for $l = \frac{1}{2}$. The function $b: \Lambda + [-\frac{1}{2}, \frac{1}{2}] \to C$ which is -1 on all intervals $[2^k - \frac{1}{2}, 2^k + \frac{1}{2}]$ and 1 on all other intervals of $\Lambda + [-\frac{1}{2}, \frac{1}{2}]$ is not the restriction to $\Lambda + [-\frac{1}{2}, \frac{1}{2}]$ of a uniformly continuous function; hence this function does not belong to $B(\Lambda + [-\frac{1}{2}, \frac{1}{2}])$.

The idea of the proof is to discretize the problem, replacing $\mathbf{R}$ by $n^{-1}\mathbf{Z}$ for a sufficiently large n; n is an integer large enough for $2/n + 2l < d_\Lambda$ and $1/n < 2l_1$ to hold, where l_1 is defined to be the l of Lemma 3. Each interval $[\lambda - l_1, \lambda + l_1]$ contains at least one point λ' in $n^{-1}\mathbf{Z}$; we arbitrarily choose one such point and let Λ' be the set of all such λ'. Then Λ' is also a topological Sidon set. For each λ in Λ, let F_λ be the finite set $[\lambda - l - 1/n,\ \lambda + l + 1/n] \cap n^{-1}\mathbf{Z} - \lambda'$. Then all the F_λ are clearly contained in the same finite subset F of $n^{-1}\mathbf{Z}$. The set $\Lambda' + F$ is a finite union of translates of Λ'; hence, by Theorem IV, $\Lambda' + F$ is a topological Sidon set. The set Λ'' of all $\lambda' + F_\lambda$, $\lambda' \in \Lambda'$ is contained in $\Lambda' + F$ and is a topological Sidon set. It should be noticed that Λ'' is the intersection of $n^{-1}\mathbf{Z}$ and the enlarged set $\Lambda + [-l - n^{-1}, l + n^{-1}]$.

The function b can be extended to a locally constant function still denoted by b on the union of the intervals $[\lambda - l - 1/n, \lambda + l + 1/n]$ $\lambda \in \Lambda$. Consider now the restriction b'' of this new b to Λ''; b'' is the restriction to Λ'' of an element β of $B(\mathbf{R})$, whose norm does not exceed $A\|b''\|_\infty$. Let $\gamma: \mathbf{R} \to \mathbf{C}$ be defined as follows:

(a) if $t \in n^{-1}\mathbf{Z}$, $\gamma(t) = \beta(t)$;

(b) on each interval $[n^{-1}j, n^{-1}(j + 1)]$, γ is linear.

If we can prove that γ belongs to $B(\mathbf{R})$ and that $\|\gamma\|_{B(\mathbf{R})} = \|\beta\|_{B(\mathbf{R})}$ the proof is complete: on elements t of $n^{-1}\mathbf{Z}$ belonging to the same interval $[\lambda - l - 1/n, \lambda + l + 1/n]$, b'' is equal to the constant $b(\lambda)$ and so is γ.

The norm in $B(\mathbf{R})$ of a function $\varphi : \mathbf{R} \to \mathbf{C}$ is the same as the norm of φ_a defined by $\varphi_a(t) = \varphi(at)$. The following lemma now ends the proof:

LEMMA 5: *Let $\varphi : \mathbf{R} \to \mathbf{C}$ be any element of $B(\mathbf{R})$ and let $\psi : \mathbf{R} \to \mathbf{C}$ be defined by the two conditions: $\varphi(j) = \psi(j)$ for all $j \in \mathbf{Z}$ and ψ is linear on each interval $[j, j+1]$. Then ψ belongs to $B(\mathbf{R})$ and $\|\psi\|_{B(\mathbf{R})} \leqslant \|\varphi\|_{B(\mathbf{R})}$.*

Proof: Let S be the measure $\sum_{-\infty}^{+\infty} \delta(x - j)$. By Poisson's formula, the Fourier transform of S is S itself. Let $\Delta : \mathbf{R} \to [0, 1]$ be the function defined by $\Delta(t) = \sup(0, 1 - |t|)$. Then $\Delta = \hat{u}$, where $u \geqslant 0$ belongs to $L^1(\mathbf{R})$; Δ is a positive definite function.

Let $v(x) = \sum_{-\infty}^{+\infty} u(x + k)$ be $u * S$. The Fourier transform of v is the product $\Delta S = \delta(x)$. Hence $v = 1$.

With this notation $\psi = (S\varphi) * \Delta$ is the Fourier transform of the measure $\nu = (S * \mu)\, u$ where $\hat{\mu} = \varphi$. We compute the norm of ν:

$$\|\nu\| \leqslant \int_{\mathbf{R}} u(x) \sum_{-\infty}^{+\infty} d|\mu|\,(x - k) = \int_{\mathbf{R}} v(x)\, d\,|\mu(x)| = \|\mu\|.$$

Lemma 5 is proved and so is Theorem VI.

6. Interpolation of bounded functions defined on a topological Sidon set by the Fourier-transforms of complex bounded Radon measures on R

THEOREM VII: *Let Λ be a topological Sidon set and d_Λ the infimum of the distances between two distinct points of Λ. Then there exists an increasing function $\omega : [0, 2^{-1}d_\Lambda[\to (0, +\infty)$ with the following property: for each l in $[0, 2^{-1}d_\Lambda[$, each positive ε and each bounded function $b : \Lambda \to \mathbf{C}$, we can find a complex bounded Radon measure μ such that*

(a) *on each $[\lambda - l, \lambda + l]$, $\lambda \in \Lambda$, $\hat{\mu}(t) = b(\lambda)\, \Delta_l\,(t - \lambda)$,* where $\Delta_l(s) = \sup(0, 1 - |s|/l)$;

(b) *outside $\Lambda + [-l, l]$, $|\hat{\mu}(t)| \leqslant \varepsilon$;*

(c) $\|\mu\| \leqslant \varepsilon^{-1}\omega(l) \sup_\Lambda |b|$.

This result is striking because, with an error not exceeding ε, the extrapolation of $b : \Lambda \to \mathbf{C}$ is completely defined on the whole real line.

The proof of Theorem VII is identical to that of Theorem III: the only difference is the definition of μ_ω and of f. Here μ_ω is the measure defined by Theorem VI such that $\hat{\mu}_\omega(t) = \omega_j$ on $[\lambda_j - l, \lambda_j + l]$, and f in $L^1(\mathbf{R})$ is defined by $\hat{f}(t) = \sup(0, 1 - l^{-1}|t|)$.

7. Estimates for trigonometric sums whose frequencies belong to a topological Sidon set

THEOREM VIII: *Let Λ be a topological Sidon set of real numbers and K any compact interval $[a, b]$, $a < b$, of real numbers. Then there exists a constant $C > 0$ such that for every trigonometric sum $P(t) = \sum_{\lambda \in \Lambda} c_\lambda \exp 2\pi i \lambda t$ whose frequencies belong to Λ, we have*

$$\sum_{\lambda \in \Lambda} |c_\lambda| \leqslant C \sup_{[a,b]} |P|.$$

In the proof of Theorem VIII, some additive properties of topological Sidon sets will be needed.

7.1. *Additive properties of Sidon sets*

We give here some additive properties of topological Sidon sets; for better results, see [10] or [49].

PROPOSITION 2: *Let Λ be a topological Sidon set of real numbers, written as an increasing sequence $\{\lambda_j;\ -\infty < j < +\infty\}$ of real numbers. Then an integer $m \geqslant 1$ can be found such that* $\lim_{|j| \to +\infty} \lambda_{j+m} - \lambda_j = +\infty.$

Remark: For $m = 1$, this property may fail. The set Λ of all 2^k, $k \geqslant 0$, and all $2^k + 1$, $k \geqslant 0$ is a topological Sidon set and for an infinite set of j, $\lambda_{j+1} - \lambda_j = 1$.

The proof of Proposition 2 is based on the following lemma:

LEMMA 6: *Let Λ be a topological Sidon set. Assume that, for all trigonometric sums $P(t) = \sum_{\lambda \in \Lambda} c_\lambda \exp 2\pi i \lambda t$, $\sum_{\lambda \in \Lambda} |c_\lambda| \leqslant C \sup_{\mathbf{R}} |P|$. Assume that m is a positive integer, A a finite set of 2^m elements, B a finite set of m elements and that all the $m2^m$ sums $a + b$, $a \in A$, $b \in B$ are distinct. Finally assume that Λ contains $A + B$. Then $m \leqslant C^2$.*

Proof of Lemma 6: Let Ω be $\{-1, 1\}^m$, already used in Section 3. Elements ω of Ω are sequences $(\omega_1, \ldots, \omega_m)$ of $+$ or -1. We can put Ω in a (1-1) correspondence with A; let $\omega \to a(\omega)$ be this correspondence. Consider the trigonometric sums $P_\omega(t) = \sum_1^m \omega_j \exp 2\pi i b_j t$, where $B = \{b_1, \ldots, b_m\}$ and $P(t) = \sum_{\omega \in \Omega} P_\omega(t) \exp 2\pi i a(\omega) t$.

Then $P(t) = \sum_{\nu \in A+B} c(\nu) \exp 2\pi i \nu t$, where $c(\nu) = \pm 1$. Hence $m2^m = \sum_{\nu \in A+B} |c(\nu)| \leqslant C \sup_{\mathbf{R}} |P|$. We shall prove that $\sup_{\mathbf{R}} |P| \leqslant \sqrt{m}\, 2^m$, which gives Lemma 6.

We have $|P(t)| \leqslant \sum_{\omega\in\Omega} |P_\omega(t)| \leqslant 2^{m/2} (\sum_{\omega\in\Omega} |P_\omega(t)|^2)^{1/2}$ by Schwarz's inequality. Parseval's relation in $L^2(\Omega)$ then gives $\sum_{\omega\in\Omega}|P_\omega(t)|^2 = m2^m$.

Proof of Proposition 2: Assume that there exist an integer $m \geqslant 1$, a $T \geqslant 0$ and an infinite set J of integers such that $0 \leqslant \lambda_{j+m} - \lambda_j \leqslant T$. Replacing J by an infinite subset, also denoted by J, it may be assumed that each difference $\lambda_{j+k} - \lambda_j$ tends to a limit μ_k, $1 \leqslant k \leqslant m$. If $d_\Lambda = \inf_{-\infty<j<+\infty} \lambda_{j+1} - \lambda_j$ we have $\mu_{k+1} - \mu_k \geqslant d_\Lambda > 0$. Hence the set B of μ_k, $1 \leqslant k \leqslant m$, has m elements. The set of all sums $\lambda_j + \mu_k, j \in J, 1 \leqslant k \leqslant m$, is adjacent to Λ. Hence there is an integer j_0 such that, if J' denotes the set of all j in J greater than j_0, and Λ' the set of all sums $\lambda_j + \mu_k$, $j \in J'$, $1 \leqslant k \leqslant m$, we have $\sum_{\lambda\in\Lambda'} |c_\lambda| \leqslant 2C \sup_{\mathbf{R}} |P|$ for all trigonometric sums $P(t) = \sum_{\lambda\in\Lambda'} c_\lambda \exp 2\pi i\lambda t$; this is a consequence of the stability of topological Sidon sets; C is defined by the condition that $\sum_{\lambda\in\Lambda} |c_\lambda| \leqslant C \sup_K |P|$ for all $P(t) = \sum_{\lambda\in\Lambda} c_\lambda \exp 2\pi i\lambda t$.

Select 2^m points among the λ_j, $j \in J'$, such that the corresponding $\lambda_j + B$ are pairwise disjoint; let A be the set of these λ_j and apply Lemma 6. We get $m \leqslant 4C^2$ which ends the proof.

COROLLARY 1: *For each topological Sidon set* Λ, *there exists an integer* $m \geqslant 1$ *such that, for every* $T > 0$, *a partition* $\Lambda = M_0 \bigcup_{j\in J}\Lambda_j$ *can be achieved with the following properties*

(a) M_0 *is finite;*

(b) $\operatorname{Card}\Lambda_j \leqslant m$;

(c) *the different* $\Lambda_j + [-T, T], j \in J$, *are pairwise disjoint.*

Proof: We can clearly restrict our attention to $\Lambda \cap [0, +\infty[$. Making the corresponding partition for $\Lambda \cap]-\infty, 0]$ and enlarging M_0, we obtain the corollary. Assume that all λ in Λ are positive. The integer m is defined by Proposition 2. There exists an integer k_0 such that $k \geqslant k_0$ implies $\lambda_{k+m} - \lambda_k \geqslant (2m + 1) T$. Let J be the set of all $j > k_0$ such that $\lambda_j - \lambda_{j-1} \geqslant (2 + 1/m) T$. For each j in J, let j' be the integer in J which immediately follows j; then $j' \leqslant j + m$ and putting $\Lambda_j = \{\lambda_j, \lambda_{j+1}, \ldots, \lambda_{j'-1}\}$ we get the required partition of Λ.

7.2. *A lemma due to Zygmund*

PROPOSITION 3: *Let a be a positive element of* $L^1(\mathbf{R})$, *n an integer* $\geqslant 2$, *and d a positive real number. Let E be the set of all* (n, n) *matrices*

$$H = [\hat{a}\,(x_j - x_k)]_{1 \leqslant j,k \leqslant n}$$

for all increasing sequences $(x_j)_{1 \leqslant j \leqslant n}$ *of real numbers such that* $x_{j+1} - x_j \geqslant d$, $(1 \leqslant j \leqslant n-1)$. *Then there exists a positive* $\eta = \eta(a, n, d)$ *such that for every H in E,* $\det H \geqslant \eta > 0$.

The proof depends on the following lemma:

LEMMA 7: *Let f be an element of* $L^2(\mathbf{R})$, *and let* $(x_j)_{1 \leqslant j \leqslant n}$ *and d be as in Proposition* 3. *There is a positive* $\varepsilon = \varepsilon(f, n, d)$ *such that for every sequence* $(c_j)_{1 \leqslant j \leqslant n}$ *of complex numbers* $\|\sum_1^n c_j f(x - x_j)\|_2 \geqslant \varepsilon (\sum_1^n |c_j|^2)^{1/2}$.

Proof of Lemma 7: Assume that a sequence $c_{j,k}$, $1 \leqslant j \leqslant n$, $k \geqslant 1$ can be found such that $\sum_1^n |c_{j,k}|^2 = 1$ while for some $x_{j,k}$, $1 \leqslant j \leqslant n$, such that $x_{j+1,k} - x_{j,k} \geqslant d$, we have $\|g_k\|_2 = \|\sum_1^n c_{j,k} f(x - x_{j,k})\|_2 \to 0$ as $k \to +\infty$. We can choose a subsequence such that $c_{j,k} \to d_j$ as $k \to +\infty$; we have $\sum_1^n |d_j|^2 = 1$ and there is an index l, $1 \leqslant l \leqslant n$ such that $d_l \neq 0$. Since the L^2 norm is translation invariant, it may be assumed that $x_{l,k} \to 0$ and $x_{j,k} \to \xi_j$, where $-\infty \leqslant \xi_j \leqslant +\infty$.

The sum $g_k = \sum_1^n c_{j,k} f(x - x_{j,k})$ can be divided into three terms. The first one, called g_k^1 involves those $x_{j,k}$ which tend to $-\infty$ as k tends to infinity. The second one called g_k^2 involves those $x_{j,k}$ which have a finite limit and the third one, g_k^3, those $x_{j,k}$ which tend to $+\infty$. Then

$$\lim_{k \to +\infty} \|g_k\|_2^2 = \lim_{k \to +\infty} \|g_k^1\|_2^2 + \|g_k^2\|_2^2 + \|g_k^3\|_2^2,$$

which implies $\|g_k^2\|_2 \to 0$ as $k \to +\infty$. Hence $\sum_1^{n\prime} d_j f(x - \xi_j) = 0$, where the dash means that this sum is only taken over finite ξ_j.

Taking the Fourier transform, we obtain $(\sum_1^{n\prime} d_j \exp -2\pi i \xi_j t) \hat{f}(t) = 0$. Hence $\sum_1^{n\prime} d_j \exp -2\pi i \xi_j t$ vanishes on a set of positive Lebesgue measure and all $d_j = 0$. This contradicts $d_l \neq 0$ and Lemma 7 is proved.

Another equivalent version of Lemma 7 is the following. Let $a \geqslant 0$ be in $L^1(\mathbf{R})$. Assume that a is not the zero function. Then for each integer $n \geqslant 2$ and each $d > 0$ there is an $\varepsilon > 0$ such that for each trigonometric sum $P(t) = \sum_1^n c_j \exp 2\pi i \lambda_j t$, whose frequencies $\lambda_1 < \lambda_2 < \cdots < \lambda_n$ satisfy $\lambda_{j+1} - \lambda_j \geqslant d$, $(1 \leqslant j \leqslant n-1)$, we have

$$\int_{\mathbf{R}} |P(t)|^2 a(t)\, dt \geqslant \varepsilon \sum_1^n |c_j|^2.$$

Proof of Proposition 3: With the preceding notation, let $X = (c_1, \ldots, c_n)$; then $\bar{X} H^t X = \sum_{1 \leqslant j,k \leqslant n} \bar{c}_j c_k \hat{a}(\lambda_j - \lambda_k) = \int_{\mathbf{R}} |P(t)|^2 a(t)\, dt \geqslant \varepsilon \|X\|_2^2$. Hence H is Hermitian and all the eigenvalues of H are real and positive; let $l_1 \leqslant l_2 \leqslant \cdots \leqslant l_n$ be these eigenvalues and X_j, $1 \leqslant j \leqslant n$, the corresponding eigenvectors. We have $\bar{X}_j H^t X_j = l_j \|X_j\|_2^2 \geqslant \varepsilon \|X_j\|_2^2$; hence $l_j \geqslant \varepsilon$ and $\det H = l_1 \cdots l_n \geqslant \varepsilon^n$.

7.3. A first step in the proof of Theorem VIII

First of all, a weak form of Theorem VIII given by Proposition 4 below will be proved:

PROPOSITION 4: *Let Λ be a topological Sidon set and l a positive real number. Then there exists a finite subset F of Λ such that for all trigonometric sums P whose frequencies belong to $\Lambda \backslash F$,* $\sup_{[-l,l]} |P|$ *and* $\sup_{\mathbf{R}} |P|$ *are equivalent norms.*

We shall prove later that we can choose $F = \emptyset$.

Proof of Proposition 4: Let $a(t)$ be the function $\sup(0, 1 - |t|/l)$. The idea of the proof is the following: for any $b : \Lambda \to \mathbf{C}$ bounded by 1, we construct a measure $\nu = a\mu$ such that $\|\nu\| \leqslant C$ and $|\hat{\nu}(\lambda) - b(\lambda)| \leqslant \frac{1}{2}$. Then ν is carried by $[-l, l]$ and Lemma 2 provides the proof. In fact the inequalities $|\hat{\nu}(\lambda) - b(\lambda)| \leqslant \frac{1}{2}$ will hold for $\lambda \in \Lambda \backslash M_0$, where M_0 is a fixed finite subset of Λ and these inequalities will be obtained by means of a somewhat technical choice of μ which we now describe.

Roughly speaking, to get μ, we write $\hat{\nu} = \hat{\mu} * \hat{a}$; $\hat{\mu}$ will be concentrated on a small neighbourhood of Λ and a suitable partition of Λ enables to separate the different terms of $\hat{\mu} * \hat{a}$.

(1) *A partition of Λ*. Let $\hat{a}$ be the Fourier transform of a, ε a positive real number whose exact value will be given at the very end of the proof and $T > 0$ a real number sufficiently large that $\int_{|t|>T} |\hat{a}(t)|\, dt \leqslant \varepsilon^2$ (in fact $\hat{a} \geqslant 0$).

We apply Proposition 2 to get a decomposition of Λ, $\Lambda = M_0 \bigcup_{j \in \mathbf{Z}} \Lambda_j$, where $\operatorname{Card} \Lambda_j \leqslant m$ and where the $\Lambda_j + [-T, T]$ are pairwise disjoint; M_0 is a finite subset of Λ.

(2) *A function Δ' used in the definition of μ*. Let $d = d_\Lambda$ be the infimum of the distances between distinct points of Λ and let $\Delta' : \mathbf{R} \to [0, +\infty)$ be defined by $\Delta'(x) = k \sup(0, 1 - 3|x|/d)$, where $k > 0$ is a constant such that $\int_{\mathbf{R}} \Delta'(x)\, \hat{a}(x)\, dx = 1$.

(3) *A measure μ*. Let $c : \Lambda \to \mathbf{C}$ be a bounded function whose connection with $b : \Lambda \to \mathbf{C}$ will be explained later. The measure μ has the following three properties:

(a) $\|\mu\| \leqslant C_1 \varepsilon^{-1} \sup_{\Lambda} |c(\lambda)|$;

(b) $\hat{\mu}(x) = c(\lambda)\, \Delta'(x - \lambda)$ if $|x - \lambda| \leqslant d/3$, $\lambda \in \Lambda$;

(c) $|\hat{\mu}(x)| \leqslant \varepsilon \sup_{\Lambda} |c(\lambda)|$ otherwise.

(4) *A function $c : \Lambda \to \mathbf{C}$ and a measure $d\nu$.*

If $\operatorname{Card} \Lambda_j = 1$ and $\lambda \in \Lambda_j$, $b(\lambda) = c(\lambda)$.

If $2 \leqslant \operatorname{Card} \Lambda_j \leqslant m$, we write Λ_j as an increasing sequence $\lambda_1, \ldots, \lambda_k$ (all of whose terms depend on j), $2 \leqslant k \leqslant m$, and for $1 \leqslant s \leqslant k$, $c(\lambda_s) = x_s$ is defined to be the solution of the linear system

$$\sum_{s=1}^{k} x_s f(\lambda_r - \lambda_s) = b(\lambda_r),$$

where $b: \Lambda \to \mathbf{C}$ is the function that we attempt to interpolate in the Fourier transform of a measure carried by $[-l, l]$ and f is the convolution $\hat{a} * \Delta'$.

Proposition 3 shows that for a constant C_2 depending only on l, d and m, we have $|x_s| \leqslant C_2$, $(1 \leqslant s \leqslant k)$—we recall that $|b(\lambda)| \leqslant 1$ for all $\lambda \in \Lambda$—, hence

$\|\mu\| \leqslant C_3/\varepsilon$;

$\hat{\mu}(x) = c(\lambda)\, \Delta'(x - \lambda)$ if $|x - \lambda| \leqslant d/3$ for some $\lambda \in \Lambda$;

$|\hat{\mu}(x)| \leqslant C_2 \varepsilon$ otherwise.

We define $d\nu(t) = a(t)\, d\mu(t)$.

An estimate for $|\hat{\nu}(\lambda) - b(\lambda)|$. If $\lambda \in \Lambda_j$ and $\operatorname{Card} \Lambda_j = 1$, we write $\hat{\nu}(\lambda) = (\hat{a} * \hat{\mu})(\lambda) = \int_{\mathbf{R}} \hat{\mu}(\lambda - t)\, \hat{a}(t)\, dt$ as the sum of three terms I_1, I_2 and I_3; $I_1 = \int_{-d/3}^{d/3} \hat{\mu}(\lambda - t)\, \hat{a}(t)\, dt$, $I_2 = \int_{d/3 \leqslant |t| \leqslant T} \hat{\mu}(\lambda - t)\, \hat{a}(t)\, dt$ and $I_3 = \int_{|t| > T} \hat{\mu}(\lambda - t)\, \hat{a}(t)\, dt$.

Then $|I_2| \leqslant C_2 \varepsilon$ since $|\hat{\mu}| \leqslant \varepsilon$ on the domain of integration while $\int_{\mathbf{R}} |\hat{a}(t)|\, dt = 1$. On the other hand $|I_3| \leqslant C_3 \varepsilon$ since $|\hat{\mu}| \leqslant C_3/\varepsilon$ and $\int_{|t| > T} |\hat{a}(t)|\, dt \leqslant \varepsilon^2$. The main term $I_1 = b(\lambda) \int_{-d/3}^{d/3} \Delta'(t)\, \hat{a}(t)\, dt = b(\lambda)$. Hence $|\hat{\nu}(\lambda) - b(\lambda)| \leqslant \varepsilon (C_2 + C_3)$.

If $\lambda \in \Lambda_j$ and $2 \leqslant \operatorname{Card} \Lambda_j \leqslant m$, we put $\lambda = \lambda_{j_s}$; Ω_1 is the union of Intervals $[\lambda_r - d/3, \lambda_r + d/3]$, $1 \leqslant r \leqslant k$, Ω_2 is the complement of Ω_1 lying in $[\lambda_{j_s} - T, \lambda_{j_s} + T]$ and Ω_3 is the complement of $[\lambda_{j_s} - T, \lambda_{j_s} + T]$. We write $\hat{\nu}(\lambda) = I_1 + I_2 + I_3$, where $I_i = \int_{\Omega_i} \hat{\mu}(\lambda - t)\, \hat{a}(t)\, dt$ For the same reason as above, $|I_2| \leqslant C_2 \varepsilon$ and $|I_3| \leqslant C_3 \varepsilon$ On each interval $[\lambda_{j_r} - d/3, \lambda_{j_r} + d/3]$, $\hat{\mu}$ is $c(\lambda_{j_r})\, \Delta'(x - \lambda_{j_r})$ Hence $I_1 = \sum_{r=1}^{k} c_{\lambda_r} f(\lambda_{j_s} - \lambda_{j_r}) = b(\lambda_{j_s})$

In all cases $|\hat{\nu}(\lambda) - b(\lambda)| \leqslant C_4 \varepsilon$, where the constant C_4 depends only on l and Λ. Proposition 4 is proved if $C_4 \varepsilon \leqslant \frac{1}{2}$.

7.4. Conclusion of the proof of Theorem VIII

To prove Theorem VIII, it remains to add to $\Lambda \backslash M_0$ the finite set M_0. Since Theorem VIII is translation invariant on Λ, the following lemma can be applied:

LEMMA 8: *Let Λ be a coherent set of frequencies and K a compact set of real numbers such that $\sup_K |f|$ and $\sup_{\mathbf{R}} |f|$ are equivalent norms for all almost periodic functions f whose frequencies belong to Λ. Assume that* 0 *does not belong to Λ. For each positive ε let K_ε be the set of all real x whose distance from K does not exceed ε. Then $\sup_{K_\varepsilon} |f|$ and $\sup_{\mathbf{R}} |f|$ are equivalent norms for all almost periodic functions f whose spectra lie in $\Lambda \cup \{0\}$.*

Proof: If the lemma were false, we could find a sequence $g_k = a_k + f_k$ of such almost periodic functions with the following properties:

(a) $\sup_{K_\varepsilon} |g_k| \to 0$;

(b) $\sup_{\mathbf{R}} |g_k| = 1$.

This second condition implies $|a_k| \leqslant 1$. Taking a subsequence if necessary, it may be assumed that $a_k \to a$ as $k \to +\infty$. Then $\sup_{K_\varepsilon} |a + f_k| \to 0$ and the sequence of $(f_k)_{k \geqslant 1}$ is a Cauchy sequence in $\mathscr{C}(K)$. Since $\sup_K |f_k|$ and $\sup_{\mathbf{R}} |f_k|$ are equivalent norms, f_k converges, on the whole of $\mathbf{R}$, to an almost periodic function f such that $f = -a$ on K_ε and whose frequencies belong to Λ. For each h, $|h| \leqslant \varepsilon$, $f(x + h) - f(x) = 0$ on K. Hence $f(x + h) = f(x)$ on the whole of $\mathbf{R}$, since the spectrum of $f(x + h) - f(x)$ is contained in Λ, and f is a constant. Since $0 \notin \Lambda$ there is only one possibility: $f = 0$. Hence $a = 0$ and $\sup_{\mathbf{R}} |g_k| \to 0$. We thus have a contradiction.

B. Idempotent and semi-idempotent measures

8. Idempotent measures on l.c.a. groups

8.1. Some definitions: idempotent measures, $S(\mu)$, the coset ring

Let G be an l.c.a. group, Γ the dual group and $M(G)$ the space of all bounded complex valued Radon measures μ on G.

DEFINITION 2: *A measure μ in $M(G)$ is said to be idempotent if the convolution product $\mu * \mu$ equals μ.*

This definition is equivalent to the following one:

DEFINITION 3: *A measure μ in $M(G)$ is said to be idempotent if its Fourier transform $\hat{\mu}$ takes only the values* 0 *or* 1.

Let μ be an idempotent measure; put $S(\mu) = \{\gamma \in \Gamma; \hat{\mu}(\gamma) = 1\}$. It is equivalent to determine μ or $S(\mu)$.

To give the characterization of $S(\mu)$ for idempotent measures μ, another definition will be needed.

DEFINITION 4: *Let Ω be the coset-ring of Γ defined in the following way: Ω is the smallest family of subsets of Γ with the properties:*

(a) *Ω contains all translates of open subgroups of Γ by elements of Γ, i.e. open cosets of Γ;*

(b) *Ω is stable with respect to finite unions, finite intersections and complements.*

THEOREM IX: *Idempotent complex valued Radon measures μ on G are characterized by the property: the support $S(\mu)$ of the Fourier transform $\hat{\mu}$ of μ belongs to the coset ring of Γ.*

8.2. The trivial half of Theorem IX

PROPOSITION 5: *Each element E of Ω is an $S(\mu)$ for an idempotent measure μ.*

Let Λ be an open subgroup of Γ and H the annihilator of Λ. Then Γ/Λ is discrete; hence H, the dual group of Γ/Λ, is compact. Let m_H be the Haar measure of H normalized by $m_H(H) = 1$. Then m_H can be regarded as an element of $M(G)$ and has the following properties:

(a) if $\gamma \in \Lambda$ then $\hat{m}_H(\gamma) = 1$;

(b) if $\gamma \notin \Lambda$ then $\hat{m}_H(\gamma) = 0$.

Hence $\Lambda = S(m_H)$.

If $E = \Lambda + \gamma_0$, we define $d\mu = \chi_{\gamma_0}(x)\, dm_H(x)$; then $E = S(\mu)$. To prove Proposition 5, it suffices to show that the family Ω' of $S(\mu)$ for idempotent μ is stable with respect to finite unions, finite intersections and complements. If μ and ν are idempotent measures $S(\mu * \nu) = S(\mu) \cap S(\nu)$, $S(\mu + \nu - \mu * \nu) = S(\mu) \cup S(\nu)$ and $S(\delta - \mu) = \Gamma \backslash S(\mu)$, where δ is the unit mass concentrated in 0 and $\backslash$ denotes complementation. Hence $\Omega' \supset \Omega$.

The proof of the reverse inclusion $\Omega' \subset \Omega$ is deeper.

8.3. Reduction to the compact case

LEMMA 9: *Let μ and ν be two distinct idempotent measures on G. Then $\|\mu - \nu\| \geqslant 1$. The same result holds if the Fourier transforms of μ and ν take only integral values.*

Proof: $\|\mu - \nu\| \geqslant \|\hat{\mu} - \hat{\nu}\|_\infty = 1$.

DEFINITION 5: *The support group of μ is defined to be the smallest closed subgroup of G on which μ is concentrated.*

PROPOSITION 6: *If μ is idempotent, the support group of μ is compact.*

Proof: Let H be the support group of μ. Consider μ as a measure on H and let Γ' be the dual group of H. By definition of H, μ is not concentrated on a proper closed subgroup of H, therefore for all $\gamma \neq 0$ in Γ', $\chi_\gamma \mu \neq \mu$. Hence $\|\chi_\gamma \mu - \mu\| \geqslant 1$.

On the other hand $\|\chi_\gamma \mu - \mu\|$ tends to 0 as γ tends to zero in Γ'. These two observations imply that Γ' is discrete.

To see that $\|\chi_\gamma \mu - \mu\| \to 0$ as $\gamma \to 0$, let C, for any positive ε, be a compact subset of G defined by $\int_{G \setminus K} d|\mu| \leqslant \varepsilon/4$. Consider the neighbourhood V of 0 in Γ' defined by $\sup\limits_{K} |\chi_\gamma(x) - 1| \leqslant \varepsilon/2\|\mu_j\|$. Then, for any γ in V, $\|\chi_\gamma \mu - \mu\| \leqslant 2 \int_{G \setminus K} d|\mu| + \int_K \varepsilon/2\|\mu\| \, d|\mu| \leqslant \varepsilon$.

Assume now that Theorem IX has been proved for any compact group H. Let μ be an idempotent measure on G and H the support group of μ; let Λ be the annihilator of H and set $S_H(\mu) = \{\gamma \in \Gamma/\Lambda; \hat{\mu}(\gamma) = 1\}$, where μ is regarded as a measure on H.

If $\pi : \Gamma \to \Gamma/\Lambda$ is the canonical homomorphism whose kernel is Λ, $S(\mu) = \pi^{-1}(S_H(\mu))$. If $S_H(\mu)$ belongs to the coset ring of Γ/Λ, then $S(\mu)$ belongs to the coset ring of Γ.

8.4. *'Translations lemmas'*

We arrive now at the main part of the proof.

The main tool in the proof (Theorem X below) is a refinement of the translation lemma of Helson ([15], p. 66, lemma 3.5.1). In all of what follows it will be assumed that G is metrizable which implies that the dual group Γ is countable. On $M(G)$ the weak-star topology $\sigma(M(G), C(G))$ is defined by the duality with the space of all complex-valued continuous functions on G. The unit ball $\|\mu\| \leqslant 1$ of $M(G)$ is a metrizable compact convex set for the weak star topology.

The trivial modifications necessary when G is not metrizable are left to the reader: all sequences should be replaced by nets.

The following result can be expressed in two equivalent ways; as usual $B(\Gamma)$ denotes the Banach algebra of Fourier transforms of all complex Radon measures μ on G.

THEOREM X: *Let μ be a complex Radon measure on G, $(\gamma_k)_{k \geqslant 1}$ a sequence* *of elements of the dual group* Γ, $\gamma_k(x)$ *the value at* x *of the character* $\gamma_k : G \to \mathbf{T}$ *and* $d\mu_k(x)$ *the complex Radon measure* $\gamma_k(x)\, d\mu\,(x)$.

Let $\mu = \mu_a + \mu_s$ *be the Lebesgue decomposition of* $\mu : \mu_a = f\, dx$, *for some* $f \in L^1(G)$ *and* μ_s *is singular with respect to the Haar measure* dx *of* G.

Assume that μ_k *tends to a complex Radon measure* ν *in the weak-star topology.*

Then either $\lim \|\nu - \mu_k\| = 0$ $(k \to +\infty)$ *or the weak limit* ν *has the following two properties:*

(i) $|\nu| \leqslant |\mu_s|$ *and therefore* ν *is singular with respect to* dx;

(ii) $\|\nu\| < \|\mu\|$ *(note that the inequality is strict).*

An equivalent way of stating this result will be given. Let φ be an element of $B(\Gamma)$ and $(\gamma_k)_{k \geqslant 1}$ a sequence of elements of Γ. Assume that for each γ in Γ, $\varphi\,(\gamma - \gamma_k)$ tends to a limit $\psi(\gamma)$. Then either $\lim\limits_{k \to +\infty} \|\psi(\gamma) - \varphi\,(\gamma - \gamma_k)\|_{B(\Gamma)} = 0$ or ψ is the Fourier transform of a singular measure and $\|\psi\|_{B(\Gamma)} < \|\varphi\|_{B(\Gamma)}$.

Before proving Theorem X, a corollary must be given.

COROLLARY 1: *If the measure* μ *and the sequence* $(\gamma_k)_{k \geqslant 1}$ *satisfy the hypothesis of theorem* X, *and if the Fourier transform* $\hat{\mu}$ *of* μ *takes only integral values, then either* $\gamma_k \mu = \nu$ *for sufficiently large* k *or* $|\nu| \leqslant |\mu_s|$ *and* $\|\nu\| < \|\mu\|$.

Proof: Lemma 9 shows that if $\gamma_k \mu = \mu_k$ is a Cauchy sequence, then $\mu_k = \nu$ for sufficiently large k.

8.5. Proof of $|\nu| \leqslant |\mu_s|$: If $(\gamma_k)_{k \geqslant 1}$ belongs to a finite subset of Γ, (μ_k) belongs to a finite subset of $M(G)$ and for sufficiently large k, $\mu_k = \nu$. In this case Theorem X is proved. Otherwise there is a subsequence of the γ_k, $k \geqslant 1$, tending to infinity. The following lemma will be useful:

LEMMA 10: *If* γ_k, $k \geqslant 1$, *tends to infinity in* Γ *and* $f \in L^1(G)$, *then* $f_k(x) = \gamma_k(x) f(x)$ *tends to* 0 *in the weak star topology* $\sigma\,(M(G), C(G))$.

Proof of Lemma 10: Since $\|f_k\|_1 = \|f\|_1$ it suffices to test the weak convergence on a total subset of $C(G)$; for example we can test on all characters $\gamma(x)$, $\gamma \in \Gamma$. But in this case $\int_G f_k(x)\, \overline{\gamma(x)}\, dx = \hat{f}(\gamma - \gamma_k)$, which tends to 0 by the Riemann–Lebesgue lemma ([15], th. 1.2.4, p. 9).

To prove $|\nu| \leqslant |\mu_s|$, it suffices to show that, for any g in $C(G)$, $|\int_G g\, d\nu| \leqslant \int_G |g|\, d|\mu_s|$. But $\int_G g\, d\nu = \lim\limits_{k \to +\infty} \int_G g(x) \gamma_k(x)\, d\mu(x) = \lim\limits_{k \to +\infty} \int_G g(x) \gamma_k(x) \times f(x)\, dx + \lim\limits_{k \to +\infty} \int_G g(x)\, \gamma_k(x)\, d\mu_s\,(x) = \lim\limits_{k \to +\infty} \int_G g(x)\, \gamma_k(x)\, d\mu_s\,(x)$. Hence $|\int_G g\, d\nu| \leqslant \int_G |g|\, d|\mu_s|$.

8.6. The inequality (ii) depends on the following two interesting lemmas:

LEMMA 11: *Let μ be a complex valued Radon measure on a compact space G, θ a Borel function on G such that $|\theta| = 1$ and $\theta\mu = |\mu|$, ε a positive number in* (0, 1) *and f a complex valued continuous function on G such that* $\|f\|_\infty \leqslant 1$.

Then $\mathcal{R}e\,(\int_G f\,d\mu) \geqslant (1-\varepsilon)\,\|\mu\|$ implies $\|f-\theta\|_1 \leqslant \sqrt{2\varepsilon}\,\|\mu\|$, where the L^1 norm is taken in the space $L^1\,(d|\mu|)$ and $\mathcal{R}e$ denotes the real part.

We already know that $\|\mu\| = \sup_{\|f\|_\infty \leqslant 1} \mathcal{R}e \int_G f\,d\mu$; Lemma 11 locates precisely those f for which this supremum is approached.

Now to the proof. If $|z| \leqslant 1$, $1 - \mathcal{R}ez \geqslant |1-z|^2/2$. Hence $\mathcal{R}e \int_G f\,d\mu = \mathcal{R}e \int_G f\bar{\theta}\,d|\mu| \geqslant (1-\varepsilon)\,\|\mu\|$ implies $2^{-1}\int_G |1-\bar{\theta}f|^2\,d|\mu| \leqslant \int_G (1-\mathcal{R}e\bar{\theta}f)\,d|\mu| \leqslant \|\mu\| - (1-\varepsilon)\,\|\mu\| \leqslant \varepsilon\|\mu\|$. Schwarz's inequality gives $(\int_G |1-\bar{\theta}f|\,d|\mu|)^2 \leqslant \|\mu\| \int_G |1-\bar{\theta}f|^2\,d|\mu| \leqslant 2\varepsilon\|\mu\|^2$. Since $|\theta| = 1$, the required inequality follows.

LEMMA 12: *With the hypothesis of Lemma* 11, *let $(g_k)_{k\geqslant 1}$, be a sequence of Borel functions on G such that $|g_k| = 1$. Assume that $g_k\mu$ tends to a limit ν in the weak star topology $\sigma\,(M(G), C(G))$. Then either $\|\nu - g_k\mu\| \to 0$ as $k \to +\infty$ or $\|\nu\| < \|\mu\|$.*

Clearly $\|\nu\| \leqslant \|\mu\|$. We shall prove that $\|\nu\| = \|\mu\|$ implies the strong convergence of $g_k\mu$ to ν. For any positive ε, let f be a continuous function on G such that $\|f\|_\infty \leqslant 1$ and $\mathcal{R}e \int_G f\,d\nu \geqslant (1-\varepsilon)\,\|\mu\|$. The weak-star convergence implies that for $k \geqslant k(\varepsilon)$, $\mathcal{R}e \int_G fg_k\,d\mu \geqslant (1-2\varepsilon)\,\|\mu\|$. Lemma 11 gives $\|fg_k - \theta\|_1 \leqslant 2\sqrt{\varepsilon}\,\|\mu\|$; hence $\|f - \bar{g}_k\theta\|_1 \leqslant 2\sqrt{\varepsilon}\,\|\mu\|$ and for k and $k' \geqslant k(\varepsilon)$, $\|(\bar{g}_{k'} - \bar{g}_k)\,\theta\|_1 = \|g_{k'} - g_k\|_1 \leqslant 4\sqrt{\varepsilon}\,\|\mu\|$. The sequence $g_k\mu$ is thus a Cauchy sequence and ν is its strong limit.

Returning to Theorem X, we apply Lemma 12 to $g_k(x) = \gamma_k(x)$ to terminate the proof.

8.7. Proof of Theorem IX

From now on, a larger class $F(G)$ of measures μ will be used: $F(G)$ is the subset of $M(G)$ defined by $\hat{\mu}(\Gamma) \subset \mathbf{Z}$.

It will be proved that for each μ in $F(G)$ and each rational integer a, the set of all γ in Γ such that $\hat{\mu}(\gamma) = a$ belongs to the coset ring of Γ.

Sketch of the proof: We have to show that on a compact group G each complex valued Radon measure μ with integral Fourier coefficients can be written

$$d\mu\,(x) = \sum_1^m P_j(x)\,dm_j\,(x), \tag{6.11}$$

where m_j is the Haar measure of a compact subgroup G_j of G normalized by $\int_{G_j} dm_j = 1$ and P_j is a trigonometric sum over G_j with integral coefficients.

The first term $dM_1 = P_1(x)\, dm_1(x)$ of this sum will be determined, up to multiplication by a character, by a minimal property: among all weak limits of products $\gamma(x)\, d\mu\,(x)$, $\gamma \in \Gamma$, M_1 has the smallest norm. Putting $\mu = M_1 + M_1'$, it will be shown that $\|\mu\| = \|M_1\| + \|M_1'\|$, $\|M_1\| \geqslant 1$ and $\|M_1'\| \leqslant \|\mu\| - 1$. Repeating this descent argument, we get (6.11).

8.8. *Definition of* $\mathscr{A}$

For a μ in $F(G)$, $\mathscr{A}$ is the set of all measures $\gamma(x)\, d\mu\,(x)$ for all γ in Γ such that $\int_G \gamma(x)\, d\mu\,(x) \neq 0$. All measures in $\mathscr{A}$ have the same norm $\|\mu\|$.

Two cases may occur:

(1) $\mathscr{A}$ is finite; then $\mathscr{A} = \{\bar{\gamma}_1\, d\mu, \ldots, \bar{\gamma}_n\, d\mu\}$. Let G_0 be the support group of μ. Consider the Fourier coefficients of μ with respect to the dual group of G_0: all but n of them vanish. Hence $d\mu\,(x) = \sum_1^n p_j \gamma_j\,(x)\, dm_0\,(x)$, where m_0 is the Haar measure of G_0 normalized by $m_0(G_0) = 1$ and where the p_j are rational integers. Theorem IX is therefore proved in this case.

(2) $\mathscr{A}$ is infinite. Let $\bar{\mathscr{A}}$ be the weak star closure of $\mathscr{A}$: $\bar{\mathscr{A}}$ is compact and metrizable. Since $\int_G d\nu$ is a rational integer different from 0 for all $\nu \in \mathscr{A}$, $|\int_G d\nu| \geqslant 1$ for all $\nu \in \bar{\mathscr{A}}$; hence $\bar{\mathscr{A}}$ does not contain 0.

The corollary of Theorem X shows that, if $\mu \in F(G)$, each ν in $\bar{\mathscr{A}}\backslash\mathscr{A}$ has the following two properties: $\|\nu\| < \|\mu\|$ and ν is singular with respect to the Haar measure of G.

8.9. *An optimal measure* ν

The norm decreases as we pass to the weak limits; hence there is a ν in $\bar{\mathscr{A}}$ whose norm is minimal.

Let γ be an element of Γ and $\gamma\bar{\mathscr{A}}$ the set of all measures $\gamma(x)\, d\lambda\,(x)$ for all measures λ in $\bar{\mathscr{A}}$. Since Γ is a group, $\gamma\bar{\mathscr{A}}$ is contained in $\bar{\mathscr{A}}$.

Let $\mathscr{B}$ be the set of all measures $\sigma = \gamma\nu$ for which $\int_G d\sigma \neq 0$. Then $\mathscr{B}$ is finite, for otherwise the preceding argument would imply the existence of a $\sigma \in \bar{\mathscr{B}} \subset \bar{\mathscr{A}}$ such that $0 < \|\sigma\| < \|\nu\|$ and this is impossible by the optimal choice of ν.

We have $\mathscr{B} = \{\bar{\gamma}_1\nu, \ldots, \bar{\gamma}_n\nu\}$, where $\gamma_1, \ldots, \gamma_n \in \Gamma$.

8.10. Definition of the group G_1 and computation of ν

If $\hat{\mu}$ does not assume the value a on Γ, any translate of $\hat{\mu}$ also has this property. Since $\mathbf{Z}$ is discrete, $\hat{\nu}$ which is a pointwise limit of translates of $\hat{\mu}$, does not assume the value a. Let $p_j = \int_G \bar{\gamma}_j(x)\, d\nu(x)$, $1 \leqslant j \leqslant n$. Then each p_j is a rational integer and is a value assumed by $\hat{\mu}$.

Let G_1 be the support group of ν: regarding ν as a measure carried by G_1, all but a finite number of the Fourier coefficients of ν vanish. *Hence* $d\nu(x) = \sum_1^n p_j \gamma_j(x)\, dm_1(x)$, *where* m_1 *is the Haar measure of* G_1 *normalized by* $\int_{G_1} dm_1(x) = 1$.

8.11. Computation of the restriction M_1 of μ to G_1 in terms of ν

PROPOSITION 7: *There exists a character* γ' *in* Γ *such that* $dM_1(x) = \gamma'(x)\, d\nu(x)$.

The proof of Proposition 7 depends on the following lemma:

LEMMA 13: *If* ν *is the weak limit of the sequence* $d\mu_k(x) = \gamma_k(x)\, d\mu(x)$, *then* ν *is also the weak limit (in the topology* $\sigma(M(G_1), C(G_1))$ *of the sequence* $\gamma_k(x)\, dM_1(x)$.

Proof of Lemma 13: We write $\mu = M_1 + M_1'$, where M_1 is carried by G_1 and where $|M_1'|(G_1) = 0$ To prove Lemma 13, it suffices to show that $\gamma_k M_1' = \mu_k'$ tends to 0 in the topology $\sigma(M(G), C(G))$.

For each positive ε, there exists a continuous function $f: G \to [0, 1]$ such that $\int_G f|dM_1'| \leqslant \varepsilon$ and $f = 1$ on G_1. Then for each g in $C(G)$, $|\int_G fg\, d\mu_k'| \leqslant \varepsilon \|g\|_\infty$. Writing $g = fg + (1 - f)g$, we get, for sufficiently large k, $|\int_G (1 - f)\, g\gamma_k\, d\mu| \leqslant \varepsilon$, since $\gamma_k\, d\mu$ tends to ν carried by G_1 on which $(1 - f)\, g = 0$. Since $(1 - f)\, \mu_k = (1 - f)\, \mu_k'$, we have $|\int_G g\, d\mu_k'| \leqslant \varepsilon(1 + \|g\|_\infty)$ for some sufficiently large k. Lemma 13 is now proved.

Proof of Proposition 7: We apply Theorem X to the group G_1 and the sequence of measures $\gamma_k(x)\, dM_1(x)$, $k \geqslant 1$.

There are three cases to consider:

(a) $\|\gamma_k dM_1 - \nu\| \to 0$ as $k \to +\infty$, which implies $\|M_1 - \bar{\gamma}_k \nu\| \to 0$ as $k \to +\infty$. Since $\nu \in F(G)$, the Cauchy sequence $(\bar{\gamma}_k \nu)$ is stationary for sufficiently large k and $M_1 = \bar{\gamma}_k \nu$.

(b) the set of measures $\gamma_k\, dM_1$ is infinite and does not converge in norm. But then its weak limit is singular with respect to the Haar measure of G_1, and this is not the case here.

(c) the set of measures $\gamma_k dM_1$ is finite. Then the weak limit $\nu = \gamma_k\, dM_1$ for sufficiently large k.

In all cases $dM_1(x) = \gamma'(x)\, d\nu(x)$.

We have proved that $dM_1(x) = \sum_1^n p_j \gamma_j'(x)\, dm_1(x)$, where the p_j are rational integers and dm_1 is the Haar measure of the group G_1. It may be observed that dM_1 is a weak limit of measures $\gamma(x)\, d\mu(x)$ for $\gamma \in \Gamma$, and that $\|M_1\|$ is minimal.

8.12. At this stage, we write $M_1' = \mu - M_1$. Then $\|M_1'\| \leqslant \|\mu\| - 1$. Everything that was done with μ can be repeated with M_1'. After a finite number of steps, we get $\mu = M_1 + \cdots + M_m$, where each M_j has the same form as M_1.

The set of all γ for which $\hat{\mu}$ takes a given value a can be defined, starting with the sets $E(b, j)$, where $\hat{M}_j$ takes the value b, by means of the operations of finite union, finite intersection and complement: write all decompositions $a = b_0 + \cdots + b_m$, where $b_j \in \mathbf{Z}$ and $|b_j| \leqslant \|\mu\|$; then the set $E(a)$ of γ for which $\hat{\mu}$ takes the value a is the union, taken over all these decompositions, of the intersections $E(b_0, 0) \cap E(b_1, 1) \cap \cdots \cap E(b_m, m)$.

Now each $E(b, j)$ belongs to the coset ring of Γ and so, therefore, does $E(a)$. Hence Theorem IX is proved.

9. Semi-idempotent measures

Let Γ be a countable subgroup of $\mathbf{R}$; we consider Γ as a discrete abelian group. Let G be the compact dual group.

DEFINITION 6: *A complex radon measure μ on G is said to be semi-idempotent if for all $t > 0$ in Γ, we have $\hat{\mu}(t) = 0$ or $\hat{\mu}(t) = 1$.*

THEOREM XI: *With the above notation, there exists, for any semi-idempotent measure μ, an idempotent measure σ such that $\hat{\sigma}(t) = \hat{\mu}(t)$ for all $t > 0$ in Γ.*

As in the proof of Theorem IX, the hypothesis will be enlarged.

DEFINITION 7: *Let μ be a complex Radon measure on a compact group G whose dual Γ is a subgroup of* $\mathbf{R}$. *We write $\mu \in F^+(G)$ if $\hat{\mu}(\gamma) \in \mathbf{Z}$ for all $\gamma > 0$ in Γ.*

DEFINITION 8: *Let μ_1 and μ_2 be two complex Radon measures on G. We write $\mu_1 \sim \mu_2$ if $\hat{\mu}_1(\gamma) = \hat{\mu}_2(\gamma)$ for all positive γ in Γ.*

It will be proved that for each μ in $F^+(G)$, there is a σ in $F(G)$ such that $\sigma \sim \mu$.

If μ is semi-idempotent, the subset E of Γ, where $\hat{\sigma}$ takes the value 1 belongs to the coset-ring of Γ. Hence there is an idempotent measure σ' such that $\sigma' \sim \mu$ and Theorem XI will be proved.

From now on, it will be assumed that $\mu \in F^+(G)$. Two cases may occur.

9.1. The periodic case

Let Γ^+ be the set of all $\gamma > 0$ in Γ and $\varphi : \Gamma^+ \to \mathbf{Z}$ the restriction to Γ^+ of the Fourier transform of μ. The periodic case is, by definition, the case where a $d > 0$ can be found in Γ such that for all γ in Γ^+, $\varphi(\gamma + d) = \varphi(\gamma)$.

In this case, a measure σ is given by the following lemma:

LEMMA 14: *With the preceding assumptions,* $\lim_{n \to +\infty} [\bar{d}(x)]^n \, d\mu(x) = d\sigma(x)$ *exists; this limit is taken in the weak topology* $\sigma(M(G), C(G))$ *and has the following two properties:* $\sigma \in F(G)$ *and for each* γ *in* Γ^+, $\hat{\sigma}(\gamma) = \hat{\mu}(\gamma)$.

Lemma 14 clearly gives Theorem XI in the periodic case; $\bar{d}(x)$ is the complex conjugate of the value taken at $x \in G$ by the character $d \in \mathrm{Hom}(G, \mathbf{T})$.

Proof of Lemma 14: Put $d\mu_n(x) = [\bar{d}(x)]^n \, d\mu(x)$; then $\|\mu_n\| = \|\mu\|$ and to show that μ_n has a weak-star limit σ, it suffices to prove that $\hat{\mu}_n(\gamma)$ has a limit for each γ in Γ.

If $\gamma > 0$, $\hat{\mu}_n(\gamma) = \varphi(\gamma + nd) = \varphi(\gamma)$ for all $n \geqslant 0$.

If $\gamma < 0$, there is an integer m such that $\gamma + md > 0$, as Γ is a subgroup of $\mathbf{R}$. Hence for $n \geqslant m$, $\hat{\mu}_n(\gamma) = \varphi(\gamma + nd)$ is constant.

9.2. The aperiodic case

If the hypothesis of Lemma 14 is not satisfied, $\|\gamma'(x) \, d\mu(x) - \gamma''(x) \, d\mu(x)\| \geqslant 1$ for distinct points γ' and γ'' in Γ^+.

We define $\mathscr{A}$ to be the set of all measures $d\nu(x) = \bar{\gamma}(x) \, d\mu(x)$ for all γ in Γ^+ such that $\int_G d\nu(x) \neq 0$; then $\int_G d\nu(x) \in \mathbf{Z}$. If no such γ exists, it is plain that $\mu \sim 0$ and Theorem XI is proved.

Two other cases may occur:

(a) $\mathscr{A}$ *is finite:* Since for distinct γ in Γ^+, the corresponding measures $d\nu$ are different, $\mathscr{A}$ is finite if there exist an integer $n \geqslant 1$, n distinct elements $\gamma_1, \ldots, \gamma_n$ of Γ^+ and n non-zero rational integers $p_1, \ldots, p_n$ such that: if $\gamma = \gamma_j$, $1 \leqslant j \leqslant n$, $\hat{\mu}(\gamma) = p_j$ and if $\gamma \neq \gamma_j$ for all j, $1 \leqslant j \leqslant n$, $\hat{\mu}(\gamma) = 0$. Then $\mu \sim \sum_1^n p_j \bar{\gamma}_j(x) \, dx = \sigma$ and Theorem XI is proved.

(b) $\mathscr{A}$ *is in finite:* Let $\bar{\mathscr{A}}$ be the weak star closure of $\mathscr{A}$. Then $\mathscr{A}$ is metrizable and compact and $0 \notin \bar{\mathscr{A}}$.

9.3. An optimal measure ν and a number l

The measure ν is an element of $\bar{\mathscr{A}}$ whose norm is minimal. There is a sequence $(\gamma_k)_{k\geqslant 1}$ of elements of Γ^+ such that $d\mu_k(x) = \bar{\gamma}_k(x)\, d\mu(x)$ tends to ν in the weak-star topology. Passing to a subsequence if necessary it may be assumed that $\gamma_k \to l \geqslant 0$ in the usual topology of $\mathbf{R}$ (l may be $+\infty$).

LEMMA 15: *Let $\mathscr{B}$ be the set of all measures $d\varrho(x) = \bar{\gamma}(x)\, d\nu(x)$ for all $\gamma > -l$ such that $\gamma \in \Gamma$ and $\int_G d\varrho(x) \neq 0$. Then the weak-star closure $\bar{\mathscr{B}}$ of $\mathscr{B}$ is contained in $\bar{\mathscr{A}}$.*

Proof: It suffices to show that $\mathscr{B}$ is contained in $\bar{\mathscr{A}}$. Writing $d\nu(x) = \lim_{k\to+\infty} \bar{\gamma}_k(x)\, d\mu(x)$ and $d\varrho_k(x) = (\bar{\gamma} + \bar{\gamma}_k)(x)\, d\mu(x)$, we get $d\varrho = \lim_{k\to+\infty} d\varrho_k$. Since $\gamma > -l$ and $\lim_{k\to+\infty} \gamma_k = l$ for the usual topology of $\mathbf{R}$, we have $\gamma + \gamma_k > 0$ for sufficiently large k and so the measure $d\varrho_k$ belongs to the set of all measures $\bar{\gamma}'(x)\, d\mu(x)$ for positive γ' in Γ. To prove that $d\varrho_k \in \mathscr{A}$ it suffices to prove that $\int_G d\varrho_k \neq 0$ if k is sufficienly large. But $\int_G d\varrho_k = 0$ for an infinite set of k would imply $\int_G d\varrho(x) = 0$, which is not true. Lemma 15 is therefore proved.

9.4. Computation of ν

LEMMA 16: *There exist a closed subgroup G_1 of G and a trigonometric sum P_1 with integral coefficients such that $d\nu(x) = P_1 dm_1(x) + d\nu'(x)$, where m_1 is the Haar measure of G_1 and the Fourier transform of ν' vanishes on $\Gamma \cap (-l, +\infty)$.*

Three cases may occur in the proof:

(1) *The periodic case:* There exists a $d > 0$ in Γ such that for each $\gamma > -l$, $\hat{\nu}(\gamma + d) = \hat{\nu}(\gamma)$. We define a measure ν_1 by $d\nu_1(x) = \lim\, [\bar{d}(x)]^n\, d\nu(x)$; then $\nu_1 \in F(G)$, $\hat{\nu}_1(\gamma) = \hat{\nu}(\gamma)$ for all $\gamma > -l$ in Γ and the product $\bar{\gamma}\nu_1$ belongs to $\bar{\mathscr{A}}$ for all γ in Γ such that $\int_G \bar{\gamma}(x)\, d\nu_1(x) \neq 0$, since $\gamma + nd > 0$ for n sufficiently large. Since the norms decrease in the passage to weak limits $\|\nu_1\| \leqslant \|\nu\|$; $\nu_1 = 0$ implies that $\hat{\nu}(\gamma) = 0$ for all $\gamma > -l$ (in this case $\hat{\mu}(\gamma) = 0$ for all $\gamma > 0$). By the optimal choice of ν, $\|\nu_1\| = \|\nu\|$ and the norm of each non-zero weak-star limit ν_2 of $\gamma\nu_1$, $\gamma \in \Gamma$, also has this property.

By the proof of Theorem X, $d\nu_1(x) = P_1(x)\, dm_1(x)$, for otherwise $d\nu_1(x)$ would be a sum $P_1(x)\, dm_1(x) + P_2(x)\, dm_2(x) + \cdots + P_n(x)\, dm_n(x)$ and $P_1(x)\, dm_1(x)$ would be a weak-star limit of products $\gamma\nu_1$ such that $0 < \|P_1\, dm_1\| < \|\nu_1\| - 1$.

From now on, in the periodic case our optimal ν will be replaced by the optimal $\gamma\nu_1$ for a $\gamma \in \Gamma$ such that $\int_G \gamma\nu_1 \neq 0$.

We can then write, in the periodic case, $d\nu(x) = P_1(x)\, dm_1(x)$.

(2) *The aperiodic case; $\mathscr{B}$ is infinite:* Since we are in the aperiodic case, $\|\gamma'\nu - \gamma''\nu\| \geqslant 1$ if $-l < \gamma' < \gamma''$. Let $\gamma_k\nu$, $k \geqslant 1$, be an infinite sequence of distinct measures in $\mathscr{B}$ and let ϱ be the weak-star limit of an infinite subsequence of these $\gamma_k\nu$. Then $\varrho \in \bar{\mathscr{A}}$, $\varrho \neq 0$ and, by Theorem X, $\|\varrho\| < \|\nu\|$ which contradicts the optimal choice of ν. This case therefore does not occur.

(3) *The aperiodic case; $\mathscr{B}$ is finite:* This means that, for all $\gamma > -l$, $\hat{\nu}(\gamma) = 0$ with the exception of n values of γ, γ_j, for which $p_j = \hat{\nu}(\gamma_j) \neq 0$ $(1 \leqslant j \leqslant n)$. We can then write $d\nu(x) = (\sum_1^n p_j\gamma_j(x))\, dx + d\nu'(x)$, where $\hat{\nu}'(\gamma) = 0$ for all $\gamma > -l$ in Γ.

9.5. *Computation of μ in terms of ν*

From now on let G_1 be the support group of the measure $d\nu$ (in the aperiodic case $G_1 = G$).

LEMMA 17: *With the preceding notation, if P_1 is not identically* 0, *$d\nu$ is not singular with respect to the Haar measure dm_1 of G_1.*

Proof: In the periodic case, we have replaced ν by $\nu_1 = P_1\, dm_1$, which is absolutely continuous with respect to the Haar measure of G_1.

In the aperiodic case we have $\hat{\nu}(\gamma) = 0$ for all $\gamma > -l$ in Γ with the exception of $\gamma = \gamma_j$, $1 \leqslant j \leqslant n$, for which $\hat{\nu}(\gamma_j) = p_j$. Then the following proposition will be applied ([15], th. 8.2.3, p. 200).

PROPOSITION 8: *Let Γ be a subgroup of* **R**, *G the compact dual group and σ a singular complex measure on G such that $\hat{\sigma}(\gamma) = 0$ for all $\gamma > 0$. Then $\int_G d\sigma = 0$.*

Proof: Let E be the closure, in the Hilbert space $L^2(d\sigma)$, of the vector space of all trigonometric sums $P(x) = \sum_{\gamma<0} a_\gamma\gamma(x)$. Assume for a moment that 1 belongs to E. This means that we can find a sequence $P_n(x) = \sum_{\gamma<0} a_n(\gamma)\,\gamma(x)$, such that $\lim_{n\to+\infty} \int_G |1 + P_n|^2\, d|\sigma| = 0$. Hence, by Schwarz's inequality, $\lim_{n\to+\infty} \int_G (1 + P_n)\, d\sigma = 0$. But $\int_G P_n\, d\sigma = 0$ by hypothesis; hence $\int_G d\sigma = 0$.

To prove that 1 belongs to E, let φ be the orthogonal projection of 1 on E. Then $1 - \varphi$ is orthogonal to all elements ψ of E; applying this remark to $\psi(x) = \gamma(x)\,(1 - \varphi(x))$ for $\gamma < 0$, we get $\int_G \gamma(x)\, |1 - \varphi(x)|^2\, d|\sigma|(x) = 0$. By complex conjugation this is also true for $\gamma > 0$. Hence $|1 - \varphi|^2$

$d|\sigma| = c dx$; since σ is singular, $c = 0$ and $\varphi = 1$. Proposition 8 is thus proved.

In our context, we can order the real numbers γ_j, $(1 \leqslant j \leqslant n)$ in such a way that $\gamma_n > \gamma_j$ for all other j. Then, if $d\sigma(x) = \overline{\gamma_n(x)}\, d\nu(x)$ is singular, $\hat{\sigma}(\gamma) = 0$ for all $\gamma > 0$ and $\int_G d\sigma = p_n \neq 0$. This is impossible and so ν is not singular with respect to dx. Lemma 17 is therefore proved.

9.6. *Final stage of the computation of μ*

Let M_1 be the restriction of μ to G_1 and let $\mu = M_1 + M_1'$. If ν is the weak-star limit of the sequence $\bar{\gamma}_k\mu$, then ν, carried by G_1, is also the weak-star limit of the sequence $\bar{\gamma}_k M_1$, $k \geqslant 1$ (Lemma 13).

It will be proved that $\bar{\gamma}_k M_1 = \nu$ for sufficiently large k; hence by Lemma 16, $M_1 = \lim_{k\to+\infty} \gamma_k \nu \sim \sigma_1 = \lim_{k\to+\infty} \gamma_k \nu_1$ $(\nu_1 = P_1\, dm_1)$; we have replaced γ_k by a subsequence if necessary, to ensure that the latter limit exists. Note that $\sigma_1 \in F(G)$. Then $M_1 \in F^+(G)$, $M_1' \in F^+(G)$ and since $\|M_1'\| \leqslant \|\mu\| - 1$, we reach the required conclusion $\mu \sim \sigma_1 + \cdots + \sigma_n = \sigma \in F(G)$ after a finite number of steps.

To show that $\nu = \bar{\gamma} M_1, \gamma > -l$, Theorem X will be applied. Two cases may occur.

(1) If $\|\bar{\gamma}_k M_1 - \nu\| \to 0$ as $k \to +\infty$, we have $\|M_1 - \gamma_k\nu\| \to 0$; in the periodic case $\nu = \nu_1 = P_1(x)\, dm_1$ and $\gamma_k\nu$, $k \geqslant 1$, cannot be a Cauchy sequence without being constant. Hence $M_1 = \gamma_k\nu$ for sufficiently large k. In the aperiodic case $\|\gamma''\nu - \gamma'\nu\| \geqslant 1$ for $\gamma' \neq \gamma''$, and the conclusion is the same.

(2) If $\nu = \lim_{k\to+\infty} \bar{\gamma}_k M_1$ in the weak-star topology but not in the strong topology, ν is singular with respect to the Haar measure of G_1. We have seen that this is not the case.

9.7. *A particular case*

PROPOSITION 9: *Let $(\lambda_k)_{k\geqslant 1}$ be an increasing sequence of positive real numbers such that $\lambda_k - k \to 0$ as $k \to +\infty$. Let Γ be the subgroup of* **R** *generated by the λ_k, $k \geqslant 1$, and let G be the compact dual group.*

Assume that a complex Radon measure μ on G has the following properties: $\hat{\mu}(\lambda_k) = 1$ for all $k \geqslant 1$ and $\hat{\mu}(t) = 0$ if $t \geqslant 0$ does not belong to the set Λ of λ_k, $k \geqslant 1$.

Then $\lambda_k = k$ for sufficiently large k.

Proof: The notation of the proof of Theorem XI will be used: $\mathscr{A}$ is the set of all measures $\bar{\lambda}_k(x)\, d\mu\,(x)$; if $\mathscr{A}$ is finite, $\lambda_{k+1} - \lambda_k$ assumes a finite number of values and since $\lambda_k - k$ tends to 0, $\lambda_{k+1} - \lambda_k = 1$ for sufficiently large k, which implies $\lambda_k = k$ for sufficiently large k. If $\mathscr{A}$ is infinite, the optimal measure ν is a weak limit of a subsequence of $\bar{\lambda}_k(x)\, d\mu\,(x)$. If $x = \lambda_{k+j} - \lambda_k$ for some fixed j and sufficiently large k in the subsequence, $\hat{\nu}(x) = 1$. Otherwise $\hat{\nu} = 0$. Hence the support $S(\nu)$ of $\hat{\nu}$ is contained in $\mathbf{Z}$ and ν is idempotent.

Let G_1 be the support-group ov ν and let M_1 be the restriction of μ to G_1. Then $dM_1\,(x) = \lambda_k(x)\, d\nu\,(x)$ for infinitely many k. Hence $S(M_1)$, the support of $\hat{M}_1$, is contained in $\mathbf{Z}$. Let $M_1' = \mu - M_1$. Then $S(M_1') \subset \Lambda \cup \mathbf{Z}$.

Repeating the preceding argument, we obtain a decomposition $\mu \sim M_1 + \cdots + M_m$ of μ, where the spectrum of each M_j is contained in $\mathbf{Z}$ for $1 \leqslant j \leqslant m - 1$ and the spectrum of M_m is finite. Hence $\lambda_k = k$ for sufficiently large k.

10. Behaviour at infinity of special a-periodic trigonometric series

THEOREM XII: *Let $(\lambda_k)_{k \geqslant 1}$ be an increasing sequence of positive real numbers such that $\lambda_k - k \to 0$ as $k \to +\infty$. Let Λ be the set of all λ_k, $k \geqslant 1$ and $\mathscr{C}_\Lambda$ the space of all mean periodic functions whose spectra are contained in Λ.*

If $\lambda_k \neq k$ for an infinite number of $k \geqslant 1$, there is a mean-periodic function f whose spectrum lies in Λ which is not bounded.

Proof: If all mean periodic functions in $\mathscr{C}_\Lambda$ are bounded, Λ is a coherent set of frequencies. Hence a positive l can be found such that the Banach algebras $B(\Lambda + [-l, l])$ and $B(\Lambda \times [-l, l])$ are isomorphic (see Chapter IV, Theorem X). For each $\varepsilon > 0$, let Δ_ε be defined by $\Delta_\varepsilon(t) = \sup\,(0, 1 - |t|/\varepsilon)$. Then Δ_ε belongs to $A(\mathbf{R})$ and $\|\Delta_\varepsilon\|_{A(\mathbf{R})} = 1$. Applying Theorem X of Chapter IV, we find a constant C and, for each ε in $(0, l)$, a complex Radon measure μ_ε on $\mathbf{R}$, such that $\|\mu_\varepsilon\| \leqslant C$ and $\hat{\mu}_\varepsilon\,(\lambda_k + t) = \Delta_\varepsilon(t)$ if $k \geqslant 1$ and $|t| \leqslant l$.

Replacing l by a smaller number if necessary, it may be assumed that $l \leqslant \frac{1}{2}$. Now assuming this, let $T_l(t)$ be the 1-periodic function defined on $[-\frac{1}{2}, \frac{1}{2}]$ by $T_l(t) = 2\Delta_{2l/3}(t) - \Delta_{l/3}(t)$; then $T_l(t) = 1$ on $[-l/3, l/3]$ and $T(t) = 0$ outside $[-2l/3, 2l/3]$. The Fourier coefficients of T_l are $0(k^{-2})$ and so T_l is the Fourier-transform of a discrete measure σ carried by $\mathbf{Z}$.

For $k \geqslant k_0$, $|\lambda_k - k| \leqslant l/6$. Choose ε in $(0, l/6)$; then $\nu_\varepsilon = \mu_\varepsilon * \sigma$ has the following properties

(i) $\|\nu_\varepsilon\| \leqslant C_2$, where C_2 is independent of ε;

(ii) if $t \geqslant \lambda_{k_0}$ and $|t - \lambda_k| \leqslant \varepsilon$, then $\hat{\mu}_\varepsilon(t) = \Delta_\varepsilon(t)$;

(iii) for all other $t \geqslant \lambda_{k_0}$, $\hat{\mu}_\varepsilon(t) = 0$.

Let Γ be the discrete group generated by Λ. Let $j: \Gamma \to \mathbf{R}$ be the canonical injection and $h: \mathbf{R} \to G$ the dual map. Then h is injective and $h(\mathbf{R})$ is dense in G. When $\mathbf{R}$ is thought of as a dense subgroup of G, the bounded measures ν_ε become complex Radon measures on G. In the weak-star topology $(\sigma(M(G), C(G))$, a subsequence of the ν_ε tends to a measure ν with the following properties: (i) $\|\nu\| \leqslant C_2$, (ii) $\hat{\nu}(t) = 0$ for $t \geqslant \lambda_{k_0}$, $t \neq \lambda_k$, and (iii) $\hat{\nu}(\lambda_k) = 1$ if $k \geqslant k_0$.

Proposition 9 shows that $\lambda_k = k$ for sufficiently large k, and Theorem XII is thus proved.

Let f be an unbounded mean periodic function whose spectrum lies in Λ. We write $f(t) \sim \sum_{k \geqslant 1} a_k \exp 2\pi i \lambda_k t$ where it may be assumed that all the λ_k are different from zero. Then $\sum_{k \geqslant 1} |a_k|^2 < +\infty$ and the primitive g of f has the absolutely convergent series $\sum_{k \geqslant 1} (a_k/2\pi i \lambda_k) \exp 2\pi i \lambda_k t$; hence g is almost periodic and continuously differentiable; the spectrum of g lies in Λ but the derivative of g is not bounded.

11. Notes

Helson sets are the compact versions of Sidon sets and are defined by $A(E) = C(E)$. The family of Helson sets is stable with respect to finite unions [94].

Theorem IV is a remarkable discovery of Drury [40] and Theorem VIII was found by Mme Deschamps-Gondim [38].

A simple proof of P. J. Cohen's theorem (Theorem IX) was obtained by Itô and Amemiya [53] and the characterization of semi-idempotent measures can be found in [52].

CHAPTER VII

PISOT NUMBERS AND SPECTRAL SYNTHESIS

Let $\theta > 2$ be a Pisot number and E the compact set of real numbers of Cantor type constructed with the dissection ration θ^{-1}. Then for each bounded continuous function $\varphi : \mathbf{R} \to \mathbf{C}$ whose spectrum lies in E, there exists a sequence φ_k, $k \geqslant 1$, of finite trigonometric sums whose frequencies belong to E such that $\varphi_k \to \varphi$ uniformly on each compact subset of the line and $\sup_{\mathbf{R}} |\varphi_k| \leqslant \sup_{\mathbf{R}} |\varphi|$. This is a strong form of spectral synthesis.

In Sections 1–4 classical facts about spectral synthesis are recalled. In Sections 6 and 7 a fundamental theorem is proved, and in Section 8 a p-adic version of the preceding results is given.

1. Spectral synthesis and structure of closed ideals of a group algebra

1.1. The Banach algebra $A(\mathbf{R})$

Let $A(\mathbf{R})$ be the Banach algebra of all continuous functions $f: \mathbf{R} \to \mathbf{C}$, which are the Fourier-transforms of functions in $L^1(\mathbf{R})$; $A(\mathbf{R})$ is normed by $\|f\|_{A(\mathbf{R})} = \int_{-\infty}^{+\infty} |\hat{f}(t)|\, dt$.

Clearly $\|f\|_\infty = \sup_{\mathbf{R}} |f| \leqslant \|f\|_{A(\mathbf{R})}$ and $\|f\|_{A(\mathbf{R})}$ is translation-invariant.

PROPOSITION 1: *Each compactly supported continuously differentiable function $f: \mathbf{R} \to \mathbf{C}$ belongs to $A(\mathbf{R})$.*

Proof: The hypothesis implies that f and f' belong to $L^2(\mathbf{R})$. Let F be the Fourier-transform of f; then the Fourier-transform of f' is $-2\pi i x F$ and, by Parseval's relation, F and xF belong to L^2. Schwarz' inequality gives $\int_{-1}^{1} |F(x)|\, dx \leqslant \sqrt{2}\, (\int_{-1}^{1} |F|^2\, dx)^{1/2} < +\infty$ and $\int_{|x| \geqslant 1} |F(x)|\, dx \leqslant (\int_{|x|\geqslant 1} x^{-2}\, dx)^{1/2} (\int_{|x|\geqslant 1} x^2 |F|^2\, dx)^{1/2} \leqslant +\infty$. Hence F belongs to $L^1(\mathbf{R})$. We have got

$$\|f\|_{A(\mathbf{R})} \leqslant \sqrt{2}\, \|f\|_2 + \sqrt{2}\, \|f'\|_2. \tag{7.1}$$

The following result shows how approximate units can be found in $A(\mathbf{R})$ ($A(\mathbf{R})$ does not have a unit).

THEOREM I: *Let f and g be two functions in* $A(\mathbf{R})$.

If $f(0) = 1, \quad \lim_{n \to +\infty} \|g(x) - f(n^{-1}x)\, g(x)\|_{A(\mathbf{R})} = 0.$

If $g(0) = 0, \quad \lim_{n \to +\infty} \|f(nx)\, g(x)\|_{A(\mathbf{R})} = 0.$

Let F and G be two functions in $L^1(\mathbf{R})$ whose Fourier transforms are f and g and for each $n \geqslant 1$, let $F_n = nF(nx)$ and $\tilde{F}_n = n^{-1}F(n^{-1}x)$. Then $\|F_n\|_1 = \|F\|_1$. We have to prove that

$$\text{if} \quad \int_{-\infty}^{+\infty} F(x)\, dx = 1, \quad \|G - F_n * G\|_1 \to 0 \quad \text{as} \quad n \to +\infty \tag{7.2}$$

$$\text{if} \quad \int_{-\infty}^{+\infty} G(x)\, dx = 0, \quad \|\tilde{F}_n * G\|_1 \to 0 \quad \text{as} \quad n \to +\infty. \tag{7.3}$$

In each case if we replace F by F' and G by G' such that $\|F - F'\|_1 \leqslant \varepsilon$, $\|G - G'\|_1 \leqslant \varepsilon$ and if we put $F'_n = nF'(nx)$, we get $\|F_n - F'_n\| \leqslant \varepsilon$ and, for all $n \neq 0$, $\|G - F_n * G - G' + F'_n * G'\|_1 \leqslant \varepsilon + \varepsilon\|F\|_1 + \varepsilon\|G\|_1 + \varepsilon^2$ while $\|\tilde{F}_n * G - \tilde{F}'_n * G'\|_1 \leqslant \varepsilon\|F\|_1 + \varepsilon\|G_1\| + \varepsilon^2$. Hence it suffices to prove the result for F and G belonging to a dense subspace of $L^1(\mathbf{R})$: we choose the continuous functions with compact support.

If the supports of F and G are contained in $[-T, T]$, the support of $G - F_n * G$ is contained in $[-2T, 2T]$ and, putting $\varepsilon_n = \sup_{|x-y| \leqslant n^{-1}T} |G(x) - G(y)|$, we get $|G(x) - (F_n * G)(x)| = |\int_{-T/n}^{T/n} G(x)\, F_n(y)\, dy - \int_{-T/n}^{T/n} G(x - y)\, F_n(y)\, dy| \leqslant \varepsilon_n\|F_1\|$. Hence $\|G - F_n * G\|_1 \to 0$ for these continuous compactly supported F and G and the proof of (7.2) is complete.

Making a simple change of variables, we see that $\|\tilde{F}_n * G\|_1 = \|F * G_n\|_1$ and returning to the case of continuous compactly supported F and G, we get, with the preceding notation,

$$|(F * G_n)(x)| = \left| \int_{-T/n}^{T/n} F(x - y)\, G_n(y)\, dy - \int_{-T/n}^{T/n} F(x)\, G_n(y)\, dy \right|$$
$$\leqslant \varepsilon_n\|G\|_1,$$

and (7.3) is proved.

If E is a closed set of real numbers, the restriction algebra $A(E)$ is defined by the following rule: a continuous function $f: E \to \mathbf{C}$ belongs to $A(E)$ if f

is the restriction to E of a function g of $A(\mathbf{R})$, and the norm of f in $A(E)$ is the infimum, taken over all such extensions g, of $\|g\|_{A(\mathbf{R})}$.

PROPOSITION 2: *If f is a function in $A(\mathbf{R})$ vanishing at 0, the norm of f in the restriction algebra $A[-\varepsilon, \varepsilon]$ tends to 0 with ε.*

Proof: Let $\alpha : \mathbf{R} \to [0, 1]$ be a function in $A(\mathbf{R})$, which is 1 on $[-1, 1]$. Then, with $\varepsilon = n^{-1}$ and the notation of Theorem I, $\|f\|_{A[-\varepsilon,\varepsilon]} \leqslant \|f(x)\,\alpha\,(nx)\|_{A(\mathbf{R})} \to 0$ as $n \to +\infty$. Since for each f in $A(\mathbf{R})$, $\|f\|_{A[-\varepsilon,\varepsilon]}$ decreases with ε, the general case follows.

PROPOSITION 3: *Let f be a function in $A(\mathbf{R})$ and x a real number such that $f(x) \neq 0$. There exists a function h in $A(\mathbf{R})$ such that $f(x)\,h(x) = 1$ on a neighbourhood of x.*

Proof: By translation we reduce the problem to the case $x = 0$ and we may assume that $f(0) = 1$. If $\varepsilon < 1$, using the notation of Proposition 2, we get $\|1 - f\|_{A[-\varepsilon,\varepsilon]} = \|\alpha - f\|_{A[-\varepsilon,\varepsilon]} \to 0$ with ε.

For sufficiently small ε, let g be a function in $A(\mathbf{R})$ such that $\|g\|_{A(\mathbf{R})} \leqslant \frac{1}{2}$ and $1 - f = g$ on $[-\varepsilon, \varepsilon]$. Put $h = \sum_0^\infty g^k$; this series is convergent in $A(\mathbf{R})$ and also for the uniform norms. Hence $hf = 1$ on $[-\varepsilon, \varepsilon]$.

PROPOSITION 4: *Let K be a compact subset of $\mathbf{R}$ and f a function in $A(K)$ which never vanishes on K. Then $1/f$ belongs to $A(K)$.*

Proof: Let $\mathscr{S}$ be the algebra of all complex valued functions all of whose derivatives, including the function itself, decay rapidly at infinity; $\mathscr{S}$ is invariant under Fourier transform and is contained in $A(\mathbf{R})$.

For each x in K there exists a neighbourhood $V(x)$ of x and a function g_x in $A(\mathbf{R})$ such that $fg_x = 1$ on $V(x)$. Standard arguments on partitions of unity show that there exist an integer n, n points $x_1, \ldots, x_n$ of K and n regular functions $\alpha_1, \ldots, \alpha_n$ in $\mathscr{S}$ such that, writing V_j for $V(x)$ and g_j for g_x if $x = x_j$, each α_j is supported by V_j and $\alpha_1 + \cdots + \alpha_n = 1$ on K.

We have: $1 = \sum_1^n \alpha_j = \sum_1^n \alpha_j g_j f$ on K and $1/f = \sum_1^n a_j g_j \in A(\mathbf{R})$.

1.2. The ideals $I(E)$ and $J(E)$

Among the closed ideals of $A(\mathbf{R})$,a first type can be defined in the following way: let E be a closed set of real numbers and $I(E)$ be the set of all f in $A(\mathbf{R})$ which vanish on E; $I(E)$ is plainly a closed ideal of $A(\mathbf{R})$.

Let $J_0(E)$ be the set of all f in $A(\mathbf{R})$ which vanish on a neighbourhood of E where the neighbourhood depends on f. Clearly $J_0(E)$ is an ideal of $A(\mathbf{R})$ and so is the closure $J(E)$ of $J_0(E)$ in $A(\mathbf{R})$.

DEFINITION 1: *The zero set of an ideal I of $A(\mathbf{R})$ is the largest closed set on which all the functions of I vanish.*

THEOREM II: *Let I be a closed ideal of $A(\mathbf{R})$ and E the zero set of I. Then $J(E) \subset I \subset I(E)$.*

The second inclusion being obvious, it suffices to prove the first one. The proof depends on two lemmas:

LEMMA 1: *Let I be an ideal of $A(\mathbf{R})$, E the zero set of I and K a compact set of real numbers disjoint from E. Then there exists a function f in I which is identically 1 on K.*

Proof: For each x in K, there is a function f_x in I such that $f_x(x) = 1$, since otherwise x would belong to the zero set of I. Since I is an ideal, $g_x = f_x\bar{f}_x$ belongs to I; $g_x \geqslant 0$ and there is a neighbourhood $V(x)$ of x such that $g_x \geqslant \frac{1}{2}$ on $V(x)$. Let $V_1, \ldots, V_n$ be a covering of K with n of these $V(x)$ and let f be the corresponding sum $\sum_1^n g_j$ where we write g_j for g_x and V_j for $V(x)$ if $x = x_j$. Then $f \in I$, and $f(x) \geqslant \frac{1}{2}$ for all x in K. Proposition 4 shows that there exists a function h in $A(\mathbf{R})$ such that $fh = 1$ on K. Hence fh is the required element of I.

LEMMA 2: *Let I be an ideal of $A(\mathbf{R})$. The compactly supported functions of I are dense in I.*

Proof: If α is any compactly supported function $A(\mathbf{R})$ such that $\alpha(0) = 1$, then, for each f in I, $f_n = f(x)\,\alpha\,(n^{-1}x)$ belongs to I and tends to f.

Proof of Theorem II: Since I is closed, it suffices to show that $J_0(E) \subset I$. Lemma 2 shows that we can further restrict our attention to compactly supported functions g in $J_0(E)$. Let K be the support of such a g; K does not intersect E and by Lemma 1 there exists a function f in I which equals 1 on K. Then $g = gf$ belongs to I and the proof is complete.

DEFINITION 2: *A closed set E of real numbers is called a set of synthesis if one of the following two equivalent statements is true:*

(a) *each function f in $A(\mathbf{R})$ which vanishes on E is the limit, in the norm of $A(\mathbf{R})$ of a sequence f_k of functions in $A(\mathbf{R})$ which vanish on neighbourhoods Ω_k of E;*

(b) *$I(E)$ is the unique closed ideal of $A(\mathbf{R})$ whose zero set is E.*

Theorem II shows that (a) $\Leftrightarrow$ (b).

2. Spectral synthesis and atomization of distributions

2.1. We begin with a technical lemma:

LEMMA 3: *For any closed set E of real numbers, $\mathcal{S} \cap J_0(E)$ is dense in $J(E)$.*

By Lemma 2 it suffices to show that each function f in $A(\mathbf{R})$ whose support is a compact set K disjoint from E may be approached by a sequence f_k of functions in $\mathcal{S}$. Since $\mathcal{S}$ is dense in $A(\mathbf{R})$, we can find a sequence g_k in $\mathcal{S}$ whose limit is f. Let α be any function in $\mathcal{S}$ which equals 1 on K and which is 0 on a neighbourhood of E. Then the sequence $f_k = \alpha g_k$, $k \geqslant 1$, tends to f as required.

DEFINITION 3: *Let $PM(\mathbf{R})$ be the space of all distributions S on $\mathbf{R}$ whose Fourier transforms $\hat{S}$ belong to $L^\infty(\mathbf{R})$. We put $\|S\|_{PM} = \|\hat{S}\|_\infty$ and this norm is called the pseudo-measure norm. The duality between $PM(\mathbf{R})$ and $A(\mathbf{R})$ is defined by*

$$\langle S, \hat{F} \rangle = \langle \hat{S}, F \rangle = \int_{-\infty}^{+\infty} \hat{S}(t)\, F(t)\, dt, \quad F \in L^1(\mathbf{R}), \quad \hat{S} \in L^\infty(\mathbf{R}).$$

For any closed set E of real numbers, let $PM(E)$ be the space of all S in $PM(\mathbf{R})$ which are supported by E.

LEMMA 4: *For each S in $PM(E)$ and each f in $J(E)$, $\langle S, f \rangle = 0$.*

By continuity we may restrict our attention to $f \in \mathcal{S} \cap J_0(E)$, and in this case $\langle S, f \rangle = 0$ by the definition of the support of a distribution.

2.2. The weak-star topology $\sigma(PM(\mathbf{R}), A(\mathbf{R}))$

In what follows $\sigma(PM(\mathbf{R}), A(\mathbf{R}))$ will denote the weak-star topology defined by the duality between $A(\mathbf{R})$ and $PM(\mathbf{R})$. A base of neighbourhoods of 0 for $\sigma(PM(\mathbf{R}), A(\mathbf{R}))$ is given by the sets $V(F, \varepsilon)$ defined as follows: for each $\varepsilon > 0$ and each finite subset F of $A(\mathbf{R})$, $V(F, \varepsilon)$ is the set of all S in $PM(\mathbf{R})$ such that $\sup_F |\langle S, f \rangle| \leqslant \varepsilon$.

Since $A(\mathbf{R})$ contains a contable dense subset, the restriction of the weak-star topology to the unit ball B of $PM(\mathbf{R})$ is metrizable and B is compact for $\sigma(PM(\mathbf{R}), A(\mathbf{R}))$.

2.3. A description of PM (E)

THEOREM III: *Let E be a closed set of real numbers. The following three properties of E are equivalent:*

(a) *E is a set of synthesis;*

(b) *for each S in PM (E) and each f in $A(\mathbf{R})$ vanishing on E, $\langle S, f\rangle = 0$;*

(c) *each distribution S carried by E whose Fourier transform is bounded is the weak star limit of a net $(\mu_\alpha)_{\alpha\in A}$ of measures carried by finite subsets of E.*

Proof: To get the equivalence between (a) and (b), we consider the ideal $J(E)$ as a closed linear subspace of $I(E)$. The Hahn–Banach theorem shows that $I(E) = J(E)$ if and only if (b) holds.

The equivalence between (b) and (c) is also given by the Hahn–Banach theorem applied to $PM(E)$ topologized by the weak-star topology $\sigma(PM(E), A(\mathbf{R}))$; let $M_f(E)$ be the linear subspace of all measures carried by finite subsets of E. Then $M_f(E)$ is dense in $PM(E)$ if and only if (b) is true.

2.4. The dual of a restriction algebra

DEFINITION 4: *For a closed set E of real numbers, we define $N(E)$ to be the subspace of all S in PM (E) such that $\langle S, f\rangle = 0$ for all f in $I(E)$.*

PROPOSITION 5: *The dual space of the restriction algebra $A(E)$ is $N(E)$.*

The proof is trivial.

COROLLARY 1: *A closed set E of real numbers is a set of synthesis if and only if the dual of $A(E)$ is $PM(E)$.*

3. A strong form of spectral synthesis

In certain cases that we shall now describe, the property of spectral synthesis gives a very concrete description of the space of bounded continuous functions $\varphi: \mathbf{R} \to \mathbf{C}$, whose spectra lie in a compact set of real numbers.

3.1. We begin with a very simple remark. If E is a closed set of real numbers and if a sequence φ_k, $k \geqslant 1$, of trigonometric sums whose frequencies belong to E satisfies the two conditions

(a) $\sup\limits_{k\geqslant 1} \sup\limits_{\mathbf{R}} |\varphi_k| < +\infty$;

(b) $\varphi_k \to \varphi$ uniformly on each compact subset of $\mathbf{R}$, then φ is bounded and continuous and the spectrum of φ also lies in E. In fact let S be the Fourier-transform of φ and let F be any function in $L^1(\mathbf{R})$ whose Fourier-transform f vanishes on E. By Lebesgue's dominated convergence theorem

$$\lim_{k \to +\infty} \int_{-\infty}^{+\infty} \varphi_k(-x)\, F(x)\, dx = \int_{-\infty}^{+\infty} \varphi(-x)\, F(x)\, dx = \langle S, f \rangle$$

but $\varphi_k(x)$ is a finite sum $\sum_{\lambda \in E} a(k, \lambda) \exp 2\pi i \lambda x$, and

$$\int_{-\infty}^{+\infty} \varphi_k(-x)\, F(x)\, dx = \sum_{\lambda \in E} a(k, \lambda)\, \hat{F}(\lambda) = 0 \quad \text{since } f \text{ vanishes on } E.$$

Hence $\langle S, f \rangle = 0$ and the spectrum of φ, which by definition is the support of its Fourier-transform S, is contained in E.

3.2. Is it possible that, conversely, any bounded continuous $\varphi : \mathbf{R} \to \mathbf{C}$ whose spectrum lies in E can be built up as the limit, in the sense of uniform convergence on compact sets, of a uniformly bounded sequence of finite trigonometric sums whose frequencies lie in E?

THEOREM IV: *Let E be a closed set of real numbers. The following three statements are equivalent:*

(a) *for a constant $C > 0$ and each bounded continuous function $\varphi : \mathbf{R} \to \mathbf{C}$ whose spectrum lies in E there exists a sequence φ_k, $k \geqslant 1$, of finite trigonometric sums whose frequencies belong to E such that*

$$\sup_{\mathbf{R}} |\varphi_k| \leqslant C \sup_{\mathbf{R}} |\varphi|$$

and $\varphi_k(x) \to \varphi(x)$ uniformly on each compact subset of $\mathbf{R}$; (7.4)

(b) *for a constant $C > 0$ and each S in PM (E), there exists a sequence S_k, $k \geqslant 0$, of measures supported by finite subsets of E and such that, for each f in $A(\mathbf{R})$*

$$\|S_k\|_{PM} \leqslant C \|S\|_{PM} \quad \text{and} \quad \langle S_k, f \rangle \to \langle S, f \rangle \quad \text{as} \quad k \to +\infty;$$

(c) *each S in PM (E) is the weak star limit of a sequence S_k of measures carried by finite subsets of E.*

Moreover, the same constant C may be chosen in (a) and (b).

3.3. This theorem is surprising since (a) is very precise (for uncountable compact sets we cannot get better approximations) and (c) seems to be fairly general. However, the proof is very simple and uses only standard functional analysis.

(c) ⇒ (b): Let B be the unit ball $\|S\|_{PM} \leqslant 1$ of $PM(E)$ and, for each $\varrho > 0$, let ϱB be the homothetical ball $\|S\|_{PM} \leqslant \varrho$.

The Banach–Steinhaus theorem shows that, if $\langle S_k, f\rangle \to \langle S, f\rangle$ for each f in $A(\mathbf{R})$, then $\sup_{k \geqslant 1} \|S_k\|_{PM} < +\infty$.

Since $A(\mathbf{R})$ contains a countable dense subset, the topology $\sigma(PM(\mathbf{R}), A(\mathbf{R}))$ restricted to B is metrizable, and B is compact in this weak-star topology.

Let $B_1 \subset B$, be the set of all complex measures μ carried by finite subsets of E whose pseudo-measure norms do not exceed 1; let C be the weak-star closure of B_1: C is clearly contained in B.

By (c) and our first remark, B is the union of the increasing sequence $D_n = B \cap nC$, $n \geqslant 1$, of compact subsets of B. Since $D_n - D_n \subset D_{2n}$ and B is compact, Baire's theorem shows that $D_n = B$ for n sufficiently large and (b) is proved.

(b) ⇒ (a): For each real t, let $\varphi_t : E \to \mathbf{C}$ be defined by $\varphi_t(x) = \exp 2\pi i t x$. The mapping $\Phi : \mathbf{R} \to A(E)$ defined by $\Phi(t) = \varphi_t$ is continuous as (7.1) shows and, for each $T > 0$, the subset K of all φ_t in $A(E)$, for which $|t| \leqslant T$, is compact We apply the well-known result.

Proposition 6: *Let X be a Banach space and L_k, $k \geqslant 1$, a sequence of continuous linear forms on X. The following three statements are equivalent:*

(i) *for each x in X, $L_k(x)$ has a limit as $k \to +\infty$;*

(ii) *there is a constant $C > 0$ such that $\|L_k\| \leqslant C$ and $L_k(x)$ has a limit for all x in a total subset of X;*

(iii) *$L_k(x)$ converges uniformly on compact subsets of X.*

(a) ⇒ (*c*): By Lebesgue's dominated convergence theorem, (7.4) implies $\int_{\mathbf{R}} \varphi_k(x) f(x)\,dx \to \int_{\mathbf{R}} \varphi(x) f(x)\,dx$ for each f in $L^1(\mathbf{R})$, which is (c).

3.4. We say that E is a set of strong synthesis if E possesses one of the three equivalent properties of Theorem IV. Varopoulos [94] gave an example of a set of synthesis which is not a set of strong synthesis, and Malliavin [61] gave an example of a closed set E of real numbers which is not a set of synthesis; we shall give another example in Chapter VIII.

A countable compact set E is always a set of strong synthesis. Any bounded continuous $\varphi : \mathbf{R} \to \mathbf{C}$ whose spectrum lies in E is, in this case, an almost periodic function and the approximation of φ by finite trigonometric sums whose frequencies lie in E holds in the norm of $L^\infty(\mathbf{R})$.

If E is not countable, E carries a diffuse bounded Radon measure μ, i.e. μ has no discrete part; we put $\varphi(x) = \hat{\mu}(-x)$. By Wiener's theorem ([24],

Ch. XVI, th. 4.19) the square mean $\lim_{T \to +\infty} 1/2T \int_{-T}^{T} |\hat{\mu}|^2 \, dx$ is 0 and $\hat{\mu}$ cannot be almost periodic. Hence, if E is not countable, (a) cannot be improved to uniform convergence on the whole of $\mathbf{R}$.

4. Spectral synthesis and weighted approximation

Using weight functions, a strong form of spectral synthesis can be defined where the approximation of the function by trigonometric sums holds in the strong sense.

4.1. $L^1(\mathbf{R}, \omega)$, $L^\infty(\mathbf{R}; \omega)$, $C(\mathbf{R}; \omega)$, $C_0(\mathbf{R}; \omega)$

In what follows $\omega : \mathbf{R} \to [1, +\infty[$ is an even function which is increasing and concave on $[0, +\infty[$, and we assume that $\omega(0) = 1$. For example $\omega(x) = 1 + |x|^\alpha$, $0 \leqslant \alpha \leqslant 1$, is such a function.

We define $L^1(\mathbf{R}; \omega)$ to be the subspace of all f in $L^1(\mathbf{R})$ satisfying $\int_{-\infty}^{+\infty} |f(x)| \, \omega(x) \, dx < +\infty$; this integral is equal to the norm of f in $L^1(\mathbf{R}; \omega)$. If multiplication is defined to be convolution, $L^1(\mathbf{R}; \omega)$ becomes a Banach algebra and $A(\mathbf{R}; \omega)$ denotes the function algebra of all Fourier transforms of elements of $L^1(\mathbf{R}; \omega)$.

The dual space of $L^1(\mathbf{R}; \omega)$ is the Banach space of all complex valued Lebesgue-measurable functions φ on the line such that

$$|\varphi(x)| \leqslant C\omega(x) \quad \text{almost everywhere.} \tag{7.5}$$

The norm of φ in $L^\infty(\mathbf{R}; \omega)$ is the infimum of all the $C > 0$ which can appear in the right hand side of (7.5).

$C(\mathbf{R}; \omega)$ is the subspace of all continuous functions φ in $L^\infty(\mathbf{R}; \omega)$ and $C_0(\mathbf{R}, \omega)$ is the subspace of all φ in $C(\mathbf{R}; \omega)$ such that

$$\varphi(x) = o(\omega(x)) \quad \text{as} \quad |x| \to +\infty. \tag{7.6}$$

4.2. Strong synthesis for weighted norms

DEFINITION 5: *The spectrum of a function φ in $C_0(\mathbf{R}; \omega)$ is the support of the Fourier transform S of φ.*

Note that φ is a tempered distribution.

DEFINITION 6: *A closed set E of real numbers has the property $S(\omega)$ if each continuous function $\varphi : \mathbf{R} \to \mathbf{C}$ which satisfies* (7.6) *and whose spectrum lies in E is the limit, in the norm topology of $C_0(\mathbf{R}; \omega)$, of a sequence of trigonometric sums whose frequencies belong to E.*

PROPOSITION 7: *If E has the property $S(\omega)$ for all weights ω, E is a set of synthesis.*

Proof: Let f be an element of $L^1(\mathbf{R})$; it is quite easy to construct a weight function ω such that $\lim_{|x|\to+\infty} \omega(x) = +\infty$ and $f \in L^1(\mathbf{R}; \omega)$. Let φ be any bounded continuous function whose spectrum lies in E; then $\varphi \in C_0(\mathbf{R}; \omega)$, and $S(\omega)$ gives a sequence

$$\varphi_k(x) = \sum_{\lambda \in E} a(k, \lambda) \exp 2\pi i \lambda x \tag{7.7}$$

such that

$$\sup_{\mathbf{R}} \left| \frac{\varphi(x) - \varphi_k(x)}{\omega(x)} \right| \to 0 \quad \text{as} \quad k \to +\infty. \tag{7.8}$$

Since $f \in L^1(\mathbf{R}; \omega)$, (7.8) implies $\int_{-\infty}^{+\infty} \varphi_k(-x) f(x)\, dx \to \int_{-\infty}^{+\infty} \varphi(-x) f(x)\, dx$.

If $\hat{f}$ vanishes on E we get $\int_{-\infty}^{+\infty} \varphi(-x) f(x)\, dx = 0$ and E is a set of synthesis by (b) in Theorem III.

5. Pisot numbers and spectral synthesis

5.1. *The results*

Let $\theta > 2$ be a *Pisot number and E the compact set of all sums* $\sum_1^\infty \varepsilon_k \theta^{-k}$, $\varepsilon_k = 0$ *or* 1; *E is a Cantor set constructed with the dissection ratio* θ^{-1}.

The following theorems show that very nice forms of spectral synthesis hold for E.

THEOREM V: *Let θ be a Pisot number and let $\varphi : \mathbf{R} \to \mathbf{C}$ be any bounded continuous function whose spectrum lies in E. There exists a sequence φ_k, $k \geqslant 1$, of finite trigonometric sums such that*

(a) *for each $k \geqslant 1$, the frequencies of φ_k belong to E;*

(b) *$\varphi_k(x) \to \varphi(x)$ uniformly on each compact subset of* $\mathbf{R}$;

(c) $\sup_{\mathbf{R}} |\varphi_k| \to \sup_{\mathbf{R}} |\varphi|$.

The proof will in fact show further, that the frequencies of φ_k belong to the finite set of the 2^k sums $\sum_1^k \varepsilon_j \theta^{-j}$, $\varepsilon_j = 0$ or 1.

THEOREM VI: *Let θ be a Pisot number, let $\omega : \mathbf{R} \to [1, +\infty[$ be an even unction which is increasing and concave on $[0, +\infty[$ and let $\varphi : \mathbf{R} \to \mathbf{C}$ be a continuous function satisfying $\varphi(x) = o(\omega(x))$ as $|x| \to +\infty$. There exists a sequence φ_k, $k \geqslant 1$, of finite trigonometric sums such that, for each $k \geqslant 1$,*

(a) *the frequencies of φ_k belong to E;*

(b) $\sup_{\mathbf{R}} (\varphi(x) - \varphi_k(x))/\omega(x) \to 0$ *as* $k \to +\infty$.

Theorem VI simultaneously gives synthesis and uniqueness: if ω is bounded and if φ is a complex valued continuous function whose spectrum lies in E and which vanishes at infinity, (b) implies that $\sup_{\mathbf{R}} |\varphi(x) - \varphi_k(x)| \to 0$ as $k \to +\infty$; hence φ is almost periodic and this implies $\varphi = 0$. Since we have already proved this property of uniqueness in Chapter III, we may assume that $\omega(x) \to +\infty$ with x

Theorem V does not give a method of obtaining φ_k in terms of φ. In the following statement φ_k depends linearly on φ but we lose a little at (c).

THEOREM VII: *Let θ be a Pisot number. There exist a positive constant C and a sequence L_k, $k \geqslant 1$, of linear mappings such that for any bounded continuous $\varphi : \mathbf{R} \to \mathbf{C}$ whose spectrum lies in E,*

(a) *the frequencies of $L_k(\varphi)$ belong to a finite subset F_k of E;*

(b) $|(L_k\varphi)(x) - \varphi(x)| \leqslant C\theta^{-k}|x| \sup_{\mathbf{R}} |\varphi|$;

(c) $\sup_{\mathbf{R}} |L_k\varphi| \leqslant C \sup_{\mathbf{R}} |\varphi|$.

In fact, a little more will be proved: there is a finite subset F of E such that F_k is the finite set of all sums $\sum_1^k \varepsilon_j\theta^{-j} + \theta^{-k}\lambda$, $\varepsilon_j = 0$ or 1, $\lambda \in F$; we note further that F_k is contained in E.

5.2. *Sketch of the proofs*

Theorem VII will be an easy consequence of the atomizing properties of harmonious sets found in Chapter IV. By simple modifications in the proof we shall get Theorem VI. To obtain Theorem V, we shall give in Section 7 a characterization of the restriction algebra $A(E)$ in the style of Bochner's theorem ([15], p. 32, th. 1.9.1).

5.3. *Proof of Theorem VII*

Let Λ be the set of all finite sums $\sum_{k\geqslant 0} \varepsilon_k\theta^k$, $\varepsilon_k = 0$ or 1 and, for each $k \geqslant 1$, let Λ_k be the subset of the 2^k sums $\sum_1^k \varepsilon_j\theta^j$, $\varepsilon_j = 0$ or 1.

For each rational integer p, let τ_p be the isometry of $L^\infty(\mathbf{R})$ which carries φ into the function defined by $x \to \varphi(\theta^p x)$. If the spectrum of φ lies in E and if $p = k \geqslant 1$, the spectrum of $\tau_k(\varphi)$ lies in $\theta^k E = \Lambda_k + E \subset \Lambda + E$. The dilation $x \to \theta^k x$ is suggested by the right-hand side of (b).

We recall Theorem XI of Chapter IV: let Λ be an harmonious set of real numbers and E a compact set of real numbers. There exist a finite subset F of E, a linear mapping L which maps bounded continuous functions $\varphi : \mathbf{R} \to \mathbf{C}$ whose spectra lie in $\Lambda + E$ into almost periodic functions

$\psi = L(\varphi)$ whose spectra lie in $\Lambda + F$ and a positive constant C such that:

$$\text{for each } \lambda \text{ in } \Lambda, \text{ the spectrum of } \psi \text{ lies in } \lambda + F \text{ whenever the spectrum of } \varphi \text{ lies in } \lambda + E. \quad (7.9)$$

$$|\psi(x) - \varphi(x)| \leqslant C|x| \sup_{\mathbf{R}} |\varphi| \quad (7.10)$$

$$\sup_{\mathbf{R}} |\psi| \leqslant C \sup_{\mathbf{R}} |\varphi| \quad (7.11)$$

For a function w in the class $\mathscr{S}$ of Schwartz, (7.11) can be improved to

$$\sup_{\mathbf{R}} |\psi| \leqslant \sup_{\mathbf{R}} |w\varphi| \quad \text{and} \quad (7.10) \text{ to } |\psi(x) - \varphi(x)| \leqslant |x| \sup_{\mathbf{R}} |w\varphi|. \quad (7.12)$$

Starting with a φ whose spectrum lies in E, the spectrum of $\tau_k\varphi$ lies in $\Lambda_k + E \subset \Lambda + E$. If θ is a Pisot number, Λ is harmonious. We define $L_k(\varphi)$ to be $(\tau_{-k} \circ L \circ \tau_k)(\varphi)$; (a) shows that the spectrum of $(L \circ \tau_k)(\varphi)$ is contained in $\Lambda_k + F$; hence the spectrum of $L_k(\varphi)$ is contained in $\theta^{-k}\Lambda_k + \theta^{-k}F = F_k$. Statements (7.10) and (7.11) give the corresponding statements in Theorem VII.

5.4. *Proof of Theorem VI*

A first step in the direction of Theorem VI is the following Proposition 8.

PROPOSITION 8: *For each Pisot number $\theta > 2$ there exist a constant $C > 0$ and a sequence L_k, $k \geqslant 1$, of linear mappings with the following properties:*

(a) *for each continuous function φ in $L^\infty(\mathbf{R}; \omega)$ whose spectrum lies in E, $L_k(\varphi)$ is a trigonometric sum whose frequencies belong to E;*

(b) *$L_k(f) \to f$ uniformly on each compact set of real numbers;*

(c) $\sup_{\mathbf{R}} |(L_k(f))/\omega| \leqslant C \sup_{\mathbf{R}} |f/\omega|$.

We first show why Proposition 8 implies Theorem VI. If $\varphi(x) = o(\omega(x))$, as $|x| \to +\infty$, there exists a weight function ϖ in the sense of Section 4.1 such that $\varphi(x) = 0(\varpi(x))$ and $\varpi(x) = o(\omega(x))$, as $|x| \to +\infty$. By Proposition 8, $|(L_k f)(x)| \leqslant C\varpi(x)$ for each real x; this inequality and (b) together imply Theorem VI.

5.5. *Proof of Proposition 8*

Some properties of the weight function ω will be needed. We put $\omega(x) = 1 + \alpha(x)$; on $[0, +\infty[$, α is increasing, concave and $\alpha(0) = 0$. It follows that $\alpha(x)/x$ is decreasing on $[0, +\infty[$ and

$$0 \leqslant s \leqslant 1 \text{ and } t \geqslant 0 \text{ imply } \alpha(st) \geqslant s\alpha(t) \quad (7.13)$$

$$s \geqslant 1 \text{ and } t \geqslant 0 \text{ imply } \alpha(st) \leqslant s\alpha(t) \quad (7.14)$$

$$s \geqslant 0 \text{ and } t \geqslant 0 \text{ imply } \alpha(s + t) \leqslant \alpha(s) + \alpha(t). \quad (7.15)$$

LEMMA 5: *Let E be a compact set of real numbers. There is a constant C_1 such that, for each continuous function $\varphi : \mathbf{R} \to \mathbf{C}$ in $L^\infty(\mathbf{R}; \omega)$ whose spectrum lies in E*

$$|\varphi(x) - \varphi(0)| \leqslant C_1|x| \sup_{\mathbf{R}} |\varphi/\omega|. \tag{7.16}$$

Proof of the lemma: By (7.14) we have $1 \leqslant \omega(x) \leqslant 1 + C|x|$; hence if $F \in L^1(\mathbf{R})$ and $xF \in L^1(\mathbf{R})$, F belongs to $L^1(\mathbf{R}; \omega)$. Taking the Fourier transforms we see that if $f \in A(\mathbf{R})$ and f', the derivative of f, also belongs to $A(\mathbf{R})$, then $f \in A(\mathbf{R}; \omega)$ and $\|f\|_{A(\mathbf{R};\omega)} \leqslant \|f\|_{A(\mathbf{R})} + C\|f'\|_{A(\mathbf{R})}$. Let $\alpha : \mathbf{R} \to [0, 1]$ be an indefinitely differentiable function which is 1 on a neighbourhood of E and 0 outside a compact subset of $\mathbf{R}$. Put $f_x(t) = (\exp 2\pi i t x - 1)\,\alpha(t)$. Using (7.1) it is easy to check that $\|f_x\|_{A(\mathbf{R})} \leqslant C|x|$ and $\|f_x'\|_{A(\mathbf{R})} \leqslant C|x|$. Let S be the Fourier-transform of φ; then $|\varphi(x) - \varphi(0)| = |\langle S, f_x\rangle| \leqslant \sup_{\mathbf{R}} |\varphi/\omega| \, \|f_x\|_{A(\mathbf{R};\omega)} \leqslant C_1|x| \sup_{\mathbf{R}} |f/\omega|$.

Returning to the proof of Proposition 8, we have, if $|x| \leqslant 1$, $|\varphi(x) - \varphi(0)| \leqslant C_1|x| \, \|\varphi\|$. If $|x| \geqslant 1$, $\alpha(x)$ and $\omega(x)$ are equivalent; hence for all real x, $|\varphi(x) - \varphi(0)| \leqslant C_2\alpha(x) \, \|\varphi\|$, $\|\varphi\| = \sup_{\mathbf{R}} |\varphi/\omega|$.

If Proposition 8 is proved for $\varphi(x) - \varphi(0)$, the result also follows for $\varphi(x)$. From now on, it will be assumed that, for some constant C_2 and each real x

$$|\varphi(x)| \leqslant C_2\alpha(x) \, \|\varphi\|.$$

We define φ_k by $\varphi_k(x) = \varphi(\theta^k x)$ and a trigonometric sum P_k by $P_k(\theta^k x) = (L[\varphi_k])(x)$; L is defined by (7.9)—(7.12).

The frequencies of P_k belong to E as we have already observed in Section 5.3.

If $|x| \geqslant 1$, $|\varphi_k(x)| = |\varphi(\theta^k x)| \leqslant C_2\alpha(\theta^k x) \, \|\varphi\| \leqslant C_2|x| \, \alpha(\theta^k) \, \|\varphi\|$.

If $|x| \leqslant 1$, $|\varphi_k(x)| = |\varphi(\theta^k x)| \leqslant C_2\alpha(\theta^k x) \, \|\varphi\| \leqslant C_2\alpha(\theta^k) \, \|\varphi\|$.

Since w in (7.12) decays rapidly at infinity, $\sup_{\mathbf{R}} |w\varphi_k| \leqslant C_3\alpha(\theta^k) \, \|\varphi\|$ and (7.12) gives $|\varphi_k(x) - (L\varphi_k)(x)| \leqslant C_3|x| \, \alpha(\theta^k) \, \|\varphi\|$, which, for $|x| \leqslant 1$, does not exceed $C_3\alpha(\theta^k x) \, \|\varphi\|$. Hence $|(L\varphi_k)(x)| \leqslant C_5\alpha(\theta^k x)$ for $|x| \leqslant 1$.

For all real x, (7.12) gives $|(L\varphi_k)(x)| \leqslant C_4\alpha(\theta^k) \, \|\varphi\|$. A fortiori, $|(L\varphi_k)(x)| \leqslant C_5\alpha(\theta^k x) \, \|\varphi\|$ for all real x. Finally $|P_k(x) - \varphi(x)| \leqslant C_3|x| \, \alpha(\theta^k)/\theta^{-k} \, \|\varphi\|$ and $|P_k(x)| \leqslant \alpha(x) \, \|\varphi\|$, which proves Proposition 8.

6. Bochner's property for a Banach algebra

6.1. Let A be a semi-simple self-adjoint regular Banach algebra with a unit. If E is the compact spectrum of A, A can be regarded, by Gelfand's representation, as a subalgebra of the algebra $C(E)$ of all continuous complex valued functions on E. The norm of f in A is greater than $\|f\|_\infty$, the sup. norm over E.

For each compact subset K of E, $A(K)$ is the restriction algebra to K of all functions in A; $\|f\|_{A(K)} = \inf \|g\|$ for those g in A which equal f on K. If $K = E$, $A(E)$ is the whole algebra A.

6.2. Definition 7: *We say that Bochner's property holds for A if there exists an increasing sequence F_k, $k \geqslant 1$, of finite subsets of E such that the following implication is true*

$$f \in C(E) \quad \text{and} \quad \lim_{k \to +\infty} \|f\|_{A(F_k)} < +\infty \quad \text{imply} \tag{7.17}$$

$$f \in A(E) \quad \text{and} \quad \lim_{k \to +\infty} \|f\|_{A(F_k)} = \|f\|_{A(E)}. \tag{7.18}$$

This means, very curiously, that the property of belonging to A for a continuous function $f: E \to \mathbf{C}$ can be verified by looking at f on finite sets.

Example 1: Let $E = [0, 1]$ and let A be the Banach algebra of all complex-valued continuously differentiable functions f on $[0, 1]$; $\|f\|_A = \sup_{[0,1]} |f| + \sup_{[0,1]} |f'|$. Let B be the Banach algebra of all f satisfying a Lipschitz condition $|f(x) - f(y)| \leqslant C\,|x - y|$ on $[0, 1]$; we put $\|f\|_B = \sup_{[0,1]} |f| + \inf C$.

Then for each finite subset F of $[0, 1]$ and each $f: F \to \mathbf{C}$ we have $\|f\|_{A(F)} = \|f\|_{B(F)}$. Since A and B are distinct, Bochner's property cannot hold for A.

Example 2: If $E = [0, 1]$ and A is the Banach algebra of all restrictions to E of 2π-periodic absolutely convergent Fourier series, then Bochner's property holds for A, as Theorem III shows.

Example 3: More generally let G be any metrizable compact abelian group and let $A(G)$ be the Banach algebra of all continuous functions $f: G \to \mathbf{C}$ whose Fourier series are absolutely convergent. Then Bochner's property holds for $A(G)$. In the same context, Bochner's property holds for the algebra $A_p(G)$, $1 < p < +\infty$ defined by Herz [56, 57].

6.3. Dual forms of Bochner's property

Let A^* be the dual space of A normed by the dual norm. Since A is regular and self-adjoint, the support of each S in A^* can be defined. From now on elements S of A^* are called pseudomeasures and $\|S\|_{PM}$ is the dual norm.

For a finite subset F of E, let $I(F)$ be the ideal of all f in A vanishing on F. An element S of A^* which is orthogonal to $I(F)$ is a measure μ carried by F.

LEMMA 6: *Let F be a finite subset of E and B the set of all complex measures μ carried by F whose norm in A^* does not exceed one.*

Then, for each complex valued function f on F,

$$\|f\|_{A(F)} = \sup_{B} \left| \int_F f \, d\mu \right|. \tag{7.19}$$

Proof: Apply the metric form of the Hahn–Banach theorem.

Another easy lemma will be needed:

LEMMA 7: *Let X be a separable Banach space, X^* the strong dual of X, Y a linear subspace of X^* and $C > 0$ a constant. Let B be the set of all y in Y whose norm in X^* does not exceed one. Then the following two properties of Y are equivalent:*

(a) $\|x\| \leqslant C \sup_{y \in B} |\langle x, y\rangle|$ (*if* $C = 1$, $\|x\| = \sup_{y \in B} |\langle x, y\rangle|$);

(b) *for each x^* in X^* there exists a sequence y_k, $k \geqslant 1$, of elements of Y such that $\|y_k\| \leqslant C\|x^*\|$ and, for every x in X,*

$$\langle x, y_k\rangle \to \langle x, x^*\rangle \quad \text{as} \quad k \to +\infty.$$

An immediate consequence of Lemmas 6 and 7 is

PROPOSITION 9: *Let A be a regular self-adjoint semi-simple Banach algebra with a unit, and let E be the spectrum of A. Then A has Bochner's property if and only if there exists a sequence $(F_k)_{k \geqslant 1}$ of finite subsets of E such that for each f in C(E), the following two properties are equivalent:*

for each $k \geqslant 1$ and each complex measure μ_k carried by F_k,

$$|\textstyle\int f \, d\mu_k| \leqslant \|\mu_k\|_{A^*} \tag{7.20}$$

$$f \in A \quad \text{and} \quad \|f\|_A \leqslant 1. \tag{7.21}$$

In this case, for each S in A^* there exists a sequence $(\mu_k)_{k\geqslant 1}$ of measures such that

$$\mu_k \text{ is carried by } F_k \tag{7.22}$$

$$\|\mu_k\|_{A^*} \leqslant \|S\|_{A^*} \tag{7.23}$$

$$\text{For every } f \text{ in } A,\ \langle \mu_k, f\rangle \to \langle S, f\rangle \quad \text{as} \quad k \to +\infty. \tag{7.24}$$

7. Bochner's property and harmonic analysis

7.1. THEOREM VIII: *Let $\theta > 1$ be a Pisot number, $N \geqslant 1$ an integer and $E = E(N, \theta)$ the compact set of all sums $\sum_{k\geqslant 1} \varepsilon_k \theta^{-k}, \varepsilon_k = 0, 1, \ldots, N$. For each $k \geqslant 1$, let F_k be the finite subset of E consisting of the $(N+1)^k$ sums $\sum_1^k \varepsilon_j \theta^{-j}$, $\varepsilon_j = 0, 1, \ldots, N$.*

Then Bochner's property holds for E and F_k, $k \geqslant 1$. More precisely

$$f \in C(E) \quad \textit{and} \quad \sup_{k\geqslant 1} \|f\|_{A(F_k)} < +\infty \quad \textit{implies} \tag{7.25}$$

$$f \in A(E) \quad \textit{and} \quad \lim_{k\to+\infty} \|f\|_{A(F_k)} = \|f\|_{A(E)}. \tag{7.26}$$

The compact set E is a set of synthesis and each distribution S carried by E whose Fourier-transform is bounded is the weak star limit of a sequence S_k, $k \geqslant 1$, of measures carried by F_k and such that $\|S_k\|_{PM} \leqslant \|S\|_{PM}$.

For each bounded continuous $\varphi: \mathbf{R} \to \mathbf{C}$ whose spectrum lies in E, there exists a sequence φ_k, $k \geqslant 1$, of trigonometric sums such that

(a) *for each $k \geqslant 1$, the frequencies of φ_k belong to F_k;*

(b) $\sup_{\mathbf{R}} |\varphi_k| \leqslant \sup_{\mathbf{R}} |\varphi|$;

(c) $\varphi_k(x) \to \varphi(x)$ *uniformly on each compact set of real numbers.*

If $\theta = 2$ and $N = 1$, E is the interval $[0, 1]$ and we get the original form of Bochner's theorem.

The last part of Theorem VIII is a consequence of Bochner's property and of Theorem IV.

The proof of Theorem VIII is a generalization of an argument given by Y. Katznelson when $\theta = 3$. The idea is to define a compact metrizable abelian group G which reflects the algebraic properties of θ and to consider $\mathbf{R}$ as a dense subgroup of G.

7.2. *A compact group G*

For each $k \geqslant 1$, let Γ_k be the subgroup $\theta^{-k}\mathbf{Z} + \cdots + \theta^{-k-n+1}\mathbf{Z}$. Since θ is an algebraic integer, $\Gamma_k \subset \Gamma_{k+1}$. We define the subgroup Γ of $\mathbf{R}$ to be the union of all Γ_k, $k \geqslant 1$; equally, we can say that Γ is the inductive limit of Γ_k, $k \geqslant 1$.

We consider Γ with the discrete topology and let G be the compact dual group. Since Γ is countable, G is metrizable. Let $j : \Gamma \to \mathbf{R}$ be the canonical injection and $h : \mathbf{R} \to G$ the dual map. Since Γ is dense in $\mathbf{R}$, h is injective; since j is injective, $h(\mathbf{R})$ is dense in G.

PROPOSITION 10: *Let* $P(X) = X^n + a_1X^{n-1} + \cdots + a_n$, $a_j \in \mathbf{Z}$ *for* $1 \leqslant j \leqslant n$, *be the minimal polynomial of* θ. *Then the compact metrizable abelian group G can be regarded as a closed subgroup of the product group* $\mathbf{T}^{\mathbf{N}}$; *G consists of all* $\mathbf{x} = (\exp 2\pi i\varphi_1, \ldots, \exp 2\pi i\varphi_k, \ldots)$ *in* $\mathbf{T}^{\mathbf{N}}$ *such that, for all* $k \geqslant 1$, $\varphi_k + a_1\varphi_{k+1} + \cdots + a_n\varphi_{k+n} \equiv 0 \pmod 1$, *and* $h : \mathbf{R} \to G$ *is then defined by* $h(t) = (\exp 2\pi i\theta^{-1}t, \ldots, \exp 2\pi i\theta^{-k}t, \ldots)$.

Proof: If $\chi \in \mathrm{Hom}\,(\Gamma, \mathbf{T})$, let us put $\chi(\theta^{-k}) = \exp 2\pi i\varphi_k$, $k \geqslant 1$. For each $k \geqslant 1$, we have $\theta^{-k} + a_1\theta^{-k-1} + \cdots + a_n\theta^{-k-n} = 0$. Hence

$$\varphi_k + a_1\varphi_{k+1} + \cdots + a_n\varphi_{k+n} \equiv 0 \pmod 1 \tag{7.27}$$

and the character χ is well defined on Γ by the sequence of φ_k, $k \geqslant 1$.

Conversely, let $(\varphi_k)_{k\geqslant 1}$ be a sequence of real numbers satisfying each congruence (7.27) and let us try to define a character χ on Γ such that, for each $k \geqslant 1$, $\chi(\theta^{-k}) = \exp 2\pi i\varphi_k$.

Let $\chi_k \in \mathrm{Hom}\,(\Gamma_k, \mathbf{T})$ be defined by

$$\chi_k(\theta^{-j}) = \exp 2\pi i\varphi_j, \quad k \leqslant j \leqslant k + n - 1.$$

Since $\theta^{-k}, \ldots, \theta^{-k-n+1}$ is a basis over $\mathbf{Z}$ of the free $\mathbf{Z}$-module Γ_k, this definition is consistent.

It will be proved that

$$\chi_{k+1}|_{\Gamma_k} = \chi_k \tag{7.28}$$

and we can then define $\chi \in \mathrm{Hom}\,(\Gamma, \mathbf{T})$ by $\chi|_{\Gamma_k} = \chi_k$.

To get (7.28) we can restrict our attention to the generators θ^{-j}, $k \leqslant j \leqslant n - 1$, of the free $\mathbf{Z}$-module Γ_k; θ^{-k} is written in Γ_{k+1} as a sum $-a_1\theta^{-k-1} - \cdots - a_n\theta^{-k-n}$ and then $\chi_k(\theta^{-k}) = \exp 2\pi i\varphi_k$ while $\chi_{k+1}(\theta^{-k}) = \exp -2\pi i\,(a_1\varphi_{k+1} + \cdots + a_n\varphi_{k+n})$. The required equality is implied by the congruence (7.27). For $k + 1 \leqslant j \leqslant k + n - 1$, each θ^{-j} in Γ_k is written θ^{-j} in Γ_{k+1} and $\chi_{k+1}(\theta^{-j}) = \chi_k(\theta^{-j})$.

If $\chi : \mathbf{R} \to \mathbf{T}$ is a continuous character defined by $\chi(x) = \exp 2\pi ixt$ ($t \in \mathbf{R}$), $h(t)$ is the restricted character $\chi|_\Gamma \in \operatorname{Hom}(\Gamma, \mathbf{T})$. Clearly $h(t) = (\exp 2\pi i\theta^{-k}t)_{k\geq 1}$.

7.3. Extension to G of Fourier transforms of measures supported by F_k

It may be noted that each F_k, $k \geqslant 1$, is contained in Γ and so therefore is the union $\bigcup_{k\geqslant 1} F_k$ of all the F_k. For any measure μ carried by F_k, $\hat{\mu} : \mathbf{R} \to \mathbf{C}$ is a finite trigonometric sum whose frequencies belong to Γ (to be more precise, the frequencies of $\hat{\mu}$ belong to $-F_k$). Hence there is a unique finite trigonometric sum $P : G \to \mathbf{C}$ whose frequencies and Fourier coefficients are those of $\hat{\mu}$ such that

$$\hat{\mu} = P \circ h. \tag{7.29}$$

Since $h(\mathbf{R})$ is dense in G, we have

$$\|\mu\|_{PM} = \sup_{\mathbf{R}} |\hat{\mu}| = \sup_{G} |P|. \tag{7.30}$$

We denote by $\mathscr{E}$ the vector space of all $P : G \to \mathbf{C}$ defined by (7.29); $\mathscr{E}$ is normed by (7.30).

LEMMA 8: *Let $p \geqslant 1$, $q \geqslant 1$ be two integers and $\mu_{p,q}$ the probability measure carried by $\theta^{-p}F_q$ and defined by the convolution product*

$$\mu_{p,q} = \nu_{p+1} * \cdots * \nu_{p+q} \tag{7.31}$$

where ν_j is the sum of a mass 1/2 put in 0 and of a mass 1/2 put in θ^{-j}.

For any continuous $f : E \to \mathbf{C}$, $\lim_{p\to+\infty} \int_E f\, d\mu_{p,q} = f(0)$ uniformly in $q \geqslant 1$.

Proof: $\theta^{-p}F_q \subset \theta^{-p}E$ tends to 0 as p tends to infinity; hence $\mu_p \to \delta(x)$ in the weak-star topology $\sigma(M(E), C(E))$, uniformly on $q \geqslant 1$.

Roughly speaking, the following lemma states that, using a suitable sequence of elements of $\mathscr{E}$, we can separate $G_1 = h(\mathbf{R})$ from its complement G_2 in G.

For each real t, we have

$$\hat{\mu}_{p,q}(t) = 2^{-q}(1 + \exp -2\pi it\theta^{-p-1}) \cdots (1 + \exp -2\pi it\theta^{-p-q}).$$

The continuous extension $M_{p,q}$ of $\hat{\mu}_{p,q}$ to G is defined by

$$M_{p,q}(\mathbf{x}) = 2^{-q}(1 + \exp -2\pi i\varphi_{p+1}) \cdots (1 + \exp -2\pi i\varphi_{p+q})$$

whence $|M_{p,q}(\mathbf{x})| = |\cos \pi\varphi_{r+1}| \cdots |\cos \pi\varphi_{p+q}|$.

LEMMA 9: *For each fixed* $p \geqslant 1$ *and each* $\mathbf{x} \in G_2 = G \backslash h(\mathbf{R})$, $\lim_{q \to +\infty} M_{p,q}(\mathbf{x}) = 0$.

If the limit of the decreasing sequence $|M_{p,q}(\mathbf{x})|$, $q \geqslant 1$, is not 0, $|\cos \pi\varphi_{p,q}| \to 1$ and $\varphi_k = p_k + r_k$, $p_k \in \mathbf{Z}$, $-\frac{1}{2} \leqslant r_k < \frac{1}{2}$, $\lim_{k \to +\infty} r_k = 0$. But for each $k \geqslant 1$, we have $\varphi_k + a_1\varphi_{k+1} + \cdots + a_n\varphi_{k+n} \equiv 0 \pmod 1$. Putting $t_k = r_k + a_1 r_{k+1} + \cdots + a_n r_{k+n}$, we get $t_k \in \mathbf{Z}$ and $\lim t_k = 0$. This implies $t_k = 0$ for $k \geqslant k_0$.

Let $\theta_2, \ldots, \theta_n$ be the conjugates (over $\mathbf{Q}$) of θ. The linear recurrence relations $r_k + a_1 r_{k+1} + \cdots + a_n r_{k+n} = 0$ $(k \geqslant k_0)$ imply $r_k = t\theta^{-k} + t_2\theta_2^{-k} + \cdots + t_n\theta_n^{-k}$ for $k \geqslant k_0$, a suitable real t and suitable complex numbers $t_2, \ldots, t_n$.

By hypothesis θ is a Pisot number: $|\theta_2| < 1, \ldots, |\theta_n| < 1$. Since r_k tends to 0, the only possibility is $r_k = t\theta^{-k}$ (Chapter I, Section 6.3) or $\varphi_k \equiv t\theta^{-k} \pmod 1$ for $k \geqslant k_0$. But, for each $k \geqslant 0$, $\theta^{-k} + a_1\theta^{-k-1} + \cdots + a_n\theta^{-k-n} = 0$ and $\varphi_k + a_1\varphi_{k+1} + \cdots + a_n\varphi_{k+n} \equiv 0 \pmod 1$. By a decreasing induction on $k \geqslant 0$, we get $\varphi_k \equiv t\theta^{-k} \pmod 1$ for each $k \geqslant 0$ and so $\mathbf{x}$ belongs to $h(\mathbf{R})$.

7.4. *Conclusion of the proof of Theorem VIII*

Let $C = \lim_{k \to +\infty} \|f\|_{A(F_k)}$ and let L be the linear form defined on $\mathscr{E}$ by the following rule: if μ is a complex measure carried by one of the F_k and if the continuous function $P: G \to \mathbf{C}$ is determined by

$$P \circ h = \hat{\mu}, \quad \text{then} \quad L(P) = \int_{F_k} f\, d\mu. \tag{7.32}$$

Proposition 9 shows that $|\int_{F_k} f\, d\mu| \leqslant \|f\|_{A(F_k)} \sup_{\mathbf{R}} |\hat{\mu}| \leqslant C \sup_{G} |P|$. Hence L is continuous and the norm of L does not exceed C.

By the metric form of the Hahn–Banach theorem, L can be extended to the whole of $C(G)$ without changing its norm: hence there is a complex Radon measure σ, $\|\sigma\| \leqslant C$, such that $L(P) = \int_G P\, d\sigma$. Therefore (7.32) can be rewritten

$$\int_{\mathbf{R}} f\, d\mu = \int_G P\, d\sigma \tag{7.33}$$

for any complex measure μ carried by one of the F_k, where $P: G \to \mathbf{C}$ is the continuous function defined by $P \circ h = \hat{\mu}$.

The real line is the union of an increasing sequence of compact sets; since h is continuous, $h(\mathbf{R}) = G_1$ is a Borel subset of G and so is $G_2 = G \backslash G_1$. Let σ_1 be the restriction of σ to G_1 and σ_2 the restriction of σ to G_2. If it can be shown that $\sigma_2 = 0$, the proof ends in a few lines. Under the action of h, the Borel field of $\mathbf{R}$ is carried onto the restriction to G_1 of the Borel field of G. Hence each complex bounded Radon measure σ_1 on G_1 can be written $\sigma_1 = h(\tau)$ where τ is a complex bounded Radon measure on $\mathbf{R}$ and $\|\sigma_1\| = \|\tau\|$. This implies $\int_{\mathbf{R}} f\, d\mu = \int_G P\, d\sigma = \int_G P\, d\sigma_1 = \int_{\mathbf{R}} \hat{\mu}\, d\tau = \int_{\mathbf{R}} \hat{\tau}\, d\mu$. The equality between the first and the last term implies $f = \hat{\tau}$ on each F_k, $k \geqslant 1$. Since the union of all the F_k is dense in E, $f = \hat{\tau}$ on E. Hence $\|f\|_{B(E)} \leqslant C$. But E is compact; $A(E)$ and $B(E)$ are isometric (Chapter IV, Proposition 5) and Theorem VIII is proved.

7.5. $\sigma_2 = 0$. We return to the definition of the number $C = \lim\limits_{k \to +\infty} \|f\|_{A(F_k)}$ and we apply Proposition 9: for each positive ε, there is a $k \geqslant 1$ and measure μ carried by F_k such that $\sup\limits_{\mathbf{R}} |\hat{\mu}| = 1$ and $|\int_{F_k} f\, d\mu| \geqslant (1 - \varepsilon)\, C$. For each $q \geqslant 1$ and each $p \geqslant k$, $F_k + \theta^{-p} F_q \subset F_{p+q}$ and Lemma 8 shows that for sufficiently large p and each $q \geqslant 1$

$$\left| \int_E f\, d(\mu * \mu_{p,q}) \right| \geqslant (1 - 2\varepsilon)\, C$$

($\mu_{p,q}$ is defined by (7.31)). From now on p is fixed. Since $\mu * \mu_{p,q}$ is carried by F_{p+q}, our fundamental relation (7.33) can be applied to get

$$\left| \int_G P(\mathbf{x})\, M_{p,q}(\mathbf{x})\, d\sigma\, (\mathbf{x}) \right| \geqslant (1 - 2\varepsilon)\, C,$$

where $\hat{\mu} = P \circ h$ and $\hat{\mu}_{p,q} = M_{p,q} \circ h$.

Writing $\int_G PM_{p,q}\, d\sigma = \int_{G_1} PM_{p,q}\, d\sigma_1 + \int_{G_2} PM_{p,q}\, d\sigma_2 = I + J$, we observe that $|I| \leqslant \|\sigma_1\|$, since $\sup\limits_{G} |P| = \sup\limits_{\mathbf{R}} |\hat{\mu}| = 1$ and $\sup\limits_{G} |M_{p,q}| = 1$, and that $|J| \leqslant \int_{G_2} |M_{p,q}|\, d|\sigma_2|$. Hence, $(1 - 2\varepsilon)\, C \leqslant \|\sigma_1\| + \int_{G_2} |M_{p,q}|\, d|\sigma_2|$. Let p be fixed and let q tend to infinity. Lebesgue's dominated convergence theorem and Lemma 9 show that the last integral tends to 0. Hence $(1 - 2\varepsilon)\, C \leqslant \|\sigma_1\|$ for each positive ε, which implies $C \leqslant \|\sigma_1\|$. But $\|\sigma_1\| + \|\sigma_2\| = \|\sigma\| \leqslant C$; we get $\|\sigma_2\| \leqslant 0$ and so $\sigma_2 = 0$. Theorem VIII is now completely proved.

8. The p-adic case

8.1. Before giving any precise results, some definitions should be recalled. Let $\mathbf{Q}_p$ be the p-adic field and A_p the subring of all mp^n in $\mathbf{Q}_p$ ($m \in \mathbf{Z}$, $n \in \mathbf{Z}$ and m and p are relatively prime). The p-adic norm is defined on A_p by $|mp^n|_p = p^{-n}$ and A_p is dense in $\mathbf{Q}_p$ for this norm.

A Pisot–Chabauty p-adic number θ is defined by the following four properties:

(a) $|\theta|_p > 1$;

(b) θ is an algebraic integer over the ring A_p; let n be the degree of θ and let $P(X) = X^n + c_1X^{n-1} + \cdots + c_n$ ($c_j \in A_p$ for $1 \leqslant j \leqslant n$) be the minimal polynomial of θ;

(c) if Ω_p is the algebraic closure of $\mathbf{Q}_p$, all the other roots z of P in Ω_p satisfy $|z|_p \leqslant 1$;

(d) let $\tau_1, \ldots, \tau_n$ be the complex roots of P; then $|\tau_1| < 1, \ldots, |\tau_n| < 1$.

8.2. THEOREM IX: *Let θ be a Pisot–Chabauty number and $E(\theta)$ the compact set of all p-adic numbers $x = \sum_1^\infty \varepsilon_k\theta^{-k}$, $\varepsilon_k = 0$ or 1 Then $E(\theta)$ is a set of synthesis. Moreover for each bounded continuous $\varphi : \mathbf{Q}_p \to \mathbf{C}$ whose spectrum lies in $E(\theta)$, a sequence $(\varphi_k)_{k \geqslant 1}$ of trigonometric sums can be found with the following properties:*

$$\text{the frequencies of } \varphi_k \text{ are finite sums } \sum_1^k \varepsilon_j\theta^{-j},\ \varepsilon_j = 0 \text{ or } 1 \tag{7.34}$$

$$\sup_{\mathbf{Q}_p} |\varphi_k| \leqslant \sup_{\mathbf{Q}_p} |\varphi| \tag{7.35}$$

$$\varphi_k \to \varphi \text{ uniformly on each compact subset of } \mathbf{Q}_p \tag{7.36}$$

8.3. As in the real case, the proof of Theorem IX can be divided into two parts. First of all, a weak version of Theorem IX is given where φ_k depends linearly on φ and where (7.35) is altered; this is the following proposition:

PROPOSITION 11: *Let θ be a Pisot–Chabauty number. There exist a finite subset F of $E(\theta)$, a positive constant C and a sequence L_k, $k \geqslant 1$, of linear maps such that, for each bounded continuous $\varphi : \mathbf{Q}_p \to \mathbf{C}$ whose spectrum lies in $E(\theta)$,* *the frequencies of the trigonometric sum $L_k(\varphi)$ are finite sums*

$$\sum_1^k \varepsilon_j \theta^{-j} + \theta^{-k}\lambda, \quad \varepsilon_j = 0 \quad \text{or} \quad 1, \lambda \in F \tag{7.37}$$

$$|(L_k\varphi)(x) - \varphi(x)| \leqslant C\,|\theta|_p^{-k}\,|x|_p \sup_{\mathbf{Q}_p} |\varphi| \tag{7.38}$$

$$\sup_{\mathbf{Q}_p} |L_k(\varphi)| \leqslant C \sup_{\mathbf{Q}_p} |\varphi|. \tag{7.39}$$

Proposition 11 is an easy consequence of Theorem XI of Chapter IV, and can be proved exactly as in the real case.

8.4. The second state in the proof of Theorem IX is the following result, where $E = E(\theta)$ is defined as in Theorem IX.

THEOREM X: *Let $f: E \to \mathbf{C}$ be a continuous function and C a positive constant. The following two properties of f are equivalent:*

(a) *for any complex measure μ carried by a finite subset of E*

$$\left|\int_E f\,d\mu\right| \leqslant C \sup_{\mathbf{Q}_p} |\hat{\mu}|; \tag{7.40}$$

(b) *f is the restriction to E of the Fourier-transform of a function in $L^1(\mathbf{Q}_p)$ and $\|f\|_{A(E)} \leqslant C$.*

The implication (b) $\Rightarrow$ (a) is obvious.

The proof of (a) $\Rightarrow$ (b) is very similar to that of Section 7. The idea is to construct a compact group on which $\hat{\mu}$ can be extended by continuity and to apply the Riesz representation theorem to describe the linear form $\mu \to \int_E f\,d\mu$.

8.5. Definition of G

For each $k \geqslant 1$, let Γ_k be the group $\mathbf{Z}\theta^{-1} + \cdots + \mathbf{Z}\theta^{-k}$ (without topology) and let Γ be the inductive limit of the Γ_k. We regard Γ as a discrete group and denote its dual by G.

Our discrete group Γ is a quotient of the direct sum of a countable set of copies of $\mathbf{Z}$: to each finite sequence $\mathbf{p} = (p_k)_{k\geqslant 1}$ of rational integers corresponds $\sum_{k\geqslant 1} p_k\theta^{-k}$ in Γ. Hence G is a subgroup of the compact product group $\mathbf{T}^{\mathbf{N}}$ of a countable set of copies of $\mathbf{T}$.

Let $P(X) \in \mathbf{Z}[X]$ be the minimal polynomial of θ: $P(X) = a_0X^n + a_1X^{n-1} + \cdots + a_n$, where $a_0 = p^r = |\theta|_p$ $(r \geqslant 1)$.[1]

PROPOSITION 12: *Regarded as a subgroup of* $\mathbf{T}^{\mathbf{N}}$, *G is the group of all sequences* $\omega = (\omega_k)_{k\geqslant 1}$ *of complex numbers of absolute value* 1 *such that, for each* $k \geqslant 1$ *and for real* φ_k

$$\omega_k = \exp 2\pi i\varphi_k \tag{7.41}$$

and

$$a_0\varphi_k + \cdots + a_n\varphi_{k+n} \equiv 0 \pmod 1. \tag{7.42}$$

With this representation of G, the duality between G and Γ *is given by the following rule: for each* $\gamma = \sum_{k\geqslant 1} p_k\theta^{-k}$ *in* Γ *and each* $\omega = (\omega_k)_{k\geqslant 1}$ *in G*

$$\chi_\gamma(\omega) = \exp 2\pi i \sum_{k\geqslant 1} p_k\omega_k. \tag{7.43}$$

Proof of Proposition 12: Let G' be the subgroup of $\mathbf{T}^{\mathbf{N}}$ defined by (7.41) and (7.42). We describe an injective mapping $i: G \to G'$. Each element χ of G is a character on Γ; putting $i(\chi) = (\omega_k)_{k\geqslant 1}$ where $\omega_k = \chi(\theta^{-k})$ we get (7.42) from $a_0\theta^{-k} + \cdots + a_n\theta^{-k-n} = 0$ $(k \geqslant 1)$. Clearly i is continuous.

Let $\bar{\Gamma}$ be the closure of Γ in $\mathbf{Q}_p$: $\bar{\Gamma} = p^r\mathbf{Z}_p$. Let $j^*: \Gamma \to \bar{\Gamma}$ be the identity mapping. Then j^* is the dual of $j: A_p/p^r\mathbf{Z} \to G$ and $i \circ j(x) = (\exp 2\pi i x\theta^{-k})_{k\geqslant 1}$ for each x in A_p (mod. $p^r\mathbf{Z}$).

To prove the surjectivity of i, from which it will follow that i is a homeomorphism, it suffices to show that $I: A_p \to G'$ defined by $I(x) = (\exp 2\pi i x\theta^{-k})_{k\geqslant 1}$ has a dense range. The following lemmas conclude the proof:

LEMMA 10: *Let* $m \geqslant 1$ *be an integer. For each positive* ε *and each sequence* $(\varphi_k)_{1\leqslant k\leqslant m}$ *of numbers in* A_p *satisfying* (7.42), *there exist an x in* A_p *and m p-adic integers* p_k, $1 \leqslant k \leqslant m$ *such that* $x\theta^{-k} + p_k$ *belongs to* A_p *and*

$$|\varphi_k - x\theta^{-k} - p_k| \leqslant \varepsilon, \ (1 \leqslant k \leqslant m)^2 \tag{7.44}$$

LEMMA 11: *Let* $l \geqslant 1$ *be an integer and let* ω *be an element of* $\mathbf{T}^{\mathbf{N}}$ *satisfying* (7.41) *and* (7.42) *and such that* $\varphi_k = 0$ *for* $l + 1 \leqslant k \leqslant l + n$. *Then there is an x in* $p^{-lr-r}\mathbf{Z}$ *such that*

$$x\theta^{-k} \equiv \varphi_k \pmod 1 \quad \text{for} \quad 1 \leqslant k \leqslant l + n. \tag{7.45}$$

---

[1] $a_0 = p^r$, $r \geqslant 0$, since P is minimal and all $a_j/a_0 \in A_p$. Let $\theta_2, \ldots, \theta_n$ be the other roots of P in Ω_p. Then $|\theta|_p > 1$ and $|\theta_2|_p \leqslant 1, \ldots, |\theta_n|_p \leqslant 1$ imply $|\theta|_p = |\theta + \theta_2 + \cdots + \theta_n|_p = |a_1/a_0|_p \leqslant p^r$, If $|\theta|_p < p^r$, the other symmetric relations between the roots of P give $|a_j/a_0|_p < p^r$ for $1 \leqslant j \leqslant n$; p divides $a_0, \ldots, a_n$ and P is not minimal.

[2] Here $|t|$ denotes the absolute value of the real number $t \in A_p$.

We begin with the proof of Lemma 11. Let G_l be the group of all $\omega \in \mathbf{T}^{l+n}$ defined by (7.42) such that $\varphi_k = 0$ for $k \geqslant l+1$. Computing successively $\varphi_l, \varphi_{l-1}, \ldots$ we get p^r possible choices for φ_l (mod 1) and for each of these choices, p^r choices for φ_{l-1} and so on. Hence Card $G_l = p^{lr}$. Let $h: p^{-lr-r}\mathbf{Z}_p/p^{-r}\mathbf{Z}_p \to G_l$ be defined by $h(x) = (x\theta^{-k})_{k\geqslant 1}$, where each term is taken modulo 1. Then, the kernel of h is equal to $\{0\}$ and since Card $(p^{-(l+1)r}\,\mathbf{Z}_p/p^{-r}\mathbf{Z}_p) = p^{lr}$, h is surjective. Since x is defined modulo $p^{-r}\mathbf{Z}_p$, x may be chosen in $p^{-(l+1)r}\,\mathbf{Z}$.

To prove Lemma 10, it is convenient to consider first the case $m = n$. Let $h: A_p \to \mathbf{T}^n$ be the homomorphism defined by $h(x) = (\exp 2\pi i x\theta^{-1}, \ldots, \exp 2\pi i x\theta^{-n})$; then $h(A_p)$ is dense in $\mathbf{T}^n$. If this was not the case, a sequence $p_1, \ldots, p_n$ of rational integers could be found such that for each x in A_p, $x\,(p_1\theta^{-1} + \cdots + p_n\theta^{-n}) \equiv 0 \pmod{\mathbf{Z}_p}$. The only possibility is $p_1\theta^{-1} + \cdots + p_n\theta^{-n} = 0$; as the degree of θ is n, we get $p_1 = \cdots = p_n = 0$. Hence no proper closed subgroup of $\mathbf{T}^n$ contains $h(A_p)$.

If $m \geqslant n+1$, $m = n + l$ say. We approximate $(\varphi_{l+1}, \ldots, \varphi_{l+n})$ by $(x_1\theta^{-l-1}, \ldots, x_1\theta^{-l-n})$. Then the difference $(\varphi_k - x_1\theta^{-k})_{1\leqslant k\leqslant m}$, modulo $(\mathbf{Z}_p)^m$, is very close to an element of G_l, which can be written as a sequence $(x_2\theta^{-k})_{1\leqslant k\leqslant m}$, modulo $(\mathbf{Z}_p)^m$, and $x = x_1 + x_2$ gives the required element of A_p.

8.6. We return to the proof of Theorem X. For each $j \geqslant 1$, let ν_j be the measure assigning the mass $\frac{1}{2}$ to 0 and the mass $\frac{1}{2}$ to θ^{-j}, and for all $j \geqslant 1$ and $k \geqslant 0$ let

$$\mu_{j,k} = \nu_{j+1} * \cdots * \nu_{j+k}.$$

As in the real case, $M_{j,k}$ is the extension of $\hat{\mu}_{j,k}$ to the whole of G and we have

$$\cdot M_{j,k}(\omega)| = \prod_{j+1}^{j+k} |\cos \pi\varphi_q|.$$

LEMMA 12: *If for some fixed j,* $\lim_{k\to+\infty} M_{j,k}(\omega) \neq 0$, *then there is an x in* A_p *such that* $\omega = (x\theta^{-k})_{k\geqslant 1}$.

Proof: If this is the case, $|\cos \pi\varphi_k| \to 1$ and $\varphi_k = p_k + r_k$, where $p_k \in \mathbf{Z}$ and $r_k \to 0$. The congruences (7.42) give the corresponding ones for r_k (when $k > m$). Let $\tau_1, \ldots, \tau_n$ be the complex roots of $P(X) = 0$. Then, for $k > m$, $r_k = \lambda_1\tau_1^{-k} + \cdots + \lambda_n\tau_n^{-k}$. Since $r_k \to 0$ as $k \to +\infty$, we get $\lambda_1 = \cdots = \lambda_n = 0$ and $r_k = 0$ for $k > m$.

Lemma 11 now gives the result.

The last part of the proof is now a straightforward transcription of that given for the real case.

9. Notes

An excellent survey of the problem of synthesis for Banach algebras can be found in [14] (where sets of synthesis are called Wiener-sets).

For $\theta = 3$, Theorem V was proved by Herz in [45].

The general form of Theorem V was conjectured by Salem.

CHAPTER VIII

ULTRA-THIN SYMMETRIC SETS

We define a class of symmetric compact sets E of real numbers which are very similar to Cantor sets constructed with a dissection ratio whose inverse is a Pisot number. Such an E will be a set of uniqueness and of synthesis and all restriction algebras $A(E)$ will be isomorphic; these E will be called ultra-thin symmetric sets.

Other group algebras, like $A(D^\infty)$, can be regarded as subalgebras of $A(E)$ and, following Varopoulos and Katznelson [93, 11], this identification can be used to obtain a compact subset K of E which is not a set of synthesis.

1. Properties of ultra-thin symmetric sets

We recall the definition of symmetric sets of real numbers (already given in Chapter III, Section 2): we start with a compact interval $[a, b]$ of real numbers and a sequence $(\xi_k)_{k\geq 1}$ of real numbers belonging to the open interval $(0, \frac{1}{2})$; E is the intersection $\bigcap_{k\geq 1} E_k$ where E_k is a union of 2^k intervals of length $\xi_1 \cdots \xi_k (b - a)$ and is defined inductively by the following rule: to construct all the intervals of E_k, each compact interval I of E_{k-1} is replaced by two compact intervals I' and I'' of length $\xi_k|I|$; the left-hand end point of I' is that of I and the right-hand end point of I'' is that of I; E_0 is $[a, b]$. *E is called ultra-thin if* $\sum_1^\infty \xi_k^2 < +\infty$.

As usual, the Banach algebra $A(\mathbf{R})$ is defined by the rule: $f \in A(\mathbf{R})$ if f is the Fourier-transform of $g \in L^1(\mathbf{R})$ and $\|f\|_{A(\mathbf{R})} = \|g\|_1$. For each compact set E of real numbers, $A(E)$ is the Banach algebra of all complex valued continuous functions h defined on E which are the restrictions to E of functions f in $A(\mathbf{R})$; $\|h\|_{A(E)} = \inf \|f\|_{A(\mathbf{R})}$ over such f.

If E is contained in the open interval $(-\pi, \pi)$, and this is assumed in everything that follows, $A(E)$ is also the Banach algebra of all restrictions to E of absolutely convergent Fourier series $\sum_{-\infty}^{+\infty} c_k e^{ikx}$ where $\sum_{-\infty}^{+\infty} |c_k| < +\infty$; these two definitions of $A(E)$ lead to equivalent norms.

All ultra-thin symmetric sets are not affine copies of each other, nor even locally affine copies. However, within the scope of harmonic analysis, all ultra-thin symmetric sets are the same as the following result shows:

THEOREM I: *All ultra-thin symmetric sets E are sets of uniqueness and of synthesis and all the corresponding restriction algebras A(E) are isomorphic.*

The role played by the exponent 2 is crucial since, for each $p > 2$, we shall construct, following Salem, a sequence $(\xi_k)_{k\geqslant 1}$ of dissection ratios such that $\sum_1^\infty \xi_k^p < +\infty$ and such that the corresponding symmetric set F has the two properties: F is a set of multiplicity and for each ultra-thin symmetric set E, $A(E)$ and $A(F)$ are not isomorphic.

2. Functions whose spectra lie in ultra-thin symmetric sets

2.1. To get a more precise version of Theorem I, it is helpful to give another definition of ultra-thin symmetric sets.

DEFINITION 1: *Let $(t_k)_{k\geqslant 1}$ be a sequence of positive real numbers satisfying the two conditions:*

$$\text{for each } k \geqslant 1,\ t_k > s_{k+1} = t_{k+1} + t_{k+2} + \cdots \tag{8.1}$$

$$\text{and } \sum_{k\geqslant 1} \left(\frac{t_{k+1}}{t_k}\right)^2 < +\infty. \tag{8.2}$$

The corresponding ultra-thin symmetric set E is the compact totally disconnected set E of all sums $x = \sum_1^\infty \varepsilon_k t_k$, $\varepsilon_k = 0$ or 1.

Putting $t_k = \xi_1 \cdots \xi_{k-1}(1 - \xi_k)$, $a = 0$, $b = 1$, we see that this new definition is equivalent to the first one.

DEFINITION 2: *With the notation of Definition* 1, *we denote by*

(a) F_k *the finite set of the* 2^k *sums* $\sum_1^k \varepsilon_j t_j$, $\varepsilon_j = 0$ *or* 1;

(b) E_k *the set of the* 2^k *intervals whose origins belong to* F_k *and each of whose lengths is* s_{k+1};

(c) I_k *the interval* $[0, s_{k+1}]$.

We then have, $F_k \subset E \subset F_k + I_k$ for each $k \geqslant 1$.

We shall now describe the space of all complex valued bounded continuous functions φ whose spectra lie in E.

DEFINITION 3: *For each $k \geqslant 1$, we write φ as a perturbed trigonometric sum*

$$\varphi(x) = \sum_{\lambda\in F_k} \varphi_\lambda(x) \exp 2\pi i\lambda x = \sum \varphi_{\varepsilon_1,\ldots,\varepsilon_k}(x) \exp 2\pi i(\varepsilon_1 t_1 + \cdots + \varepsilon_k t_k)x, \tag{8.3}$$

where the spectrum of each φ_λ *lies in* I_k, $\varphi_\lambda = \varphi_{(\varepsilon_1,\ldots,\varepsilon_k)}$ *if* $\lambda = \varepsilon_1 t_1 + \cdots + \varepsilon_k t_k$ *and* $\varepsilon_k = 0$ *or* 1, *the second sum being extended over the* 2^k *such sequences of* 0's *and* 1's.

For all real numbers $x, x_1, \ldots, x_k$ and with the same summation convention, we put

$$\tilde{\varphi}_k(x, x_1, \ldots, x_k) = \sum \varphi_{(\varepsilon_1,\ldots,\varepsilon_k)}(x) \exp 2\pi i (\varepsilon_1 x_1 + \cdots + \varepsilon_k x_k) \tag{8.4}$$

and

$$\varphi_k(x) = \sum_{\lambda \in F_k} \varphi_\lambda(0) \exp 2\pi i \lambda x = \tilde{\varphi}_k(0, t_1 x, \ldots, t_k x). \tag{8.5}$$

2.2. With this notation we have

THEOREM II: *For each ultrathin symmetric set E of real numbers, there exists a constant C such that each complex valued bounded continuous function* φ *whose spectrum lies in E can be approximated in the following way:*

(i) *for all real* x, $|\varphi(x) - \varphi_k(x)| \leqslant C t_{k+1} |x| \sup_{\mathbf{R}} |\varphi|$;

(ii) $\sup_{\mathbf{R}} |\varphi_k| \leqslant \sup_{\mathbf{R}^{k+1}} |\tilde{\varphi}_k| \leqslant C \sup_{\mathbf{R}} |\varphi|$;

(iii) *on each interval I of length* C/t_k, $\sup_{\mathbf{R}} |\varphi_k| \leqslant C \sup_I |\varphi_k|$.

Roughly speaking (i) describes the quality of the approximation of the perturbed trigonometric series φ by the corresponding uniformly bounded pure trigonometric sums φ_k and (iii) shows that there exists a kind of periodicity of period C/t_k for these φ_k. Finally (ii) shows that the frequencies t_k are decreasing so rapidly that sup-norms can be estimated by replacing each $\exp 2\pi i t_k x$ by an independent $\exp 2\pi i x_k$.

The proof of Theorem II will be given in Section 4. The intuitive content of Theorem I will be described now.

A problem raised in the introduction should be recalled. A function of time is denoted by $\varphi(t)$ ('function' here means complex valued continuous function). Observations are made over a long interval $[0, L_1]$. The first feeling is that $\varphi(t)$ is periodic. Let T_1 be this period and $\nu_1 = T_1^{-1}$ the corresponding frequency. For the sake of simplicity, it will be assumed that $\varphi(t) = a(0) + a(1) \cos 2\pi\nu_1 t + a(2) \sin 2\pi\nu_2 t$. But more careful observations over longer intervals show that $a(0)$, $a(1)$ and $a(2)$ are actually three functions of time whose period T_2 could not be detected on the interval $[0, L_1]$; T_2 is very large compared with L_1 or with T_1. For the sake of simplicity, it will be assumed that

$$a(j) = a(j, 0) + a(j, 1) \cos 2\pi\nu_2 t + a(j, 2) \sin 2\pi\nu_2 t; \quad 0 \leqslant j \leqslant 2,$$

where we have $\nu_2 = 1/T_2$.

Now $\varphi(t)$ becomes an almost periodic function with nine coefficients; the frequencies of f are $\varepsilon_1\nu_1 + \varepsilon_2\nu_2$ for $\varepsilon_1, \varepsilon_2 = -1, 0$ or 1.

Better observations show that each of these nine coefficients of f itself has 'secular perturbations' whose period T_3 is very large compared with T_2 and so on.

If φ is assumed to have a 'nice' behaviour at infinity, for instance if φ is bounded, a mathematical model can be built to explain the 'projective sequence' of approximations of φ, to compute the corresponding error terms and to find the global behaviour of φ.

Let E be the set of all infinite sums $\sum_1^\infty \varepsilon_k\nu_k$, $\varepsilon_k = -1, 0$ or 1. (We assume that the sequence T_k, $k \geqslant 1$, of periods of 'secular perturbations' grows sufficiently rapidly to satisfy $\sum_1^\infty 1/T_k < +\infty$.)

Two cases have been studied in this book:

(1) We have $\sum_1^\infty (\nu_{k+1}/\nu_k)^2 < +\infty$. Then this mysterious function φ is any complex-valued bounded continuous function on the line whose spectrum lies in E. For each $k \geqslant 1$, we define the finite set to F_k consist of the 3^k sums $\varepsilon_1\nu_1 + \cdots + \varepsilon_k\nu_k$, $\varepsilon_k = -1, 0$ or 1. Then, for each $k \geqslant 1$, $\varphi(t)$ is identically $\sum_{\lambda\in F_k} \varphi_\lambda(t) \exp 2\pi i\lambda t$, where the spectrum of each φ_λ lies in $[0, \bar{\nu}_{k+1}]$, where $\bar{\nu}_{k+1} = \nu_{k+1} + \nu_{k+2} + \cdots$. We put $\varphi_k(t_0, t) = \sum_{\lambda\in F_k} \varphi_\lambda(t_0) \exp 2\pi i\lambda t$ and we get $|\varphi(t) - \varphi_k(t_0, t)| \leqslant C(|t - t_0|)\, T_{k+1}^{-1} \sup_{\mathbf{R}} |\varphi|$. Hence if $L_k = \sqrt{T_k T_{k+1}}$ and $t_0 = 0$, $|\varphi(t) - \varphi_k(0, t)| \leqslant C\sqrt{T_k/T_{k+1}} \sup_{\mathbf{R}} |\varphi|$ on $[0, L_k]$ and we cannot distinguish between $\varphi(t)$ and $\varphi_k(0, t) = \varphi_k(t)$.

A possibility exists for a local harmonic analysis of φ on $[0, L_k]$: the observation of $\varphi(t) = \sum_{\lambda\in F_k} \varphi_\lambda(t) \exp 2\pi i\lambda t$ on $[0, L_k]$ gives estimates of each perturbed coefficient $\varphi_\lambda(t)$ on $[0, L_k]$: an approximation of $\varphi_\lambda(t_0)$ is $(1/L_k)\int_{t_0}^{t_0+L_k} \varphi(t) \exp -2\pi i\lambda t\, dt$ and the error term is bounded by $C'\sqrt{T_k/T_{k+1}} \sup_{\mathbf{R}} |\varphi|$.

The global behaviour of φ can be very different from the expected behaviour of an almost periodic function. For example, let μ be a continuous measure carried by E and let $\varphi(t) = \hat{\mu}(-t)$. Then a theorem of Wiener ([24], Ch. XVI, th. 4.19) shows that $\lim_{T\to+\infty} T^{-1}\int_0^T |\varphi(t)|\, dt = 0$.

Since the derivative φ' of φ is bounded, this property implies that for each positive ε and each integer $n \geqslant 1$, there exists an interval of real numbers of length n on which $|\varphi| \leqslant \varepsilon$.

(2) θ is a Pisot number, p is a large integer and $T_k/T_{k+1} = \nu_{k+1}/\nu_k = \theta^{-p}$. Here also, our mysterious function φ is any complex-valued bounded continuous function on the line whose spectrum lies in E. All the preceding analysis of such a φ remains valid; C and C' are now two constants depending only on θ.

3. Theorem I as a corollary of Theorem II

3.1. Each ultra-thin symmetric set E is a set of synthesis

Let φ be as in Theorem II and let f be a function in $L^1(\mathbf{R})$, whose Fourier-transform vanishes on E. Then, Lebesgue's dominated convergence theorem together with (i) and (ii) gives $\lim\limits_{k\to+\infty} \int_{\mathbf{R}} \varphi_k(-x)\, f(x)dx = \int_{\mathbf{R}} \varphi(-x) f(x)\, dx$.

But, by an easy computation, $\int_{\mathbf{R}} \varphi_k(-x) f(x)\, dx = \sum_{\lambda\in F_k} \varphi_\lambda(0)\hat{f}(\lambda) = 0$. Hence $\int_{\mathbf{R}} \varphi(-x) f(x)\, dx = 0$.

3.2. Each ultra-thin symmetric set is a set of uniqueness

Let $\varphi : \mathbf{R} \to \mathbf{C}$ be a complex-valued continuous function vanishing at infinity whose spectrum lies in E. Since $t_{k+1}/t_k \to 0$ and $t_k \to 0$, there exists an increasing sequence $(x_k)_{k\geqslant 1}$ of positive real numbers such that $t_k x_k \to +\infty$, while $t_{k+1}x_k \to 0$. Let I_k be the interval $[x_k, x_k + C/t_k]$ and let $\omega_k = \sup\limits_{I_k} |\varphi|$; by the assumption on φ, $\omega_k \to 0$ as $k \to +\infty$.

By (i), $\sup\limits_{I_k} |\varphi| \geqslant -Ct_{k+1}(x_k + C/t_k)\, \|\varphi\|_\infty + \sup\limits_{I_k} |\varphi_k|$. Hence $\sup\limits_{I_k} |\varphi_k| \leqslant Ct_{k+1}(x_k + C/t_k)\, \|\varphi\|_\infty + \omega_k$ which tends to 0. (iii) now shows that $\sup\limits_{\mathbf{R}} |\varphi_k|$ also tends to 0 and so the limit φ of φ_k is identically 0.

3.3. All Banach algebras A(E) are isomorphic

Before beginning the proof, we describe a dense subalgebra of $A(E)$. For each $k \geqslant 1$, let A_k be the subspace of all f in $A(E)$, which are constant on each interval $\lambda + I_k$, $\lambda \in F_k$ (Definition 2) and let A_∞ be the union of all the A_k. *Then A_∞ is dense in $A(E)$.*

To prove this remark, it helps to regard E as a compact subset of $(-\pi, \pi)$; since E is a set of synthesis, the dual space $PM(E)$ of $A(E)$ is the Banach space of all distributions S carried by E such that $\sup_{\mathbf{R}} |\hat{S}(n)| < +\infty$.

If A_∞ is not dense in $A(E)$, there exists an $S \neq 0$ in $PM(E)$ such that $\langle S, f\rangle = 0$ for each f in A_∞. If $f = 1$, this gives $\hat{S}(0) = 0$ and integrating the Fourier series $\sum_{\mathbf{Z}} \hat{S}(n)\, e^{inx}$ of S, we obtain the Fourier series $\sum_{-\infty}^{+\infty} \hat{S}(n)\, e^{inx}/in$ of a function F in $L^2(-\infty, +\infty)$; the distributional derivative of F is S. Hence F is constant on each interval of the complement of E in $[-\pi, \pi]$ (since S is carried by E). For each x and x' in $(-\pi, \pi)\backslash E$ the restriction to E of the characteristic function of $[x, x']$ belongs to A_∞. Hence $F(x') = F(x)$ and F is constant on the complement of E; since E has Lebesgue measure 0, $F = C$, where this equality holds in the sense of distributions, and $S = 0$.

We return now to the proof of the isomorphism. Let E' be another ultra-thin symmetric set defined by a sequence $(t_k')_{k\geqslant 1}$ satisfying (8.2) and let $H: E' \to E$ be the homeomorphism defined by $H(\sum_1^\infty \varepsilon_k t_k') = \sum_1^\infty \varepsilon_k t_k$ for all $\varepsilon_k = 0$ or 1.

PROPOSITION 1: *For each $f \in A(E)$, $f \circ H$ belongs to $A(E')$ and for some constant C depending only on E and E'*

$$\|f \circ H\|_{A(E')} \leqslant C\, \|f\|_{A(E)}. \tag{8.6}$$

This proposition gives the required isomorphism since interchanging E' and E yields the reverse inequality.

In proving (8.6) we may assume that f belongs to A_∞; the general case then follows from the density of A_∞ in $A(E)$. To compute norms in $A(E')$, we test on the unit ball of the dual $PM(E')$: let φ' be any complex valued continuous function whose spectrum lies in E' and such that $\|\varphi'\|_\infty \leqslant 1$, let S' be the Fourier transform of φ', let φ_k' be defined in terms of φ' by (8.5) and let S_k' be the Fourier transform of φ_k'; S_k' is a measure carried by F_k'.

If $f \in A_k$, $f \circ H \in A_k'$ and $\langle f \circ H, S'\rangle = \langle f \circ H, S_k'\rangle = \langle f, H(S_k')\rangle$. The measure $S_k = H(S_k')$ is the Fourier transform of a trigonometric sum φ_k; if $\varphi_k' = \sum c_{(\varepsilon_1,\ldots,\varepsilon_k)} \exp 2\pi i\, (\varepsilon_1 t_1' + \cdots + \varepsilon_k t_k')\, x$, then $\varphi_k = \sum c_{(\varepsilon_1,\ldots,\varepsilon_k)} \exp 2\pi i\, (\varepsilon_1 t_1 + \cdots + \varepsilon_k t_k)\, x$ and (ii) in Theorem II applied to the ultra-thin set E' and the function φ' shows that $\|\varphi_k\|_\infty \leqslant \sup_{\mathbf{R}^k} |\sum c_{(\varepsilon_1,\ldots,\varepsilon_k)} \exp 2\pi i\, (\varepsilon_1 x_1 + \cdots + \varepsilon_k x_k)| \leqslant C$. Then $|\langle f \circ H, S'\rangle| = \langle f, H(S_k')\rangle \leqslant \|f\|_{A(E)}\, \|\varphi_k\|_\infty \leqslant C \|f\|_{A(E)}$. Taking the sup over φ', we get (8.6)

Using exactly the same argument, the following result (used in Section 7) can be obtained.

PROPOSITION 2: *For each ultra-thin symmetric set E of real numbers there exists a constant C such that, for each sequence $\theta_1, \ldots, \theta_n$ of n real numbers, the function $f: E \to \mathbf{C}$ defined by*

$$f(\sum_1^\infty \varepsilon_k t_k) = \exp 2\pi i (\sum_1^n \varepsilon_k \theta_k) \tag{8.7}$$

belongs to $A(E)$ with a norm not exceeding C.

4. Proof of Theorem II: reduction to a problem on intervals

4.1. THEOREM III: *Let $\alpha, t_1, \ldots, t_n$ be an increasing sequence of $n+1$ positive real numbers and let $q > 1$ be such that $t_1 \geqslant q\alpha$, $t_2 \geqslant q(t_1 + \alpha)$, $\ldots, t_n \geqslant q(t_{n-1} + \cdots + t_1 + \alpha)$; C will denote a constant depending only on q. Let Λ_n be the set of the 2^n sums $\sum_1^n \varepsilon_k t_k$, $\varepsilon_k = 0$ or 1 and let $a_\lambda(x)$, $\lambda \in \Lambda_n$, be a sequence indexed by Λ_n of complex valued bounded continuous functions whose spectra lie in $[0, \alpha]$. We put*

$$f(x) = \sum_{\lambda \in \Lambda_n} a_\lambda(x) \exp 2\pi i (\sum_1^n \varepsilon_k t_k) x \tag{8.8}$$

$$\tilde{f}(x, x_1, \ldots, x_n) = \sum_{\lambda \in \Lambda_n} a_\lambda(x) \exp 2\pi i (\varepsilon_1 x_1 + \cdots + \varepsilon_n x_n). \tag{8.9}$$

Then

$$\sup_{\mathbf{R}} |f| \leqslant \sup_{\mathbf{R}^{n+1}} |\tilde{f}| \leqslant (1 + C\alpha^2/t_1^2) \cdots (1 + C t_{n-1}^2/t_n^2) \sup_{\mathbf{R}} |f|. \tag{8.10}$$

4.2. Two lemmas: high frequencies and independence

Let U be the union of all intervals $[\lambda, \lambda + \alpha]$, $\lambda \in \Lambda_n$, and let $PF(U)$ be the set of all distributions S carried by U whose Fourier-transforms vanish at infinity; we put $\|S\| = \sup_{x \in \mathbf{R}} |\hat{S}(x)|$.

Then, as we observed in Chapter IV, Section 14, the dual of $PF(U)$ is $A(U)$, whose dual is $PM(U)$. Using this fact, the proof of (8.10) can be reduced to the case where f tends to 0 at infinity.

The following two lemmas will be needed:

LEMMA 1: *Let* $q > 1$, $\omega > 0$, $0 < l \leqslant 1/q\omega$, let *I be the interval* $[0, l]$ *and let* $h: \mathbf{R} \to \mathbf{T}$ *be the continuous homomorphism defined by* $h(t) = \exp 2\pi i\omega t$. *There is a constant C depending only on q such that, for each pseudo-function S carried by I*

$$\|h(S)\|_{PM(\mathbf{T})} \leqslant \|S\|_{PM(\mathbf{R})} \leqslant (1 + Cl^2\omega^2)\, \|h(S)\|_{PM(\mathbf{T})}.$$

Roughly speaking, Lemma 1 asserts that the local isomorphism between **R** and **T** defines almost an isometry for pseudo-measure norms of distributions carried by a small interval.

Proof of Lemma 1: Let S be a pseudo-function carried by I and let $\varphi(x) = \hat{S}(-x)$. We have to prove that $\sup_{\mathbf{R}} |\varphi| \leqslant (1 + Cl^2\omega^2) \sup_{\omega\mathbf{Z}} |\varphi|$; let x_0 be a point where $|\varphi(x_0)| = \sup_{\mathbf{R}} |\varphi|$: such a point exists since φ is continuous and vanishes at infinity. Then, by Taylor's formula, $\varphi(x) = \varphi(x_0) + (x - x_0) A + 2^{-1} (x - x_0)^2 B(x)$, where $A = \varphi'(x_0)$, $|B| \leqslant \sup_{\mathbf{R}} |\varphi''(x)|$. By Bernstein's inequality (Chapter V, Section 4) $\sup_{\mathbf{R}} |\varphi''| \leqslant 4\pi^2 l^2 \sup_{\mathbf{R}} |\varphi| = 4\pi^2 l^2 |\varphi(x_0)|$. Since $|\varphi(x_0)|$ is the maximum of all $|\varphi(x)|$, the minimum of $|\varphi(x_0) + (x - x_0) A|$ is obtained when $x = x_0$ and $A/\varphi(x_0)$ is a pure imaginary (see Chapter I, Section 4.3, Lemma 7). We get $|\varphi(x)| \geqslant |\varphi(x_0)| (1 - 2\pi^2 l^2 |x - x_0|^2)$. There is an $x \in \omega\mathbf{Z}$ such that $|x - x_0| \leqslant \omega/2$ which implies $\sup_{\omega\mathbf{Z}} |\varphi| \geqslant (1 - \pi^2 l^2 \omega^2)/2) \sup_{\mathbf{R}} |\varphi|$.[1]

LEMMA 2: *Let* $q > 1$, $0 < l < ql \leqslant \nu$, *and let* $a_0(x)$ *and* $a_1(x)$ *be two functions vanishing at infinity whose spectra lie in* $[0, l]$. *Then*

$$\sup_{\mathbf{R}} |a_0(x) + e^{2\pi i\nu x} a_1(x)| \leqslant \sup_{\mathbf{R}^2} |a_0(x) + e^{2\pi i y} a_1(x)|$$
$$\leqslant (1 + c\, l^2/\nu^2) \sup_{\mathbf{R}} |a_0(x) + e^{2\pi i\nu x} a_1(x)|.$$

Roughly speaking, the effect of the high frequency ν is to make $e^{2\pi i\nu x}$ independent of the low-frequency terms $a_0(x)$ and $a_1(x)$ (this observation has a very concrete interpretation: the position of the second hand of a watch gives very little information about the position of the hour hand).

Proof: Let θ be a fixed real number and put $x = (\theta + k)/\nu$, $k \in \mathbf{Z}$. We have $\sup_{\mathbf{R}} |a_0(x) + e^{2\pi i\nu x} a_1(x)| \geqslant \sup_{k\in\mathbf{Z}} |a_0(\theta\nu^{-1} + e^{2\pi i\theta} a_1(\theta\nu^{-1} + k\nu^{-1})|$.

---

[1] If $l\omega \in [\pi^{-1}, q^{-1}]$, the proof of Theorem II, Chapter III, shows that $\|S\|_{PM(\mathbf{R})} \leqslant C_q \|h(S)\|_{PM(\mathbf{T})}$.

Let $b : \mathbf{R} \to \mathbf{C}$ be defined by $b(t) = a_0(\theta\nu^{-1} + t) + e^{2\pi i\theta} a_1(\theta\nu^{-1} + t)$; the spectrum of b lies in $[0, l]$ and Lemma 1 gives

$$(1 + C\, l^2/\nu^2) \sup_{\mathbf{Z}\nu^{-1}} |b| \geqslant \sup_{\mathbf{R}} |b| = \sup_{\mathbf{R}} |a_0(t) + e^{2\pi i\theta} a_1(t)|.$$

Finally, for each real θ, $\sup_{\mathbf{R}} |a_0(x) + e^{2\pi i\theta} a_1(x)| \leqslant (1 + C\, l^2/\nu^2) \sup_{\mathbf{R}}$ $a_0(x) + e^{2\pi\nu i x} a_1(x)|$. Taking the sup over θ we get Lemma 2.

4.3. *Proof of Theorem III*

We proceed by induction on n. Let $s_{n-1} = \alpha + t_1 + \cdots + t_{n-1}$ and for each function f whose spectrum lies in $[0, \alpha] + \Lambda_n$, let $\tilde{f}(x, x_1, \ldots, x_n)$ be the $n + 1$ dimensional extension of f given by (8.4).

We can write $f(x) = g(x) + e^{2\pi i t_n x} h(x)$, where the spectrum of g and that of h both lie in $[0, s_{n-1}]$. The inequality $s_{n-1} \leqslant (q + 1/q)\, t_{n-1}$ and Lemma 2 give

$$\sup_{\mathbf{R}^2} |g(x) + e^{iy} h(x)| \leqslant (1 + C_1\, t_{n-1}^2/t_n^2) \sup_{\mathbf{R}} |f|, \tag{8.11}$$

where $C_1 = (q + 1/q)^2\, C$ and C is defined by Lemma 2.

For each fixed y, the spectrum of $g(x) + e^{iy} h(x)$ lies in Λ_{n-1}. Hence the induction hypothesis gives

$$\sup_{\mathbf{R}^n} |\tilde{g}(x, x_1, \ldots, x_{n-1}) + e^{iy} \tilde{h}(x, x_1, \ldots, x_{n-1})|$$

$$\leqslant (1 + C_1\, t_1^2/t_2^2) \cdots (1 + C_1\, t_{n-2}^2/t_{n-1}^2) \sup_{\mathbf{R}} |g(x) + e^{iy} h(x)|. \tag{8.12}$$

Combining (8.11) and (8.12) we get Theorem III.

4.4. *Proof of Theorem II*

Theorem III trivially gives (ii) in Theorem II; the convergence of the series $\sum_{k \geqslant 1} t_{k+1}^2/t_k^2$ ensures the convergence of the infinite product $\prod_{k \geqslant 1} (1 + C\, t_{k+1}^2/t_k^2)$.

To get (i) and (iii), we need only apply the following lemma:

LEMMA 3: *Let Λ be a uniformly discrete set of real numbers: $0 < l < \inf |\lambda - \lambda'|$ for all $\lambda \neq \lambda'$ in Λ, and assume that for some constant C_1 and each finite perturbed trigonometric sum $\sum_{\lambda \in \Lambda} a_\lambda(x) \exp 2\pi i \lambda x = f(x)$, constructed with complex-valued bounded continuous functions a_λ whose* *spectra lie in* $[0, l]$,

$$\sup_{\mathbf{R}} |\sum_{\lambda\in\Lambda} a_\lambda(0) \exp 2\pi i\lambda x| \leqslant C_1 \sup_{\mathbf{R}} |f(x)|. \tag{8.13}$$

Then there exists a constant C_2 *depending only on* C_1 *such that for each pure trigonometric sum* $P(x) = \sum_{\lambda\in\Lambda} c_\lambda \exp 2\pi i\lambda x$,

$$\sup_{\mathbf{R}} |P| \leqslant 2C_1 \sup_{|x|\leqslant C_2/l} |P(x)| \tag{8.14}$$

and, assuming f *is defined as above,*

$$|\sum_{\lambda\in\Lambda} a_\lambda(0) \exp 2\pi i\lambda x - \sum_{\lambda\in\Lambda} a_\lambda(x) \exp 2\pi i\lambda x| \leqslant 2\pi l C_1 |x| \sup_{\mathbf{R}} |f|. \tag{8.15}$$

Proof of Lemma 3: Let Δ be the function which is 0 outside $[0, 1]$, linear on $[0, \frac{1}{2}]$ and $[\frac{1}{2}, 1]$ and such that $\Delta(\frac{1}{2}) = 2$; Δ is the Fourier transform of ψ. Define C_2 by the condition that $|\psi(x)| \leqslant 1/2C_1$ if $|x| \geqslant C_2$ and apply (8.14) with $a_\lambda(x) = \psi(lx)\, a_\lambda$ for all $\lambda \in \Lambda$. We get $C_1 \sup_{\mathbf{R}} |\psi P| \geqslant \sup_{\mathbf{R}} |P|$. But $\sup_{\mathbf{R}} |\psi P| \leqslant \sup_{|x|\leqslant l^{-1}C_2} |\psi P| + 1/2C_1 \|P\|_\infty$, which gives (8.14).

To get (8.15), we put $f(x, y) = \sum_{\lambda\in\Lambda} a_\lambda(y) \exp 2\pi i\lambda x$. Applying (8.13) to $f(x + y)$, where y is a parameter, we get $\sup_{\mathbf{R}^2} |f(x, y)| \leqslant C_1 \sup_{\mathbf{R}} |f(x)|$.

Let us compute $f(x, y_1) - f(x, y_2)$; for each fixed x, $f(x, y)$ is a function whose spectrum lies in $[0, l]$. By Bernstein's inequality, we get, for each fixed x, $|f(x, y_1) - f(x, y_2)| \leqslant |y_1 - y_2|\, 2\pi l \sup_{y\in\mathbf{R}} |f(x, y)|$. If we replace y_1 by 0, y_2 by x and if we take the supremum with respect to x, we get (8.15).

5. Symmetric coherent sets of frequencies

5.1. THEOREM IV: *Let* $(t_k)_{k\geqslant 1}$ *be an increasing sequence of positive real numbers such that* $t_{k+1} > t_1 + \cdots + t_k = s_k$ *and* $\sum_1^\infty (t_k/t_{k+1})^2 < +\infty$. *Let* Λ *be the set of all finite sums* $\sum_{k\geqslant 1} \varepsilon_k t_k$, $\varepsilon_k = 0$ or 1. *Then* Λ *is a coherent set of frequencies. Moreover for each positive* ε, *there exists a constant* $T(\varepsilon)$ *such that for each trigonometric sum* $P(x) = \sum_{\lambda\in\Lambda} a_\lambda \exp 2\pi i\lambda x$, $\sup_{\mathbf{R}} |P| \leqslant (1 + \varepsilon) \sup_{|t|\leqslant T(\varepsilon)} |P(t)|$.

5.2. Before giving the proof, we make some remarks concerning Lemma 3: if, for some $\varepsilon > 0$, $C_1 \leqslant 1 + \varepsilon/2$, we may choose $C_2 = C_2(\varepsilon)$, in the notation of the proof of Lemma 3, sufficiently large to ensure $|\psi(x)| \leqslant \varepsilon/2$

$+ \varepsilon$ if $|x| \geqslant C_2$, which implies

$$\sup_{|x| \leqslant C_2(\varepsilon) l^{-1}} |P(x)| \geqslant (1 - \varepsilon) \|P\|_\infty; \tag{8.16}$$

$C_2(\varepsilon)$ is independent of l and Λ.

Definition of an integer m: if C is the constant of Theorem III, an integer m can be found such that $(1 + Ct_{m-1}/t_m)(1 + Ct_m^2/t_{m+1}^2) \cdots (1 + Ct_k^2/t_{k+1}^2) \cdots < 1 + \varepsilon/2$.

This is possible since $\sum_1^\infty (t_k/t_{k+1})^2 < +\infty$. Taking a larger m if necessary, we may also assume that

$$\frac{\varepsilon}{2\pi s_{m-1}} \geqslant \frac{C_2(\varepsilon)}{\sqrt{t_{m-1} t_m}}$$

since $s_{m-1}/t_m \to 0$ as $m \to +\infty$.

Now let P be any finite sum $\sum a_{(\varepsilon_1, \ldots, \varepsilon_n)} \exp 2\pi i (\varepsilon_1 t_1 + \cdots + \varepsilon_n t_n) x$, let Λ_m be the set of all finite sums $\sum_0^{m-1} \varepsilon_k t_k$. If $n < m$, there is nothing to prove: Λ_m is finite, hence Λ_m is harmonious and for each positive ε there exists a $T_1(\varepsilon) > 0$ such that for each trigonometric sum P whose frequencies belong to Λ_m, $\sup_{\mathbf{R}} |P| \leqslant (1 + \varepsilon) \sup_{|t| \leqslant T(\varepsilon)} |P(t)|$.

If $n \geqslant m$, a trigonometric sum P_n is defined by

$$P_n(x, x_m, x_{m+1}, \ldots, x_n) = \sum_{\varepsilon_k = 0, 1} a_{(\varepsilon_1, \ldots, \varepsilon_n)} \exp 2\pi i (\varepsilon_1 t_1 x + \cdots + \varepsilon_{m-1} t_{m-1} x + \varepsilon_m x_m + \cdots + \varepsilon_n x_n)$$

and it will be proved that for a suitable $T(\varepsilon)$, $\sup_{\mathbf{R}^n} |P_n| \leqslant (1 + \varepsilon) \sup_{|t| \leqslant T(\varepsilon)} |P(t)|$, which is an improvement of Theorem IV.

Let $\alpha, \beta_m, \ldots, \beta_n$ be $n - m + 2$ real numbers such that

$$|P_n(\alpha, \beta_m, \ldots, \beta_n)| \geqslant (1 - \varepsilon) \sup_{\mathbf{R}^{n-m+2}} |P_n| = (1 - \varepsilon) M.$$

Since Λ_m is finite and since for fixed $\beta_m, \ldots, \beta_n$, the spectrum of $x \to P_n(x, \beta_m, \ldots, \beta_n)$ lies in Λ_m, α can be replaced by a γ, $|\gamma| \leqslant T_1(\varepsilon)$, such that $|P_n(\gamma, \beta_m, \ldots, \beta_n)| \geqslant (1 - \varepsilon) \sup_{\mathbf{R}} |P_n(x, \beta_m, \ldots, \beta_n)| \geqslant (1 - \varepsilon)^2 M$.

Consider the interval $[\gamma, \gamma + \varepsilon/2\pi s_{m-1}]$; its length exceeds

$$\frac{C_2(\varepsilon)}{\sqrt{t_{m-1} t_m}}$$

by our choice of m. Applying (8.16) and Theorem III to the set Λ of all finite sums $\sum_{k\geqslant m}\varepsilon_k t_k$, $\varepsilon_k = 0$ or 1 and to $l = \sqrt{t_{m-1}t_m}$, we can find an x in $[\gamma, \gamma + \varepsilon/2\pi s_{m-1}]$ such that

$$|P_m(\gamma, x, \dots, x)| \geqslant (1-\varepsilon)^2 \sup_{\mathbf{R}^{n-m+1}} |P_n(\gamma; x_m \cdots x_n)| \geqslant (1-\varepsilon)^4 M.$$

From Bernstein's inequality applied, for this fixed x, to the trigonometric sum $t \to P_n(t, x, \dots, x)$ whose spectrum lies in $[0, s_{m-1}]$, we obtain $|P_n(\gamma, x, \dots, x) - P_n(x, x, \dots, x)| \leqslant \varepsilon \sup_{t\in\mathbf{R}} |P_n(t, x, \dots, x)| \leqslant \varepsilon M$. Hence

$$|P(x)| = |P_n(x, x, \dots, x)| \geqslant ((1-\varepsilon)^4 - \varepsilon) M$$

and $|x| \leqslant T_1(\varepsilon) + \varepsilon/2\pi s_{m-1} = T(\varepsilon)$; ε no longer has the same meaning as in Theorem IV; nonetheless, this theorem is now proved.

6. High frequencies and independence; complements

6.1. The exponent 2 which appears in all our computations arises for the first time in the crucial Lemma 2 which, by iteration, gives Theorem III.

It is a fascinating problem to replace the frequency ν by a finite set $\nu_1, \dots, \nu_n$ of real numbers which are independent over $\mathbf{Q}$ and to ask the following question:

Does a constant C exist such that for each sufficiently small positive l and each sequence $a_0, a_1, \dots, a_n$ of complex valued bounded continuous functions on the line whose spectra lie in $[0, l]$,

$$\sup_{\mathbf{R}^{n+1}} |a_0(x) + a_1(x)\exp 2\pi i x_1 + \cdots + a_n(x)\exp 2\pi i x_n|$$
$$\leqslant (1 + Cl^{2/n}) \sup_{\mathbf{R}} |a_0(x) + a_1(x)\exp 2\pi i\nu_1 x + \cdots + a_n(x)\exp 2\pi i\nu_n x|? \tag{8.17}$$

The answer is very surprising:

THEOREM V: *With the preceding notation, estimate* (8.17) *is the best possible in the sense that $l^{2/n}$ cannot be replaced by a number which is $o(l^{2/n})$ when l tends to* 0*; moreover this estimate* (8.17) *holds if and only if for some constant $D > 0$, each integer $m \geqslant 1$ and each sequence $p_1, \dots, p_n$ of rational integers satisfying $|p_j| \leqslant m$,*

$$|p_1\nu_1 + \cdots + p_n\nu_n| \geqslant Dm^{-n+1}. \tag{8.18}$$

Roughly speaking, Theorem V asserts that the best estimate in (8.17) for n independent frequencies holds if and only if these frequencies are strongly independent.

COROLLARY: *Let* $\nu_1, \ldots, \nu_n$ *be n real numbers satisfying* (8.18) *and let* $(t_k)_{k\geqslant 1}$ *be a sequence of positive real numbers tending to* 0 *sufficiently rapidly that* $\sum_{k\geqslant 1} (t_{k+1}/t_k)^{2/n} < +\infty$.

Let E_n *be the set of all sums* $\sum_{k\geqslant 1} \varepsilon_k t_k$ *for* $\varepsilon_k = 0$ *or* $\nu_1, \ldots,$ *or* ν_n.

Then E_n *is a set of synthesis and a set of uniqueness and all the restriction algebras* $A(E_n)$ *are isomorphic.*

The proof of the corollary is an easy transcription of the proof of Theorem I. Now to the proof of Theorem V.

6.2. Proof of (8.18) ⇒ (8.17). As in Lemma 1, we can confine our attention to the case where the functions $a_0, \ldots, a_n$ vanish at infinity; we put

$$f(x) = a_0(x) + a_1(x) \exp 2\pi i \nu_1 x + \cdots + a_n(x) \exp 2\pi i \nu_n x$$

$$f(x, y) = a_0(x) + a_1(x) \exp 2\pi i \nu_1 y + \cdots + a_n(x) \exp 2\pi i \nu_n y$$

and $M = \sup_{\mathbf{R}^2} |f(x, y)|$.

Theorem IX of Chapter I shows that for each positive ε and each interval I of real numbers of length $C\varepsilon^{-(n-1)/2}$ and for a fixed x,

$$\sup_{y\in I} |f(x, y)| \geqslant (1 - \varepsilon) \sup_{y\in \mathbf{R}} |f(x, y)|.$$

Hence there exist two real numbers x_0 and y_0 such that $|x_0 - y_0| \leqslant C\varepsilon^{-(n-1)/2}$ and $|f(x_0, y_0)| \geqslant (1 - \varepsilon) M$. As in the proof of Lemma 1, we have

$$|f(x, y_0)| \geqslant |f(x_0, y_0) + (x - x_0) A| - 2\pi^2 l^2 (x - x_0)^2 M.$$

Multiplying $f(x_0, y_0)$ by a constant of absolute value 1 if necessary, we may assume that $f(x_0, y_0)$ is positive. Then, putting $A = a + ib$, we get

$$M \geqslant |f(x, y_0)| \geqslant f(x_0, y_0) + (x - x_0) a - 2\pi^2 l^2 (x - x_0)^2 M \quad (8.19)$$

hence $(x - x_0) a \leqslant \varepsilon M + 2\pi^2 l^2 (x - x_0)^2 M$.

By an optimal choice of $x - x_0$, $|a| \leqslant 2\pi l \sqrt{2\varepsilon}\, M$; (8.19), with this estimate of a, gives $|f(y_0, y_0)| \geqslant (1 - C_1 l^{2/n}) M$ when $\varepsilon = l^{2/n}$, which is (8.17).

Proof of (8.17) ⇒ (8.18). Let $a : \mathbf{R} \to \mathbf{C}$ be a function in the class $\mathscr{S}$ of Schwartz whose Fourier-transform is a positive indefinitely differentiable function supported by [0, 1]. Assuming that $a(0) = 1$, we get, for two positive constants α and β, $0 < \beta < 1$, $|a(x)| \leqslant 1 - \alpha x^2$ if $|x| \leqslant 1$ and $|a(x)| \leqslant \beta$ if $|x| \geqslant 1$.

Let $T = l^{-(n-1)/n}$ for sufficiently small l and assume that $|x| \geqslant mT$ (the exact value of $m \geqslant 1$ will be given in a moment); we get $|a(lx)| \leqslant 1 - \alpha m^2 l^{2/n}$. Testing (8.17) on $f(x) = a(lx)\, P(x)$, where $P(x) = c_0 + \sum_1^n c_j \exp 2\pi i \nu_j x$, we get

$$\sup\,[\sup_{|x| \leqslant mT} |P(x)|,\ (1 - \alpha m^2 l^{2/n})\, \|P\|_\infty] \geqslant (1 - Cl^{2/n})\, \|P\|_\infty .$$

If $\alpha m^2 > C$, the only possibility is $\sup_{|x| \leqslant mT} |P(x)| \geqslant (1 - Cl^{2/n})\, \|P\|_\infty$; Theorem IX of Chapter I applies now to give (8.18).

The same argument shows that $l^{2/n}$ cannot be replaced in Theorem V by any number which is $o(l^{2/n})$ as l tends to 0.

7. From the group R to the group D^∞

It is indeed remarkable that in the heart of the group algebra $A(\mathbf{R})$, dwell other group algebras. For the sake of simplicity we shall confine our attention to the product group D^∞ of a countable number of copies of $\mathbf{Z}/2\mathbf{Z}$ and it will be proved that, for each ultra-thin symmetric set E, $A(D^\infty)$ is a closed subalgebra of the quotient algebra $A(E)$.

7.1. The group D^∞ and the group algebra $A(D^\infty)$

Let D be the group with two elements 0 and 1 where addition is defined modulo 2 and let D^∞ be the product group $\prod_{k \geqslant 1} D_k$ of a countable set of copies of D. An element $\boldsymbol{\omega}$ of D^∞ is a sequence $(\omega_k)_{k \geqslant 1}$ of elements of D; addition is componentwise addition modulo 2. The topology of D^∞ is that of componentwise convergence and the Borel field of D^∞ is generated by elementary sets $E(\alpha_1, \ldots, \alpha_n)$ of the type $\omega_1 = \alpha_1, \ldots, \omega_n = \alpha_n$. The Haar measure $d\boldsymbol{\omega}$ of D^∞ is defined by assigning to $E(\alpha_1, \ldots, \alpha_n)$ the measure 2^{-n}.

For each $n \geqslant 1$, let $\chi_n : D^\infty \to \mathbf{T}$ be defined by $\chi_n(\boldsymbol{\omega}) = (-1)^{\omega_n}$; χ_n is a continuous character on D^∞ and each character $\chi : D^\infty \to \mathbf{T}$ can be written $\chi_\gamma(\boldsymbol{\omega}) = \chi_1^{\alpha_1} \cdots \chi_n^{\alpha_n}$ for a suitable $n \geqslant 1$ and a suitable sequence $\gamma = (\alpha_1, \ldots, \alpha_n)$ of 0's and 1's. The dual group Γ of D^∞ is the direct sum of a countable set of copies of D. We denote by $A(D^\infty)$ the Banach algebra of all complex valued continuous functions $f : D^\infty \to \mathbf{C}$ whose Fourier series is absolutely convergent

$$f(\boldsymbol{\omega}) = \sum_{\gamma \in \Gamma} a(\gamma)\, \chi_\gamma(\boldsymbol{\omega}), \quad \|f\|_{A(D^\infty)} = \sum_{\gamma \in \Gamma} |a(\gamma)| . \tag{8.20}$$

7.2. $A(D^\infty)$ as a subalgebra of $A(E)$

Throughout this section, E is an ultra-thin symmetric set of real numbers.

THEOREM VI: *Let $\Pi : E \to D^\infty$ be the continuous surjective mapping defined by $\Pi(x) = \boldsymbol{\omega}$ if $x = \sum_1^\infty \varepsilon_k t_k$, $\varepsilon_k = 0$ or 1 and $\boldsymbol{\omega} = (\omega_k)_{k \geqslant 1}$, $\omega_k = 0$ or 1, are connected by $\omega_k \equiv \varepsilon_{2k-1} + \varepsilon_{2k}$ (mod 2) for all $k \geqslant 1$.*

Then for any complex valued continuous function $f : D^\infty \to \mathbf{C}$ the following two assertions are equivalent:

(a) $f \in A(D^\infty)$;

(b) $f \circ \Pi \in A(E)$.

Before proving this theorem, some remarks should be made. Let H be the homeomorphism between E and D^∞ defined by $H(\sum_{k \geqslant 1} \varepsilon_k t_k) = (\varepsilon_k)_{k \geqslant 1}$ when $\varepsilon_k = 0$ or 1, $k \geqslant 1$. Then H does not induce an isomorphism between $A(E)$ and $A(D^\infty)$. Using general theorems ([15], p. 78, th. 4.1.3) on homomorphisms of group algebras it can be proved that such an isomorphism between $A(E)$ and $A(D^\infty)$ cannot exist: $A(D^\infty)$ is not isomorphic to the whole of $A(E)$ but only to a closed subalgebra of $A(E)$.

7.3. *Proof of* (a) ⇒ (b)

It suffices to show that there exists a constant C such that for each character $\chi_\gamma : D^\infty \to \mathbf{T}$, $\|\chi_\gamma \circ \Pi\|_{A(E)} \leqslant C$. If $\chi_\gamma(\boldsymbol{\omega}) = \chi_1^{\alpha_1} \cdots \chi_n^{\alpha_n}$, $g = \chi_\gamma \circ \Pi$ is defined on E by

$$g(x) = \exp i\pi\,(\beta_1 \varepsilon_1 + \cdots + \beta_{2n}\varepsilon_{2n}) \quad \text{when} \quad x = \sum_1^\infty \varepsilon_k t_k$$

and where $\beta_{2k-1} = \beta_{2k} = \alpha_k$ for $1 \leqslant k \leqslant n$.

Now Proposition 2 immediately gives the result.

7.4. *Proof of* (b) ⇒ (a)

We first of all define a linear operator $L : C(E) \to C(D^\infty)$ such that, for each $f \in C(D^\infty)$, $L(f \circ \Pi) = f$ and for each f in $A(E)$, $L(f)$ belongs to $A(D^\infty)$. These two properties give the required implication (b) ⇒ (a) and L will be used to show how problems of spectral synthesis can be transferred from E to D^∞.

Let $P : D^\infty \to D^\infty$ be the continuous homomorphism defined by $P(\boldsymbol{\varepsilon}) = \boldsymbol{\omega}$ if $\boldsymbol{\varepsilon} = (\varepsilon_k)_{k \geqslant 1}$, $\boldsymbol{\omega} = (\omega_k)_{k \geqslant 1}$ and $\omega_k \equiv \varepsilon_{2k-1} + \varepsilon_{2k}$ (mod 2) for each $k \geqslant 1$. The kernel of P is the subgroup G of D^∞ consisting fo all

$g = (\varepsilon_k)_{k \geqslant 1}$ such that $\varepsilon_{2k-1} = \varepsilon_{2k}$ $(k \geqslant 1)$; G is clearly another copy of D^∞ and the Haar measure dm of G is a Radon measure on D^∞.

For each $\varphi \in C(E)$, we define a continuous function $F: D^\infty \to \mathbf{C}$ by $F \circ H = \varphi$, where H is the canonical homeomorphism[1] between E and D^∞, and we put $L(\varphi) = \psi$ if, for each $\boldsymbol{\omega} = P(\boldsymbol{\varepsilon})$,

$$\varphi(\boldsymbol{\omega}) = \int_G F(\boldsymbol{\varepsilon} + g)\, dm(g). \tag{8.21}$$

Plainly $L(f \circ \Pi) = f$ for each $f \in C(D^\infty)$ since in this case F is constant modulo G. To show that, for each $\varphi \in A(E)$, $\psi \in A(D^\infty)$, we can confine our attention to the case where $\varphi(x) = e^{inx}$ if it can be proved that $\|\psi\|_{A(D^\infty)} \leqslant C$ for some constant C not depending on n. But, in this case, F is defined by $F(\boldsymbol{\varepsilon}) = \prod_1^\infty e^{in\varepsilon_k t_k}$ and $\psi(\boldsymbol{\omega}) = \prod_1^\infty \psi_k(\omega_k)$, where $\psi_k(0) = \frac{1}{2}(1 + e^{in(t_{2k-1}+t_{2k})})$ and $\psi_k(1) = \frac{1}{2}(e^{int_{2k-1}} + e^{int_{2k}})$. For each function $h: D \to \mathbf{C}$, $\|h\|_{A(D)} = \frac{1}{2}|h(0) + h(1)| + \frac{1}{2}|h(0) - h(1)|$. Hence $\|\psi\|_{A(D^\infty)} \leqslant \prod_1^\infty \|\psi_k\|_{A(D^\infty)} = \prod_1^\infty \|\psi_k\|_{A(D)} \leqslant 1$. This last inequality follows from the observation that for any two real numbers a and b, $|\cos a \cos b| + |\sin a \sin b| \leqslant 1$; hence $\frac{1}{4}|1 + e^{ia} + e^{ib} + e^{i(a+b)}| + \frac{1}{4}|1 - e^{ia} - e^{ib} + e^{i(a+b)}| \leqslant 1$, which implies $\|\psi_k\|_{A(D)} \leqslant 1$ for each $k \geqslant 1$.

8. Spectral synthesis fails in $A(D^\infty)$

8.1. THEOREM VII: *There exists a real function f in $A(D^\infty)$ such that the closed ideals I_n generated by f^n, $n \geqslant 1$, are all distinct.*

The ideal I_n is the closure in $A(D^\infty)$ of the set of all products $f^n h$, $h \in A(D^\infty)$; hence $I_{n+1} \subset I_n$ for all $n \geqslant 1$ and we have to show that $f^n \notin I_{n+1}$. It is sufficient to produce a pseudomeasure S_n, i.e. a distribution whose Fourier coefficients form a bounded sequence, such that

$$\langle S_n, f^n \rangle \neq 0 \quad \text{while for all } h \text{ in } A(D^\infty), \quad \langle S_n, hf^{n+1} \rangle = 0. \tag{8.22}$$

For a better understanding of the proof, we present a candidate S_1. If $f: \mathbf{R} \to \mathbf{R}$ is a regular function with a finite set F of zeros without multiplicity, a distribution $S_1 = \delta' \circ f$ is well defined by $\sum_{\lambda \in F} (f'(\lambda)|f'(\lambda)|)^{-1}$ $\delta'(x - \lambda) + \mu$, where δ' is the derivative of the unit mass concentrated in 0 and μ is a measure carried by F. Hence $\langle S_1, f \rangle = \sum_{\lambda \in \Gamma}(|f'(\lambda)|)^{-1} > 0$ while, for all regular h, $\langle S_1, hf^2 \rangle = 0$. Unfortunately S_1 is not a pseudomeasure.

---

[1] H is defined in Section 7.2.

Notice that $\delta'(x) = \int_{-\infty}^{+\infty} iu\, e^{iux} du$, where the integral is a vector valued one in the space $\mathscr{D}'$ of distributions. It is tempting to seek our distributions S_n in the shape of a formula

$$S_n = \int_{-\infty}^{+\infty} (iu)^n\, e^{iuf}\, du \tag{8.23}$$

for a suitable f in $A(D^\infty)$ such that $\|e^{iuf}\|_{PM}$ decays rapidly at infinity (we recall that the pseudo-measure norm is the supremum of the absolute values of the Fourier coefficients).

8.2. PROPOSITION 3: *If there exists an f in $A(D^\infty)$ such that, for each $N \geqslant 1$ $|u|^N \|e^{iuf}\|_{PM} \to 0$ as $|u| \to +\infty$, there is a real number a such that the closed ideals generated in $A(D^\infty)$ by the elements $(a + f)^n$ are all distinct.*

Before giving the proof, some remarks should be made. Let $\Phi : \mathbf{R} \to PM\,(D^\infty)$ be the function defined by $\Phi(u) = e^{iuf}$; Φ decays rapidly at infinity and, since f belongs to $A(D^\infty)$, all derivatives of Φ exist in the strong sense and decay rapidly at infinity. The vector valued function Φ belongs to the class $\mathscr{S}$ of Schwartz.

We first choose a so that $\langle 1, S_0 \rangle \neq 0$, where the scalar product $\langle\ \rangle$ denotes here and in everything that follows, scalar products between $A(D^\infty)$ and $PM\,(D^\infty)$; integrations with respect to u will be written explicitly. The function $\theta : \mathbf{R} \to \mathbf{C}$ defined by $\theta(u) = \langle 1, e^{iuf} \rangle$ belongs to $\mathscr{S}$ and does not vanish identically since $\theta(0) = 1$; hence its Fourier transform does not vanish identically and for some real a

$$\int_{-\infty}^{+\infty} \langle 1, e^{iuf} \rangle\, e^{iua}\, du = \int_{-\infty}^{+\infty} \langle 1, e^{iu(f+a)} \rangle\, du \neq 0.$$

At this stage, we replace f by $f + a$; notice that $\|e^{iu(f+a)}\|_{PM} = \|e^{iuf}\|_{PM}$ also decays rapidly at infinity. Let $m \geqslant 0$ be an integer, $h \in A(D^\infty)$ and consider

$$I_{n,m} = \langle hf^m, S_n \rangle = \int_{-\infty}^{+\infty} \langle hf^m, e^{iuf} \rangle\, (iu)^n\, du. \tag{8.24}$$

The function $J_m : \mathbf{R} \to \mathbf{C}$ defined by $J_m(u) = \langle hf^m, e^{iuf} \rangle$ belongs to $\mathscr{S}\,(\mathbf{R})$; its derivative is $i\,\langle hf^{m+1}, e^{iuf} \rangle$ and integrating (8.24) by parts we get $I_{n,m} = 0$ if $m > n$ and $I_{n,n} = (-1)^n n!\, \langle h, S_0 \rangle$; we thus have the relations (8.22) and the proof is complete.

8.3. *Construction of f satisfying the hypothesis of Proposition 3*

For each $k \geqslant 1$, let $\varphi_k : D^\infty \to \mathbf{R}$ be defined by $\varphi_k(\omega) = \omega_{2k-1}\omega_{2k}$; if we let D_k^2, $k \geqslant 1$, be a countable number of copies of $D \times D$, we can write $D^\infty = \prod_1^\infty D_k^2$. With this notation φ_k becomes a function defined on D_k^2. Clearly $\|\varphi_k\|_{A(D_k{}^2)} = 1$ while the pseudo-measure norm of each function $e^{it\varphi_k}$, $t \in \mathbf{R}$, can be computed explicitly: for each $f: D \times D \to \mathbf{C}$, $\|f\|_{PM}$ is the sup of $\frac{1}{4}|f(0, 0) + f(1, 0) + f(0, 1) + f(1, 1)|$, $\frac{1}{4}|f(0, 0) + f(0, 1) - f(1, 0) - f(1, 1)|$, $\frac{1}{4}|f(0, 0) - f(0, 1) + f(1, 0) - f(1, 1)|$ and $\frac{1}{4}|f(0, 0) - f(0, 1) - f(1, 0) + f(1, 1)|$.

In our case $\|e^{it\varphi_k}\|_{PM(D_k{}^2)}$ is the supremum of $\frac{1}{4}|3 + e^{it}|$ and $\frac{1}{4}|1 - e^{it}|$. For all real t, $\|e^{it\varphi_k}\|_{PM} \leqslant 1$ and

$$\text{if} \quad |t - \pi| \leqslant \pi/3 \pmod{2\pi}, \ \|e^{it\varphi_k}\|_{PM(D_k{}^2)} \leqslant \tfrac{3}{4}. \tag{8.25}$$

Let $f = \sum_1^\infty a_k\varphi_k$ for a_k defined by the following rule: if $j \geqslant 4$ and $2^j/j^2 \leqslant k < 2^{j+1}/(j+1)^2$, $a_k = (2\pi/3)\,2^{-j}$. We have $\sum_1^\infty a_k < +\infty$ and f belongs to $A(D^\infty)$. For $2^j \leqslant u < 2^{j+1}$, $\|e^{iuf}\|_{PM} \leqslant \prod_{k=1}^\infty \|e^{iua_k\varphi_k}\|_{PM}$; each pseudomeasure norm in this product can be computed in the corresponding D_k^2 and (8.25) gives $\|e^{iuf}\|_{PM} \leqslant (\frac{3}{4})^{2^j/(j+1)^2}$ and $\|e^{iuf}\|_{PM} \leqslant (\frac{3}{4})^{u/2\log^2 u}$ as $u \to +\infty$. Since f is real, the same estimate holds as $u \to -\infty$.

9. Spectral synthesis fails in $A(E)$ and $A(\mathbf{R})$

THEOREM VIII: *Each compact symmetric set of real numbers contains a compact subset which is not a set of synthesis.*

It is sufficient to show that each ultra-thin symmetric set also has this property. The proof is then based on the following proposition, where the notation is that of Theorem VI.

PROPOSITION 4: *If f in $A(D^\infty)$ does not admit spectral synthesis, neither does $f \circ \Pi$ in $A(E)$ nor does any extension $\tilde{f}$ of $f \circ \Pi$ in $A(\mathbf{R})$.*

Proof: Let $U \subset D^\infty$ be the set of zeros of f, $V \subset E$ be $\Pi^{-1}(U)$. If $\tilde{f}$ could be approximated by a sequence a_j of functions in $A(\mathbf{R})$ vanishing on neighbourhoods of the zero set of $\tilde{f}$, let b_j be the restrictions of such a_j to E and $V_j \subset E$ the zero sets of the b_j. The mapping Π is clearly open and $L(b_j) = f_j$ (see the definition of L in (8.21)) would be a sequence of functions in $A(D_\infty)$ vanishing on neighbourhoods of U and tending to f. Hence f would admit spectral synthesis.

10. Counter-examples. A random set due to Salem

10.1. THEOREM IX: *For each $p > 2$ there exists a sequence $(t_k)_{k\geqslant 1}$ of positive real numbers such that $\sum_{k\geqslant 1}(t_{k+1}/t_k)^p < +\infty$ and such that the compact set E of all sums $\sum_{k\geqslant 1}\varepsilon_k t_k$, $\varepsilon_k = 0$ or 1 has the following properties:*

(a) *E is a set of multiplicity;*

(b) *there exists a complex valued bounded continuous function $\varphi : \mathbf{R} \to \mathbf{C}$ whose spectrum lies in E and which cannot be approached, in the topology of uniform convergence on compact subsets of $\mathbf{R}$, by a bounded sequence φ_k, $k \geqslant 1$, of trigonometric sums whose frequencies belong to E (here, bounded means: $\sup_{k\geqslant 1}\sup_{t\in R}|\varphi_k(t)| < +\infty$).*

10.2. Let $0 < \alpha < \frac{1}{2}$ be such that $\alpha p > 1$. Each t_k will be chosen at random and independently of the others in the interval $I_k = [(k!)^{-\alpha}, 2(k!)^{-\alpha}]$; then $\sum_{k\geqslant 1}(t_{k+1}/t_k)^p < +\infty$ and for almost all choices of the sequence $\boldsymbol{\omega}$ of such t_k, the corresponding compact symmetric set $E(\boldsymbol{\omega})$ o all $\sum_1^\infty \varepsilon_k t_k$, $\varepsilon_k = 0$ or 1 will have the required properties.

Let Ω be the compact product space $\prod_{k\geqslant 1} I_k$, dt_k the Haar measure of I_k normalized by $\int_{I_k} dt_k = 1$ and $d\boldsymbol{\omega}$ the product Radon measure on Ω: $d\boldsymbol{\omega} = dt_1 \otimes \cdots \otimes dt_k \otimes \cdots$.

10.3. To prove that $E(\boldsymbol{\omega})$ is a set of multiplicity, we use the function $\varphi(x, \boldsymbol{\omega}) = \prod_1^\infty |\cos xt_k|$ as we did in Chapter III, Section 3.3). If α is sufficiently close to $\frac{1}{2}$, which may always be assumed, $E(\boldsymbol{\omega})$ is contained in $[0, \pi[$ and to prove that $\varphi(x, \boldsymbol{\omega})$ vanishes at infinity, it suffices to consider only integral x.

Let F_k, $k \geqslant 1$, be the finite set of integers x such that $2\pi(k!)^\alpha \leqslant x < 2\pi((k+1)!)^\alpha$ and let $f_k(\boldsymbol{\omega}) = \sup_{x\in F_k}\varphi(x, \boldsymbol{\omega})$. To prove that $\lim_{k\to+\infty} f_k(\boldsymbol{\omega}) = 0$ for almost all ω in Ω, the following lemma will be needed.

LEMMA 4: *Assume that an increasing sequence m_k of integers tending to infinity exists such that for each $X > 1$*

$$\sum_{k\geqslant 1} X^{m_k}\int_\Omega f_k^{m_k}(\boldsymbol{\omega})\, d\boldsymbol{\omega} < +\infty. \tag{8.26}$$

Then, for almost all $\boldsymbol{\omega}$ in Ω, $\lim_{k\to+\infty} f_k(\boldsymbol{\omega}) = 0$.

Proof of Lemma 4: Let $g_j(\omega) = \sum_{k\geqslant 1} j^{m_k} f_k^{m_k}(\boldsymbol{\omega})$; $g_j(\boldsymbol{\omega}) \in L^1(\Omega, d\boldsymbol{\omega})$ and so each g_j is finite almost everywhere, i.e. for all $\boldsymbol{\omega}$ not belonging to a negligible set N_j. Let N be the union of all the N_j; the measure of N is 0 and for $\omega \notin N$ and $j \geqslant 1$, $\overline{\lim}_{k\to+\infty} jf_k \leqslant 1$ which implies $\overline{\lim}_{k\to+\infty} f_k(\omega) = 0$.

To obtain the estimate (8.26), we apply the following lemma:

LEMMA 5: *There is a constant C such that, for each integer $m \geqslant 1$ and each interval I of real numbers whose length $|I|$ exceeds $\pi/2$,*

$$\frac{1}{|I|}\int_I |\cos x|^m \, dx \leqslant \frac{C}{\sqrt{m}}. \tag{8.27}$$

Proof: By periodicity and symmetry the problem can be reduced to the case where $I = [0, \pi/2]$. For a suitable $l > 0$, we have $\cos x \leqslant e^{-x^2/2}$ in $[0, l]$ and for any l, $\cos x \leqslant \cos l$ in the remainder interval $[l, \pi/2]$. Hence $\int_0^{\pi/2} \cos^m x \, dx \leqslant \int_0^\infty e^{-mx^2/2} \, dx + (\pi/2) \cos^m l \sim \sqrt{\pi/2m}$, which gives (8.27).

The integral of a product of independent random functions is the product of the integrals. Hence $\int_\Omega \varphi^m (x, \omega) \, d\omega = \prod_1^\infty \int_{I_k} |\cos x \, t_k|^m \, dt_k \leqslant \prod_1^k \int_{I_j} |\cos xt_j|^m \, dt_j$. If $x \in F_k$, each of the integrals in the last product is bounded by $Cm^{-1/2}$. Therefore $\int_\Omega \varphi^m (x, \omega) \, d\omega \leqslant C^k m^{-k/2}$.

Clearly $\int_\Omega f_k^m(\omega) \, d\omega \leqslant \sum_{x\in F_k} \int_\Omega \varphi^m (x, \omega) \, d\omega \leqslant (\text{Card } F_k) \, C^k m^{-k/2} \leqslant 2\pi \, ((k + 1)!)^\alpha \, C_k m^{-k/2}$. Choose $m = m_k = k$ and then for $X > 1$ we have $\sum_{k\geqslant 1} X^{m_k} C^k m_k^{-k/2} \, ((k + 1)!)^\alpha < +\infty$ which, by Lemma 4, finishes the proof.

To prove the remaining statements of Theorem IX, the additive properties of the random sets $E(\omega)$ must be described.

10.4. *Additive properties of the random symmetric sets $E(\omega)$*

DEFINITION 4: *Let $(t_k)_{k\geqslant 1}$ be a sequence of real numbers such that $\sum_1^\infty |t_k| < +\infty$. We say that $(t_k)_{k\geqslant 1}$ is fully independent if for any bounded sequence $(p_k)_{k\geqslant 1}$ of rational integers the following implication holds:*

$$\sum_1^\infty p_k t_k = 0 \Rightarrow p_k = 0 \quad \text{for all} \quad k \geqslant 1.$$

It is easy to show that full independence implies $\underline{\lim}\, t_{k+1}/t_k = 0$ but a **Q**-linearly independent sequence $(t_k)_{k\geqslant 1}$ of real numbers such that $\lim t_{k+1}/t_k = 0$ is not always fully independent.

The terminology of the following proposition is that of Theorem IX: $I_k = [(k!)^{-\alpha}, 2(k!)^{-\alpha}]$ and $\omega = (t_k)_{k\geqslant 1}$ is a random choice of t_k in I_k.

PROPOSITION 5: *For almost all ω in Ω, $(t_k)_{k\geqslant 1}$ is fully independent.*

Proof: Let N be the set of all $\boldsymbol{\omega}$ in Ω for which there is a bounded sequence $(p_k)_{k\geqslant 1}$ of rational integers such that $\sum_1^\infty p_k t_k = 0$; if furthermore, such a sequence $(p_k)_{k\geqslant 1}$ exists with $|p_k| \leqslant m$ for all $k \geqslant 1$, we write $\boldsymbol{\omega} \in N(m)$: N is the union of all such $N(m)$. If $\boldsymbol{\omega} \in N(m)$, there is a smallest value j of k for which $p_k \neq 0$ and we then write $\boldsymbol{\omega} \in N(m, j)$: $N(m)$ is the union $\bigcup_{j\geqslant 1} N(m,j)$. If all these $N(m,j)$ have measure zero, so does N and Proposition 5 is proved. Each $N(m,j)$ is a closed subset of Ω and the measure of $N(m,j)$ will be computed by means of Fubini's theorem. Each $\boldsymbol{\omega}$ in $N(m,j)$ can be written as a pair $(\boldsymbol{\omega}', \boldsymbol{\omega}'')$, where $\boldsymbol{\omega}' = (t_1, \ldots, t_j)$ and $\boldsymbol{\omega}'' = (t_{j+1}, \ldots, t_k, \ldots)$. If for each fixed $\boldsymbol{\omega}''$ in $\Omega'' = \prod_{k\geqslant j+1} I_k$ the measure, computed in $\prod_{1\leqslant k\leqslant j} I_k = \Omega'$, of all $\boldsymbol{\omega}'$ such that $(\boldsymbol{\omega}', \boldsymbol{\omega}'') \in N(m,j))$ is 0, so is the measure of $N(m,j)$.

But the statement $(\boldsymbol{\omega}', \boldsymbol{\omega}'') \in N(m,j)$ is one of the $2m$ statements

$$p_j t_j \in G(\boldsymbol{\omega}''),\ p_j = -m, -m+1, \ldots, -1, 1, \ldots, m \qquad (8.28)$$

where $G(\boldsymbol{\omega}'')$ is the set of all sums $\sum_{k\geqslant j+1} p_k t_k$ for all integral p_k such that $|p_k| \leqslant m$.

Since t_{k+1}/t_k tends to 0. the measure of the symmetric set $G(\boldsymbol{\omega}'')$ of real numbers is 0 and so is the measure of the set of t_j satisfying (8.28).

COROLLARY: *For $\delta > \frac{1}{2}$,* let $(s_k)_{k\geqslant 1}$ *be a fully independent sequence of real numbers chosen in* $[(k!)^{-\delta}, 2(k!)^{-\delta}]$ *and let E_1 be the ultra-thin symmetric set of all sums $\sum_{k\geqslant 1} \varepsilon_k s_k$, $\varepsilon_k = 0$ or 1. Let H be the homeomorphism from $E(\boldsymbol{\omega})$ to E_1 defined by*

$$H(\sum_{k\geqslant 1} e_k t_k) = \sum_{k\geqslant 1} \varepsilon_k s_k, \quad \varepsilon_k = 0 \quad \text{or} \quad 1.$$

Then for almost all ω in Ω, the following equality holds for each complex measure μ carried by a finite subset of $E(\boldsymbol{\omega})$

$$\|H(\mu)\|_{PM} = \|\mu\|_{PM}. \qquad (8.29)$$

Proof: Let $\mathfrak{G}$ be the additive group generated by E: $\mathfrak{G}$ is the set of all sums $\sum_{k\geqslant 1} p_k t_k$ corresponding to all bounded sequences $(p_k)_{k\geqslant 1}$ of rational integers; let $\mathfrak{G}_1$ be the additive group generated by E_1.

The full independence of the sequences $(t_k)_{k\geqslant 1}$ and $(s_k)_{k\geqslant 1}$ implies that H can be extended by $H(\sum_{k\geqslant 1} p_k t_k) = \sum_{k\geqslant 1} p_k s_k$ to an isomorphism, still denoted by H, from $\mathfrak{G}$ to $\mathfrak{G}_1$. If P is any finite trigonometric sum, $P(t) = \sum_{\lambda\in\mathfrak{G}} c(\lambda) \exp 2\pi i\lambda t$ and if $G(t) = \sum_{\lambda\in\mathfrak{G}} c(\lambda) \exp 2\pi i H(\lambda) t$, Kronecker's theorem (Chapter I, Section 5) gives $\|P\|_\infty = \|Q\|_\infty$, which is (8.29).

10.5. Last part of the proof of Theorem IX

Before we conclude the proof, we shall describe the difference between the Banach algebras $A(E)$ and $A(E_1)$. It is unknown whether a restriction algebra to a set of multiplicity can be isomorphic to a restriction algebra to a set of uniqueness. Such a theorem would prove that for almost all $\boldsymbol{\omega} \in \Omega$, $A(E(\boldsymbol{\omega}))$ and $A(E_1)$ are not isomorphic.

The method we shall use is the following: let $\mathscr{R}$ be a regular semi-simple commutative Banach algebra which will be identified, by means of Gelfand's representation, with a function algebra on the spectrum K of $\mathscr{R}$; the pseudo-measures are the elements of the dual space $\mathscr{R}^*$ of $\mathscr{R}$ and the support of such a pseudomeasure can be defined in a natural way. Let $\sigma(\mathscr{R}^*, \mathscr{R})$ be the weak-star topology. We say that $\mathscr{R}$ is of type 1 if each pseudo-measure S is the weak-star limit of a sequence S_n, $n \geqslant 1$, of pseudomeasures carried by finite subsets of K. If this is not the case, we say that $\mathscr{R}$ is of type 2. General theorems of Banach give the following information on Banach algebras of type 1: there is a constant $C > 0$ such that for each S in $\mathscr{R}^*$, a sequence S_n of finitely supported pseudomeasures can be found such that

$$\|S_n\|_{\mathscr{R}^*} \leqslant C\|S\|_{\mathscr{R}^*} \quad \text{and} \quad S_n \to S \quad \text{in} \quad \sigma(\mathscr{R}^*, \mathscr{R}). \tag{8.30}$$

We can state

PROPOSITION 6: *For almost all* $\boldsymbol{\omega}$ *in* Ω, $A(E(\boldsymbol{\omega}))$ *is of type 2.*

This at once finishes the proof of Theorem IX: in the situation of restriction algebras $A(E)$ to compact sets E of real numbers, (8.30) is equivalent to

$$\hat{S}_n(x) \to \hat{S}(x) \quad \text{uniformly on compact subsets of } \mathbf{R} \tag{8.31}$$

and

$$\sup_{\mathbf{R}} |\hat{S}_n| \leqslant C \sup_{\mathbf{R}} |\hat{S}|$$

as we have already observed in Chapter VII, Section 3.2.

The following lemma will be useful in proving Proposition 6:

LEMMA 6: *Let* $f_k : \mathbf{R} \to [0, 1]$ *be defined by*

$$f_k(x) = \cos t_1 x \cdots \cos t_{k-1} x \sin t_k x \cos t_{k+1} x \cdots \cos t_n x \cdots$$

Then for almost all $\boldsymbol{\omega}$ *in* Ω, $\sup_{\mathbf{R}} |f_k| \to 0$.

The proof is an easy modification of that given in Section 10.3: the term $\sin t_k x$ instead of $\cos t_k x$ ensures uniform convergence to 0 on compact subsets of $\mathbf{R}$.

Note that $f_k = |\hat{\mu}_k|$ for a suitable measure μ_k carried by E.

LEMMA 7: *The pseudo measure norm of $H(\mu_k)$ does not tend to* 0.

Proof: By Theorem II this norm is greater than C^{-1} sup $\cos x_1 \cdots \cos x_{k-1} \sin x_k \cos x_{k+1} \ldots$, where C is a positive constant and $(x_k)_{k\geqslant 1}$ runs over all sequences of real numbers. But this last sup is 1.

Proof of Proposition 6: Assume that $A\,(E(\omega))$, which we shall write $A(E)$, is of type 1 and that ω does not belong to the union of the two negligible sets which arise in Lemma 6 and Proposition 5 respectively.

Let $(\mu_{k,j})_{j\geqslant 1}$ be a sequence of measures carried by finite subsets of E and satisfying $\|\mu_{k,j}\|_{PM} \leqslant C\|\mu_k\|_{PM}$ and $\langle \mu_{k,j}, f\rangle \to \langle \mu_k, f\rangle$ for all locally constant functions f on E; such a sequence exists by (8.30). Then by Corollary of Proposition 5 $\|H(\mu_{k,j})\|_{PM} = \|\mu_{k,j}\|_{PM}$ and clearly $\langle H(\mu_{k,j}), f\rangle \to \langle H(\mu_k), f\rangle$ for all locally constant functions f on E_1. Hence $H(\mu_k)$ is the weak-star limit of the bounded sequence of pseudo-measures $H\,(\mu_{k,j}), j \geqslant 1$, since the locally constant functions are dense in $A(E_1)$, and since norms decrease in the passage to weak-star limits $\|H(\mu_k)\|_{PM} \leqslant \underline{\lim}_{j\to+\infty} \|H(\mu_{k,j})\| \leqslant C\|\mu_k\|_{PM}$. Lemmas 6 and 7 show that this is not the case.

11. Other random sets

Let $(t_k)_{k\geqslant 1}$ be an increasing sequence of positive real numbers such that $t_{k+1} \geqslant 3t_k$ (for all $k \geqslant 1$) and let Λ be the set of all finite sums $\sum_{k\geqslant 1} \varepsilon_k t_k$, $\varepsilon_k = 0$ or 1. Theorem IV asserts that if $\sum_{k\geqslant 1} (t_k/t_{k+1})^2 < +\infty$, Λ is a coherent set of frequencies. We shall prove that the exponent 2 cannot be replaced by a greater one and that a sequence $(t_k)_{k\geqslant 1}$ can be chosen in such a way that $\sum_{k\geqslant 1} (t_k/t_{k+1})^2 < +\infty$ while Λ is not harmonious. In both cases random methods will be used.

Let $0 < \alpha < \frac{1}{2}$, let I_k, $k \geqslant 1$, be the interval $(k!)^\alpha \leqslant x \leqslant 2(k!)^\alpha$, let dt_k be the Lebesgue measure of I_k normalized by $\int_{I_k} dt_k = 1$, let Ω be the compact product space $\prod_{k\geqslant 1} I_k$ and let $d\boldsymbol{\omega}$ be the product Radon measure on Ω defined by $d\boldsymbol{\omega} = \bigotimes_{k\geqslant 1} dt_k$.

For each $\boldsymbol{\omega} = (t_k)_{k\geqslant 1}$ in Ω, let $\Lambda(\boldsymbol{\omega})$ be the corresponding discrete set of all finite sums $\sum_{k\geqslant 1} \varepsilon_k t_k$, $\varepsilon_k = 0$ or 1.

THEOREM X: *For almost all* $\boldsymbol{\omega}$ *in* $\Omega, \Lambda(\boldsymbol{\omega})$ *is not a coherent set of frequencies.*

Proof: Let $P_k(x, \boldsymbol{\omega}) = \exp \pi i (t_1 + \cdots + t_k) x \cos \pi t_1 x \cdots \cos \pi t_k x$; the spectrum of P_k lies in Λ and Theorem X will be proved if, whenever $0 < a < b$ there exists a negligible subset $N(a, b)$ of Ω such that, $P_k(x, \boldsymbol{\omega})$ converges uniformly to 0 on $[a, b]$ for $\boldsymbol{\omega} \notin N(a, b)$; then, if $\boldsymbol{\omega}$ does not belong to the negligible union of all the $N(j^{-1}, j), j \geqslant 1, \Lambda(\boldsymbol{\omega})$ is not a coherent set of frequencies.

Let F_k be the finite set of all $x = j(kt_k)^{-1}, j \in \mathbf{N}$, lying in $[a, b]$. For each x in $[a, b]$, there is a y in F_k such that $|x - y| \leqslant (2kt_k)^{-1}$ and Bernstein's inequality (Chapter V, Section 4) shows that $|P_k(x) - P_k(y)| \leqslant Ck^{-1}$, where C does not depend on k. Hence it suffices to prove that $f_k(\boldsymbol{\omega}) = \sup_{x \in F_k} |P_k(x, \boldsymbol{\omega})|$ tends to 0 for almost all $\boldsymbol{\omega}$ in Ω. The remainder of the proof is an easy transcription of that given in Section 10.3.

If $\frac{1}{2} < \alpha < 1$ the symmetric set $\Lambda(\boldsymbol{\omega})$ constructed with $(t_k)_{k \geqslant 1}$, where $t_k \in I_k = [(k!)^\alpha, 2(k!)^\alpha]$ is a coherent set of frequencies. Moreover for each positive ε, there exists a T (depending on ε and ω) such that for each trigonometric sum P whose frequencies belong to Λ, $\sup_{\mathbf{R}} |P| \leqslant (1 + \varepsilon) \sup_{|x| \leqslant T} |P(x)|$. This property is shared by all harmonious sets. However

THEOREM XI: *For* $\frac{1}{2} < \alpha < 1$, *let* I_k *be the interval* $[(k!)^\alpha, 2(k!)^\alpha]$ *and let* $\boldsymbol{\omega} = (t_k)_{k \geqslant 1}$ *where each* t_k *is chosen at random and independently of the other* t_j *in the interval* I_k. *Let* $\Lambda(\boldsymbol{\omega})$ *be the discrete set of all finite sums* $\sum_{k \geqslant 1} \varepsilon_k t_k$, $\varepsilon_k = 0$ *or* 1. *Then for almost all choices of the sequence of* t_k, $k \geqslant 1, \Lambda(\boldsymbol{\omega})$ *is not harmonious.*

(To be more precise we recall that each t_k is equidistributed in I_k.)

This theorem is very sharp since, following the method given in Chapter II, Section 4, it can be shown that if $\sum_{k \geqslant 1} t_k/t_{k+1} < +\infty \Lambda$ is harmonious.

We shall prove that, if $0 < a < b$ there is a negligible subset $N'(a, b)$ of Ω such that, for $\boldsymbol{\omega} \notin N'(a, b)$, $\inf_{a \leqslant x < b} \sup_{\lambda \in \Lambda} \|\lambda x\| \geqslant \frac{1}{20}$. If $\boldsymbol{\omega}$ does not belong to the negligible union of all the $N'(i^{-1}, j)$, $j \geqslant 1$, Λ is not harmonious. We recall that for each real number x, $\|x\|$ is the distance from x to the nearest integer.

Assume that $\|\lambda x\| \leqslant \frac{1}{20}$ for some x in $[a, b]$ and all λ in Λ. Then $\sum_{k \geqslant 1} \|t_k x\| \leqslant \frac{1}{10}$ (see Chapter II, Section 8, Proposition 7). Let F_k be defined as in the proof of Theorem X, let $S_k(x, \boldsymbol{\omega})$ be $\|t_1 x\| + \cdots + \|t_k x\|$

and $P_k(x, \omega) = \prod_1^k (1 - \|t_j x\|)$. For each x in $[a, b]$, there is a y in F_k such that $|x - y| \leqslant (2kt_k)^{-1}$; hence $S_k(x, \omega) \leqslant \frac{1}{10}$ implies $S_k(y, \omega) \leqslant \frac{1}{5}$ for $k \geqslant k_0$. The elementary inequality $\prod_1^k (1 - \alpha_j) \geqslant 1 - \sum_1^k \alpha_j$, $(0 \leqslant \alpha_j \leqslant 1)$ gives $P_k(y, \omega) \geqslant \frac{4}{5}$ for $k \geqslant k_0$ and $y \in F_k$. But an easy modification of the argument used in Section 10.3 shows that $\sup_{F_k} P_k(y, \omega) \to 0$ for $\omega \notin N'(a, b)$. We have thus got a contradiction.

12. Notes

The main idea of modern harmonic analysis is that, when we restrict our attention to 'ultra-thin sets', all becomes clear and obvious, all Banach algebras are isomorphic...

This feeling was tested on Kronecker sets ($A(E) = C(E)$ trivially in this case) and on Helson sets ($A(E) = C(E)$).

To get more interesting restriction algebras, it is necessary to consider ultra-thin sets with a given additive structure: if K_1 and K_2 are two compact subsets of the line, $K_1 \cap K_2 = \emptyset$ and $K_1 \cup K_2$ is a Kronecker set, then $A(K_1 + K_2) = C(K_1) \hat{\otimes} C(K_2)$ — the tensor product being a projective one — as Varopoulos showed [93]. Ultra-thin symmetric sets were studied in [68] and [86].

CONCLUSION

Do connections actually exist in harmonic analysis between additive properties of a set Λ of frequencies and functional properties of corresponding functions constructed with these frequencies?

No general answer can be given.

Throughout this book, both positive and negative results are to be found.

Positive results

Relatively dense harmonious subsets of a locally compact abelian group were characterized by three equivalent properties: the first concerns Diophantine approximation, the second concerns additive properties, and the last one deals with functional properties. (The reader is referred to the Interlude.)

Rapid prediction of the size of trigonometric sums whose frequencies belong to a finite set F of real numbers is equivalent to the strong independence of F over $\mathbf{Z}$ (Theorem IX of Chapter I).

Negative results

In Chapter VIII, ultra-thin symmetric sets E and a random set F defined by Salem were studied. It may happen that E and F have the same additive structure (a homeomorphism $H: E \to F$ exists which can be extended to an isomorphism between the additive groups generated by E and F).

However, E and F do not give rise to similar functional properties: bounded continuous functions whose spectra lie in E can be nicely approximated by trigonometric sums whose frequencies belong to E; this is not the case for F.

Open problems

It is not known whether coherent sets of frequencies or Sidon sets can be characterized by additive properties.

SOME OPEN PROBLEMS ABOUT SYMMETRIC SETS

	E is a symmetric set constructed with				
	A constant dissection ratio $\xi = \theta^{-1}$			Variable dissection ratios ξ_k, $k \geqslant 1$	
	θ Pisot	θ Salem	Other θ	Random set due to Salem ($\xi_k \sim k^{-\alpha}$ and $0 < \alpha < \frac{1}{2}$)	$\Sigma_1^\infty \xi_k^2 < +\infty$
E is a set of uniqueness	yes	no	no	no	yes
E is a set of synthesis	yes	?	?	?	yes
$A(E)$ is of type 1	yes	yes	?	no	yes
Bochner's property holds	yes	yes	?	no	?

OPEN PROBLEMS

Special trigonometric series on local fields

Let K be a local field[1] of characteristic 0. A standard subset Λ of K was defined (Chapter II, Section 13) in the following way. Let k be an algebraic number field and $\pi : k \to K$ an isomorphic embedding with a dense range.

Consider the set V of all possible metrics on k (metrics are either defined by valuations or real or complex absolute values); for a $v \in V$, $|x|_v$ will denote the value of the corresponding metric at x. Let v_0 be the metric induced by that of K. Then Λ is the set of all $\pi(x)$ for all x in k such that $|x|_v \leqslant 1$ for all $v \neq v_0$ in V.

Three examples of standard sets have been thoroughly studied in this book or in classical harmonic analysis:

(a) $\Lambda = \mathbf{Z}$ and $K = \mathbf{R}$ (then $k = \mathbf{Q}$);

(b) k is a real quadratic field and Λ is the set of all quadratic integers x in k whose conjugate lies in $[-1, 1]$; note that Λ also contains 1 and -1.

(c) $K = \mathbf{Q}_p$ and Λ is the set of all mp^n in $[-1, 1]$ ($m \in \mathbf{Z}$, $n \in \mathbf{Z}$).

The last two cases are very similar: let $E(\Lambda)$ be the space of all almost periodic functions on K (identified with the dual group) whose frequencies belong to Λ. There exists a critical positive m such that for any Riemann integrable compact subset C of K

(1) if the Haar measure of C exceeds m, $\sup_C |f|$ and $\operatorname{Sup}_K |f|$ are equivalent norms on $E(\Lambda)$;

(2) if the Haar measure of C is less than m, this is not the case.

It is to be expected that this result holds for all standard sets in local fields of characteristic 0 *with the exception of the discrete subgroups of* **R** *and* **C**.

[1] It is assumed that K is not discrete.

APPENDIX

A. Additive problems

From now on, for any subsets A and B of a group G, $A + B$ will denote the set of all sums $x + y$, $x \in A$, $y \in B$ and $|A|$ will denote the Cardinal of A.

THEOREM I: *Let $p \geqslant 2$ be a prime and let G be the finite group $\mathbf{Z}/p\mathbf{Z}$ with p elements. Let A and B be two subsets of G. Then*

either $A + B = G$,

or $|A + B| \geqslant |A| + |B| - 1$.

Proof: It will be shown that $A + B \neq G$ implies $|A + B| \geqslant |A| + |B| - 1$ and the idea of the proof is to enlarge B in such a way that the complement of $A + B$ is reduced to a single element of G.

It should be noted that if B is replaced by $c + B$, $c \in G$, $A + B$ is replaced by $c + A + B$. This substitution does not affect the required conclusion. Hence if $A + B \neq G$, it may be assumed that $0 \notin A + B$.

Two cases may occur:

Case 1: *The complement of $A + B$ in G is equal to* 0.

Then $|A + B| = p - 1$. On the other hand $-A$ and B are disjoint subsets of G. Hence $|-A| + |B| \leqslant p$, which implies $|A + B| \geqslant |A| + |B| - 1$.

Case 2: *The complement C of $A + B$ in G is not equal to* 0.

LEMMA 1: *In the second case, a subset B_1 of G can be found with the following three properties:*

(1) $B \subset B_1 \subset G$;

(2) $|A + B_1| - |A + B| = |B_1| - |B| > 0$;

(3) $0 \notin A + B_1$.

Before proving this lemma, it should be explained why it implies Theorem I. Using this lemma inductively, a strictly increasing sequence $B_1 \subset B_2 \cdots$ of finite subsets of G can be constructed. Since G is finite, the process will stop. But the process only stops when the situation of case 1 has been reached. Hence an integer m and a sequence $B = B_0, B_1, \ldots, B_m$ of finite subsets of G exist such that $|A + B_{j+1}| - |A + B_j| = |B_{j+1}| - |B_j| > 0$, $(0 \leqslant j \leqslant m - 1)$ and such that the complement of $A + B_m$ is $\{0\}$. Adding all these equations, we get $|A + B_m| - |A + B| = |B_m| - |B|$ or $|A + B| = |A + B_m| - |B_m| + |B|$.

B_m satisfies the conditions of the first case. Hence $|A + B_m| \geqslant |A| + |B_m| - 1$ and so $|A + B| \geqslant |A| + |B| - 1$.

Now Lemma 1 has to be proved.

Let $n + 1$, $n \geqslant 1$, be the Cardinal of C; $C = \{0, c_1, \ldots, c_n\}$. A first observation is: *one of the sums* $c_i + c_j$, $1 \leqslant i \leqslant n$, $1 \leqslant j \leqslant n$, *belongs to* $A + B$. If this were not the case, C would be closed under addition. Since p is a prime, there are only two possibilities: $C = \{0\}$, which was our first case, and $C = G$, in which case A and B are empty and Theorem I is trivial.

We thus have a relation

(4) $c + d = a + b,\ a \in A,\ b \in B,\ c \notin A + B,\ d \notin A + B.$

From now on a relation such as (4) is chosen and b is fixed. Let B' be the set of all $b - c_j$, $1 \leqslant j \leqslant n$, for which an element a' of A exists such that $a' + b - c_j$ belongs to $C: a' + b - c_j = c_k$, $1 \leqslant k \leqslant n$. Let $B_1 = B \cup B'$.

We list some properties of B':

(5) *B' contains $b - c$*, where c is given by (4); hence B' is not empty;

(6) *B' and B are disjoint:* if $b - c_j = b'$, then $c_k = a' + b' \in A + B$, which is not the case;

(7) $0 \notin A + B'$. If $a'' + b - c_j = 0$, $c_j = a'' + b \in A + B$ which is not the case;

(8) $(A + B') \cap C = b - B'$. If $c_k \in A + B'$, then $c_k = a'' + b - c_j$, which can also be written $c_j = a'' + b - c_k$; by our definition of B', $b - c_k$ belongs to B' and c_k belongs to $b - B'$. Hence $(A + B') \cap C \subset b - B'$.

On the other hand, $b - B' \subset (A + B') \cap C$. In fact, let $x = b - c_j$ be in B'; then we can find an a' in A and a c_k in C such that $a' + b - c_j = c_k$; $c_j = b - x$ belongs to C and is a sum $a' + b - c_k$. Hence c_j also belongs to $A + B'$.

(9) *The complement of $A + B$ in $A + B_1$ is $b - B'$*; we have to show that $(A + B_1) \cap C = b - B'$. But $A + B_1 = (A + B) \cup (A + B')$ and $(A + B) \cap C$ is empty while $(A + B') \cap C = b - B'$.

(10) Hence $|A + B_1| - |A + B| = |b - B'| = |B_1| - |B|$. This last equality holds because B and B' are disjoint.

THEOREM II: *For $n \geqslant 1$, let $\mathbf{T}^n$ be the n-dimensional torus and let σ be the Haar measure of $\mathbf{T}^n$ normalized by $\sigma(\mathbf{T}^n) = 1$.*

Let A and B be two compact subsets of $\mathbf{T}^n$.

Then either $A + B = \mathbf{T}^n$ or $\sigma(A + B) \geqslant \sigma(A) + \sigma(B)$.

The idea of the proof is to reduce the problem to finite ones.

Let $(\Gamma_k)_{k \geqslant 1}$ be a sequence of finite subgroups of $\mathbf{T}^n$ such that

(a) the number of elements in Γ_k is k;

(b) if σ_k is the Haar measure of Γ_k normalized by $\sigma_k(\Gamma_k) = 1$, σ is the weak-star limit of these σ_k, $k \geqslant 1$.

We assume initially that A and B are Riemann-integrable. Then $\sigma_k(A) \to \sigma(A)$, $\sigma_k(B) \to \sigma(B)$ as $k \to +\infty$.

We have $(A + B) \cap \Gamma_k = A \cap \Gamma_k + B \cap \Gamma_k$ and for all primes p, we have either

(11) $A \cap \Gamma_p + B \cap \Gamma_p = \Gamma_p$, or

(12) $\sigma_p(A \cap \Gamma_p + B \cap \Gamma_p) \geqslant \sigma_p(A \cap \Gamma_p) + \sigma_p(B \cap \Gamma_p) - 1/p$.

If (11) holds for an infinite number of primes p, $A + B = \mathbf{T}^n$. On the other hand, if (12) holds for an infinite number of primes p, $\sigma(A + B) \geqslant \sigma(A) + \sigma(B)$.

For arbitrary compact subsets A and B, we construct a decreasing sequence A_k (resp. B_k) of Riemann-integrable compact subsets of $\mathbf{T}^n$ such that for each $\mathbf{x} \in A_k$ (resp. $\mathbf{x} \in B_k$), $d(\mathbf{x}, A) \leqslant 1/k$ (resp. $d(\mathbf{x}, B) \leqslant 1/k$).

If $A + B$ is not the whole of $\mathbf{T}^n$, $A_k + B_k$ is not the whole of $\mathbf{T}^n$ for $k \geqslant k_0$, say. Hence $\sigma(A_k + B_k) \geqslant \sigma(A_k) + \sigma(B_k)$ for $k \geqslant k_0$ and passing to the limit, $\sigma(A + B) \geqslant \sigma(A) + \sigma(B)$ which concludes the proof of Theorem II.

B. Kronecker's theorem

THEOREM III: *Let $\omega_1, \ldots, \omega_n$ be n real numbers linearly independent over $\mathbf{Q}$. For each sequence $x_1, \ldots, x_n$ of n real numbers and each positive ε, there exist a real number t and n rational integers $p_1, \ldots, p_n$ such that $|x_1 - \omega_1 t - p_1| \leqslant \varepsilon, \ldots, |x_n - \omega_n t - p_n| \leqslant \varepsilon$.*

Proof: Let Γ be the subgroup of all $\mathbf{x} = (x_1, \ldots, x_n)$ in $\mathbf{R}^n$ such that we can write $x_j = \omega_j t + p_j$ for some t in $\mathbf{R}$ and rational integers p_j, $1 \leqslant j \leqslant n$. Let H be the closure of Γ in $\mathbf{R}^n$. We will show that $H = \mathbf{R}^n$.

LEMMA 2: *Let H be a closed subgroup of $\mathbf{R}^n$. Then either $H = \mathbf{R}^n$ or there exists a linear form l on $\mathbf{R}^n$ such that $l \neq 0$ and $l(\mathbf{x}) \in \mathbf{Z}$ for all $\mathbf{x} \in H$.*

Proof of Lemma 2: If H is not the whole of $\mathbf{R}^n$, an integer $m \geqslant 0$, $m \leqslant n$ and a basis $\mathbf{e}_1, \ldots, \mathbf{e}_n$ of $\mathbf{R}^n$ can be found such that the equations of H

with respect to this basis are $x_1 = \cdots = x_m = 0$, $x_{m+1} \in \mathbf{Z}, \ldots, x_n \in \mathbf{Z}$. It is now sufficient to put $l(\mathbf{x}) = x_n$.

Returning to Theorem II, if the closure of Γ is not the whole of $\mathbf{R}^n$, a linear form $l \neq 0$ exists such that $l(\mathbf{x}) \in \mathbf{Z}$ for each $\mathbf{x}$ in Γ.

Testing on $\mathbf{x} \in \mathbf{Z}^n$, we find that $l(\mathbf{x}) = q_1x_1 + \cdots + q_nx_n$ for rational integers $q_1, \ldots, q_n$ and testing then on $\mathbf{x} = (\omega_1 t, \ldots, \omega_n t)$, we find $\omega_1 q_1 + \cdots + \omega_n q_n = 0$, which contradicts the hypothesis.

BIBLIOGRAPHY

A. Basic reference books

[1] Banach, S., *Théorie des opérations linéaires*. Monografje Matematyczne. Warszawa 1932.

[2] Bary, N.K., *A treatise on trigonometric series*, vols. I and II (translated from the Russian). New York, Pergamon Press, 1964.

[3] Borevich, Z.I. and Shafarevich, I.R., *Number theory* (translated from the Russian) New York, Academic Press, 1966.

[4] Cassels, J.W.S., *An introduction to Diophantine approximations*. Cambridge, University Press, 1957.

[5] Cassels, J.W.S., *An introduction to the geometry of numbers*. Berlin, Springer-Verlag, 1959.

[6] Edwards, R.E., *Fourier series, a modern introduction*. New York, Holt, Rinehart and Winston, 1967.

[7] Hewitt, E. and Ross, K.A., *Abstract harmonic analysis*, vols. I and II. Berlin, Springer-Verlag, 1963 and 1970.

[8] Kahane, J.-P., *Séries de Fourier absolument convergentes*. Berlin, Springer-Verlag, 1970.

[9] Kahane, J.-P., *Some random series of functions*. Boston, Heath and Co, 1968.

[10] Kahane, J.-P. and Salem, R., *Ensembles parfaits et séries trigonométriques*. Paris, Hermann, 1963.

[11] Katznelson, Y., *An introduction to harmonic analysis*. New York, Wiley and Sons, 1968.

[12] Lang, S., *Introduction to Diophantine approximations*. New York, Addison-Wesley Publ. Cy., 1966.

[13] Mann, H.B., *Addition theorems*. Interscience tracts in Pure and Applied Mathematics, number 18. 1965.

[14] Reiter, H., *Classical harmonic analysis and locally compact groups*. Oxford, Clarendon Press, 1968.

[15] Rudin, W., *Fourier analysis on groups*. Interscience tracts in Pure and Applied Mathematics, number 12. 1962.

[16] Saks, S. and Zygmund, A., *Analytic functions*. Warszawa and Wrocław 1952.

[17] Salem, R., *Algebraic numbers and Fourier analysis*. Boston, Heath, 1963.

[18] Salem, R., *Œuvres mathématiques*. Paris, Hermann, 1967.

[19] Samuel, P., *Théorie algébriques des nombres*. Paris, Hermann, 1967.

[20] Serre, J.-P., *Cours d'arithmétique*. Paris, Presses Universitaires de France, 1970.

[21] Schwartz, L., *Théorie des distributions*, vols. I and II. Paris, Hermann, 1966.

[22] Weil, A., *Basic number theory*. Berlin, Springer-Verlag, 1967.

[23] Wiener, N., *The Fourier integral and certain of its applications*. Cambridge, University Press, 1933.

[24] Zygmund, A., Trigonometric series, 2nd ed., vols. I and II. Cambridge, University Press, 1968.

B. Original papers

[25] Amara, M., Ensembles fermés de nombres algébriques, *Sém. Dubreil–Pisot*, Inst. H. Poincaré **19**, 1 (1965—66).

[26] Baker, A., Rational approximations to $\sqrt[3]{2}$ and other algebraic numbers. *Quart. J. Math.* **15** (1964), 375—383.

[27] Benzaghou, B., Algèbres de Hadamard. *Bull. Soc. Math. France*, **98**, 3 (1970), 209—252.

[28] Bertrandias, F., Caractérisation des ensembles S_q par la répartition modulo 1 en *p*-adique. *Sém. Dubreil-Pisot* **17** (1963—64), nr. 11.

[29] Bertrandias, F., Ensembles remarquables d'adèles algébriques. *Bull. Soc. Math. France*, Mém. **4** (1965).

[30] Beurling, A. and Malliavin, P., On the closure of characters and zeros of entire functions. *Acta Math.* **118** (1967), 79—93.

[31] Bonami, A., Private communication.

[32] Bonami, A., Etude des coefficients de Fourier des fonctions de L^p(G). *Ann. Inst. Fourier* (Grenoble) **2** (1970).

[33] Boyd, D. W., Transcendental numbers with badly distributed powers. *Proc. Amer. Math. Soc.* **23** (1969), 424—427.

[34] Cassels, J. W. S., On a result of Marshall Hall. *Mathematika* **3** (1956), 109—110.

[35] Chabauty, C., Sur la répartition modulo 1 de certaines suites *p*-adiques. *C. R. Acad. Sc. Paris* **231** (1950), 465—466.

[36] Decomps-Guilloux, A., Eléments algébriques dans les adèles. *Sém. Dubreil-Pisot* **4**, 1 (1970).

[37] Deschamps-Gondym, M., Compacts associés a un ensemble de Sidon. *C. R. Acad. Sc. Paris* **271** (1970), 590—592.

[38] Deschamps-Gondym, M., Sur les ensembles de Sidon topologiques. *C. R. Acad. Sc. Paris* **271** (1970), 1247—1249.

[39] Domar, Y., Sur la synthèse harmonique des courbes de R^2. *C. R. Acad. Sc. Paris* **270** (1970), 875—878.

[40] Drury, S. W., Sur les ensembles de Sidon. *C. R. Acad. Sc. Paris* **271 A** (20 juillet 1970).

[41] Ehrenpreis, L., Mean periodic functions. *Amer. J. Math.* **78** (1955), 292—328.

[42] Ferru, H., Caractérisation des nombres de Pisot complexes. *Bull. Soc. Math. France*, Mém. **19** (1969), 5—20.

[43] Ferru, H., Extension au cas de $\mathbf{R}^2$... *C. R. Acad. Sc. Paris*. (This paper contains the correction of an error in the previous one.)

[44] Hartman, S., Rolewicz, S. and Ryll-Nardzewski, C., Ultra-Kroneckerian sets. *Colloq. Math.* **16** (1967), 225–229.

[45] Herz, C., Spectral synthesis for the Cantor set. *Proc. Nat. Acad. Sc. U.S.A.* **42** (1956), 42—43.

[46] Herz, C., Spectral synthesis for the circle. *Ann. of Math.* (2) **68** (1958), 709—712.

[47] Kahane, J.-P., Lectures on mean-periodic functions. Tata Institute of Fundamental Research, Bombay (1959).

[48] Kahane, J.-P., Pseudopériodicité et séries de Fourier lacunaires. *Ann. Scient. E.N.S.* 3rd séries **79** (1962), 93—150.

[49] Kahane, J.-P., Sur les fonctions moyennes-périodiques bornées. *Ann. Inst. Fourier* (Grenoble) **7** (1957), 292–315.

[50] Kahane, J.-P. and Katznelson, Y., Contribution à deux problèmes concernant les fonctions de la classe A. *Israël. J. Math.* **1**, 2 (1963), 109—131.

[51] Katznelson, Y. and McGehee, O. C., Measures and pseudo-measures on compact subsets of the line. *Math. Scand.* **23** (1968), 57–68.

[52] Kessler, I., Semi idempotent measures on abelian groups. *Bull. Amer. Math. Soc.* **73** (1967), 25—8260.

[53] Itô, T. and Amemiya, I., A simple proof of the theorem of P.J. Cohen. *Bull. Amer. Math. Soc.* **70** (1964), 774–776.

[54] Landau, H.J., Necessary density conditions for sampling and interpolation of a certain entire function. *Acta Math.* **117** (1967), 37—52.

[55] Lohoue, N., Nombres de Pisot et synthèse harmonique dans les algèbres $A_p(R)$. *C. R. Acad. Sc. Paris* **270** (1970), 1676—1678.

[56] Lohoue, N., La synthèse des convoluteurs sur un groupe abélien localement compact. *C. R. Acad. Sc. Paris* **272** (1971).

[57] Lohoue, N., Thesis, Faculté des Sciences d'Orsay, 1972.

[58] McGehee, O.C., Sets of uniqueness and sets of multiplicity. *Israël. J. Math.* **4** (1966), 83—96.

[59] McGehee, O.C., Certain isomorphisms between quotients of a group algebra. *Pacific J. Math.* **21** (1967), 133—152.

[60] Malgrange, B., Existence et approximation des solutions des équations aux dérivées partielles et des équations de convolution. *Ann. Inst. Fourier* (Grenoble) **6** (1955), 271—355.

[61] Malliavin, P., Sur l'impossibilité de la synthèse spectrale sur la droite. *C. R. Acad. Sc. Paris* **248** (1959), 2155—2157.

[62] Malliavin, P., Impossibilité de la synthèse spectrale sur les groupes abéliens non compacts. *Publ. math. I.H.E.S. Paris* **2** (1959), 61—68.

[63] Malliavin, P., Ensembles de résolution spectrale. In: *Proc. Int. Congress Math. 1962*. Djursholm, Sweden, Institut Mittag Leffler, 1963, 368—378.

[64] Malliavin, P. et Malliavin-Brameret, M.P., Caractérisation arithmétique d'une classe d'ensembles de Helson. *C. R. Acad. Sc. Paris* **264 A** (1967), 192—193.

Malliavin, P. and Beurling, A. (See Beurling, A. and Malliavin, P.)

[65] Mela, J.-F., Sur les ensembles d'interpolation de C. Ryll-Nardzewski et de S. Hartman. *Studia Math.* **29** (1968), 167—193.

[66] Mela, J.-F., Sur certains ensembles exceptionnels en analyse de Fourier. *Ann. Inst. Fourier* (Grenoble) **18** (1969), 33—70.

[67] Mela, J.-F., Approximation Diophantienne et ensembles lacunaires. *Bull. Soc. Math. France*, Mém. **19** (1969), 26—54.

[68] Meyer, Y., Isomorphismes entre certaines algèbres de restrictions. *Ann. Inst. Fourier* (Grenoble) **18**, 2 (1968), 73—86.

[69] Meyer, Y., Algèbres de restrictions non isomorphes. *Ann. Inst. Fourier* (Grenoble) **19** (1969), 117—124.

[70] Meyer, Y., Nombres algébriques et analyse harmonique. 4th series, **3**, 1 (1970), 75—110.

[71] Meyer, Y., Nombres de Pisot, nombres de Salem et analyse harmonique. *Lecture notes in math.* **117**. Berlin, Springer-Verlag, 1970.

[72] Meyer, Y., Les nombres de Pisot et la synthèse harmonique. *Ann. Sc. E. N. S.*, 4th series, **3**, 2 (1970), 235–246.

[73] Meyer, Y. et Schreiber, J.-P., Quelques fonctions moyennes-périodiques non bornées. *Ann. Inst. Fourier* (Grenoble) **19** (1969), 231—236.

[74] Pisot, C., La répartition modulo 1 et les nombres algébriques. *Ann. Scuola Norm. Sup. Pisa* **7**, 2 (1938), 205—248.

[75] Pisot, C., Familles compactes de fractions rationnelles et ensembles fermés de nombres algébriques. *Ann. Sci. E.N.S. Paris*, **81**, 3 (1964), 165—188.

[76] Pisot, C., Quelques aspects de la théorie des entiers algébriques. Montréal, Québec, Les Presses de l'Université de Montréal, 1966.

[77] Rauzy, G., Algébricité des fonctions méromorphes prenant certaines valeurs algébriques. *Bull. Soc. Math. France* **96** (1968), 197—208.
[78] Rauzy, G., Transformations rationnelles pour lesquelles l'ensemble des nombres de Pisot-Vijayaraghavan est stable. *C. R. Acad. Sc. Paris* **268** (1969), 305—307.
[79] Rauzy, G., Caractérisation des ensembles normaux. *Bull. Soc. Math. France* **98** (1970), 401—414.
[80] Rosenthal, H. P., A characterization of restrictions of Fourier-Stieltjes transforms. *Pacific J. Math.* **28** (1969), 663—668.
[81] Rosenthal, H.P., Projections onto translation-invariant subspaces of $L^p(G)$. *Memoirs Amer. Math. Soc.* **63** (1966).
[82] Roth, K.F., Rational approximations to algebraic numbers. *Mathematika* **2** (1955), 1—20.
[83] Schmidt, W. M., Badly approximable systems of linear forms. *J. Number Theory* **1** (1969), 139—154.
[84] Schmidt, W. M., Simultaneous approximation to algebraic numbers by rationals. *Acta Math.* **125** (1970), 189—202.
[85] Schmidt, W. M., Lectures on Diophantine approximations. Dept. Math., Univ. of Colorado, Boulder, Colo., 1970.
[86] Schneider, R., Some theorems in Fourier analysis on symmetric sets. *Pacific J. Math.* **31** (1969), 175–196.
[87] Schreiber, J.-P., Une caractérisation des nombres de Pisot-Salem des corps *p*-adiques. *Bull. Soc. Math. France,* Mém. **19** (1969), 55—63.
[88] Schreiber, J.-P., Sur la notion de modèles dans les groupes abéliens localement compacts. *C. R. Acad. Sc. Paris* **272** (1971), 30—32.
[89] Schreiber, J.-P., Thesis, Faculté des Sciences d'Orsay, France, 1972.
[90] Serre, J.-P., Travaux de Baker. *Sém. Bourbaki* **22** (1969), nr 368.
[91] Varopoulos, N., Sur les ensembles parfaits et les séries trigonométriques. *C. R. Acad. Sc. Paris* **260** (1965), 3831—3834.
[92] Varopoulos, N., Sets of multiplicity in locally compact abelian groups. *Ann. Inst. Fourier* (Grenoble) **16**, 2 (1966).
[93] Varopoulos, N., Tensor algebras and harmonic analysis. *Acta Math.* **119** (1967), 51–112.
[94] Varopoulos, N., Groups of continuous functions in harmonic analysis. *Acta Math.* **125** (1970), 109—154.

INDEX